FIRST EDITION

PROCESS QUALITY

Center for the Advancement of Process Technology

Prentice Hall

Boston Columbus Indianapolis New York San Francisco Upper Saddle River
Amsterdam Cape Town Dubai London Madrid Milan Munich Paris Montreal Toronto
Delhi Mexico City Sao Paulo Sydney Hong Kong Seoul Singapore Taipei Tokyo

Editor in Chief: Vernon Anthony
Acquisitions Editor: David Ploskonka
Editorial Assistant: Nancy Kesterson
Director of Marketing: David Gesell
Senior Marketing Manager: Alicia Wozniak
Marketing Assistant: Les Roberts
Associate Managing Editor: Alexandrina Benedicto Wolf
Copyeditor: Nancy Marcello
Inhouse Production Liaison: Alicia Ritchey
Operations Specialist: Laura Weaver
Art Director: Jayne Conte
Cover Art: Center for the Advancement of Process Technology
Lead Media Project Manager: Karen Bretz
Full-Service Project Management: Lisa S. Garboski, bookworks
Composition: Aptara®, Inc.
Printer/Binder: LSC Communications, Inc.
Cover Printer: LSC Communications, Inc.
Text Font: Times Ten Roman

Credits and acknowledgments borrowed from other sources and reproduced, with permission, in this textbook appear on appropriate pages within text.

Library of Congress Cataloging-in-Publication Data

Process quality/Center for the Advancement of Process Technology.
p. cm.
Includes index.
ISBN 0-13-700409-5
1. Process control. 2. Quality control. 3. Manufacturing processes—Quality control. I. Center for the Advancement of Process Technology.
TS156.8.P7656 2010
658.5'62–dc22

2009038236

Prentice Hall
is an imprint of

PEARSON

www.pearsonhighered.com

ISBN 13: 978-0-13-700409-6
ISBN 10: 0-13-700409-5

Contents

Preface

The Process Industries Challenge

In the early 1990s, the process industries recognized that they would face a major manpower shortage due to the large number of employees retiring. Industry partnered with community colleges, technical colleges, and universities to provide training for their process technicians, recognizing that substantial savings on training and traditional hiring costs could be realized. In addition, the consistency of curriculum content and exit competencies of process technology graduates could be ensured if industry collaborated with education.

To achieve this consistency of graduates' exit competencies, the Gulf Coast Process Technology Alliance and the Center for the Advancement of Process Technology identified a core technical curriculum for the Associate's Degree in Process Technology. This core, consisting of eight technical courses, is taught in alliance member institutions throughout the United States. This textbook is intended to provide a common standard reference for the Process Quality course that serves as part of the core technical courses in the degree program.

Purpose of the Textbook

Instructors who teach the process technology core curriculum, and who are recognized in the industry for their years of experience and their depth of subject matter expertise, requested that a textbook be developed to match the standardized curriculum. Reviewers from a broad array of process industries and education institutions participated in the production of these materials so that the widest audience possible would be represented in the presentation of the content.

The textbook is intended for use in community colleges, technical colleges, universities, and corporate settings in which process technology is taught. However, educators in many disciplines will find these materials a complete reference for both theory and practical application. Students will find this textbook to be a valuable resource throughout their process technology career.

Primary Contributor

Jim Lamar holds a Bachelor's Degree in Chemical Engineering from Texas A&I University in Kingsville, Texas, as well as a Master's Degree in Quality Management from Loyola University in New Orleans. Additionally, he has completed over 1,000 hours of seminar-type training in nearly every subfunction of quality. He is certified by the American Society for Quality as a Certified Quality Auditor and a Certified Quality Engineer. He was in the initial group of Quality Systems Lead Assessors certified by the Registrar Accreditation Board in 1993. He is certified as a Six Sigma green belt and black belt and currently serves in the role of master black belt. He is also the author of two other books. With over 25 years of experience, he was able to apply his expertise to the creation of this textbook.

Organization of the Textbook

This textbook has been divided into 16 chapters to solve the quality puzzle. Chapter 1 provides an overview of the history of quality and of the forefathers of quality who helped lead us to where we are today. Chapters 2, 3, 4, 5, and 7 demonstrate the use of statistics in quality. Chapters 6, 10, 11, 13, and 14 cover possible improvement strategies. Chapter 8 explains the process of root cause analysis. Chapter 9 provides insight into dealing with customers. Chapter 12 covers management systems. Chapter 15 provides an overview of quality costs. Finally, Chapter 16 shows how all the aforementioned topics tie and work together.

Each chapter is organized in the following way:

- Learning objectives
- Key terms
- Introduction
- Key topics
- Summary
- Checking your knowledge
- Activities

The **Learning Objectives** for a chapter may cover one or more sessions in a course. For example, some chapters may take 2 weeks (or 2 sessions) to complete in the classroom setting.

The **Key Terms** are a listing of important terms and their respective definitions that students should know and understand before proceeding to the next chapter.

The **Introduction** may be a simple introductory paragraph or may introduce concepts necessary to the development of the content of the chapter itself.

Any of the **Key Topics** can have several subtopics. Although these topics and subtopics may not always follow the flow of the learning objectives all learning objectives are addressed in the chapter.

The **Summary** is a restatement of the learning outcomes of the chapter.

The **Checking Your Knowledge** questions are designed to help students to self-test on potential learning points from the chapter.

The **Activities** section contains activities that can be performed by students on their own or with other students in small groups, and activities that should be performed with instructor involvement.

Chapter Summaries

CHAPTER 1: INTRODUCTION TO PROCESS QUALITY

This chapter describes the history of the quality movement in the United States, covers the quality gurus who paved the way, and provides an introduction to key quality concepts and tools.

CHAPTER 2: VARIABILITY CONCEPTS

This chapter explains process variability, statistical trends, common and special causes, stable and unstable processes, and data distributions. By understanding the variability of the process and how that variation is distributed, past data can be used to predict future performance.

CHAPTER 3: PROCESS CAPABILITY

This chapter explains variation, various capability indexes, measurement devices, sampling, overcontrol, and statistical tools that help quantify variation. Understanding these concepts is vital to relating variation to the needs of the customer.

CHAPTER 4: VARIABLES CONTROL CHARTS

This chapter explains the purpose of control charts and their relationship to distributions of data, as well as their being a mechanism to look at data in real time. All variables control charts are based on the normal or Boltzman distribution.

CHAPTER 5: ATTRIBUTES CONTROL CHARTS

This chapter explains the purpose of attributes control charts and their use to distinguish common cause from special cause variation. Attributes control charts are based on the Poisson and binomial distributions. This chapter also includes generic rules for interpreting all types of control charts.

CHAPTER 6: OTHER BASIC QUALITY TOOLS

This chapter explains a wide variety of the most basic quality improvement tools. The understanding and use of these tools provide the key to implementing other improvement methodologies such as root cause analysis.

CHAPTER 7: DESIGNED EXPERIMENTS

This chapter explains that designed experiments are tools for gathering minimum amounts of data that provide maximum amounts of information. The interrelationship of process variables is described and explained.

CHAPTER 8: ROOT CAUSE ANALYSIS

This chapter explains that the purpose of root cause analysis is to determine the true underlying cause of a problem in order to prevent its recurrence. This chapter also explains the various root cause analysis tools and their functions.

CHAPTER 9: CUSTOMER QUALITY

This chapter explains the difference between internal and external customers, and it is only by the customer's satisfaction that a business can survive. Useful tips for how process technicians should conduct themselves in the presence of a customer are also explored.

CHAPTER 10: SIX SIGMA

This chapter explains the basic methodology behind Six Sigma and compares the tenets of Six Sigma with earlier quality efforts such as total quality management.

CHAPTER 11: TEAMS

This chapter explains the purpose of a team, the tools a team uses, and the roles technicians play on a team in the process industries.

CHAPTER 12: MANAGEMENT SYSTEMS

This chapter explains that management systems document the processes to drive continuous improvement into their processes; also covered in this chapter is a technician's role as an auditor or auditee.

CHAPTER 13: QUALITY RELIABILITY PLANNING

This chapter explains that quality reliability planning is an improvement strategy that brings the various functions within a business together to ensure a continuity of purpose and understanding.

CHAPTER 14: LEAN

This chapter explains that lean is a philosophy of excellence and describes a variety of lean tools to help companies eliminate waste from their process.

CHAPTER 15: QUALITY COSTS

This chapter explains that the language of management is money; and in order to gain management's support, it must be shown how much potential profit is being wasted.

CHAPTER 16: PUTTING THE PUZZLE TOGETHER

This chapter explains how all the concepts and tools covered in this textbook work together to improve quality.

Acknowledgments

The following organizations and their dedicated personnel voluntarily participated in the production of this textbook. Their contributions toward making this a successful project are greatly appreciated. Perhaps our gratitude for their involvement can best be expressed by this sentiment:

> The credit belongs to those people who are actually in the arena . . . who know the great enthusiasms, the great devotions to a worthy cause; who at best, know the triumph of high achievement; and who, at worst, fail while daring greatly . . . so that their place shall never be with those cold and timid souls who know neither victory nor defeat. ~Theodore Roosevelt

Process technicians both current and future will utilize the information within this textbook as a resource to understand the process industries more fully. This knowledge will strengthen these paraprofessionals by helping to make them better prepared to meet the ever challenging roles and responsibilities within their specific process industry.

PROCESS QUALITY REVIEWERS

Josephine A. Allen, Ph.D., St. James Parish Public School System Career and Technology Center, Louisiana
Kathy Brossette, Baton Rouge Community College, Louisiana
Melissa D. Lassiter Collins, BP Learning & Development, Texas
Regina Davis, The Dow Chemical Company, Freeport, Texas (Retired)
Jerry Duncan, College of the Mainland, Texas
Raymond Fisher, Fisher Consulting, Inc., Baton Rouge Community College, Louisiana
Cleve J. Fontenot, BASF Corporation—The Chemical Company, Louisiana
Henry Haney, Kenai Peninsula College, Alaska
Rhea Jones, ConocoPhillips, Oklahoma
Karen Kupsa, College of the Mainland, Texas
Carol Lamar, Independent Reviewer, Texas
Dennis Rygaard, The Dow Chemical Company, Texas (Retired)

CENTER FOR THE ADVANCEMENT OF PROCESS TECHNOLOGY STAFF

Bill Raley, Principal Investigator
Jack Berry, Director
Angelica Toupard, Associate Director of Curriculum Development
Scott Turnbough, Graphic Artist
Chris Carpenter, Web Application Developer

Cindy Cobb, Program Assistant
Charlie Dwyer, Technical Writer/Developer

This material is based upon work supported, in part, by the National Science Foundation under Grant No. DUE 0532652. Any opinions, findings, and conclusions or recommendations expressed in this material are those of the author(s) and do not necessarily reflect the views of the National Science Foundation.

A Word from the Author

Why Quality Improvement for Operators and Technicians Is Important to Me

I spent the first 6 weeks of my professional career working on shift with the operators, learning the operation. An old operator (older than me anyway) was assigned to show me the ropes. For my first lesson he showed me how to tie a hangman's noose with a small piece of rope. He then let me know in no uncertain terms that I would either learn to operate the unit or he would hang me with that noose. Despite the differences in our ages and positions in the company, that old operator and I became the best of friends, and that little section of rope has graced the walls of my office for over 25 years now as a constant reminder to me that operations is where the rubber meets the road. We either learn how to operate our plants to make a competitive profit for the company or they will close us down. If we cannot implement quality on the shop floor, we will not be successful. The same statement could be made regarding health, safety, environment, and costs. Success of the company takes place in the manufacturing unit as much as it does in the corporate board room.

I dedicate my efforts in making quality technologies available to operators and technicians to the late Don Osborne—my favorite operator. *~Jim Lamar*

CHAPTER 1

Introduction to Process Quality

"Quality is neither mind nor matter, but a certain entity independent of the other two . . . even though Quality cannot be defined, you know what it is."

~PRISIG

Objectives

Upon completion of this chapter you will be able to:

- Describe the history of the quality movement in the United States.
- Compare and contrast the different philosophies of Deming, Juran, and Crosby.
- List several of the key tools in "the quality toolbox." Explain why there are so many tools and the value of each.
- Identify key quality concepts employed in the process industries today.
- Compare and contrast the differences between statistical quality control (SQC) and statistical process control (SPC).

Key Terms

Control chart—a graphic depiction of a process that is used to distinguish common cause from special cause variation.

Quality—the performance, products, and services that consistently meet *or exceed* the expectations of the customer by doing the right job, the right way, the first time, every time.

Statistical process control (SPC)—the application of statistical techniques to the process that generates a product.

Statistical quality control (SQC)—the application of statistical techniques to the output (quality) of a process.

Total quality management (TQM)—the pulling together of many different components of quality into a single program (sometimes shown as TQC—total quality control).

Introduction

Quality is a key component of the success of any business today. Starting up a new business is difficult. Staying in business for the long haul is yet another story. Being the low-cost producer may give your business an edge in capturing a sale, but getting the customer to come back to you for the second and third sale will require that, in addition to a good price, you provide a quality product as well as quality services. The purpose of this textbook is to present you with an overview of the various components of the general concept of quality and prepare you for your role in supporting quality in the process industries.

Defining quality is a daunting task, yet this book attempts at least to describe quality in terms of what it means to those in the process industries. Part of what makes defining quality difficult is that all of the answers you receive may in fact be correct because quality is a broad concept. Let us think about quality as a jigsaw puzzle.

There are many pieces to the quality puzzle. Each piece fills in a certain gap. If any one piece is missing, the picture is incomplete and unclear. Only by filling in every single piece can one grasp the total quality picture. For the sake of illustration, the various quality topics in this textbook represent six different puzzle pieces:

1. Statistics
2. Customers
3. Management systems
4. Improvement strategies
5. Root cause analysis
6. Quality costs

As you go through each chapter, check to make sure you understand where each topic fits within the puzzle. This chapter lays the groundwork for the rest of the book by discussing the history of the quality movement, identifying key quality concepts, and promoting understanding of the technical terminology used in the field of quality. This chapter lists and briefly describes the kinds of tools that will be covered in more depth as you continue your studies through this book.

Quality Defined

Quality can be defined as conforming to requirements. Quality can also be defined as consistency. Quality is superior performance. Quality is being better than the competition. Quality is an absence of defects. Quality is meeting expectations. Quality is whatever the customer wants it to be.

Ask a dozen people on the street to define quality, and you will probably receive a dozen different answers, including some of the ones just provided. You can combine many of these concepts to define **quality** as the performance, products, and services that consistently meet or exceed the expectations of the customer by doing the right job, the right way, the first time, every time.

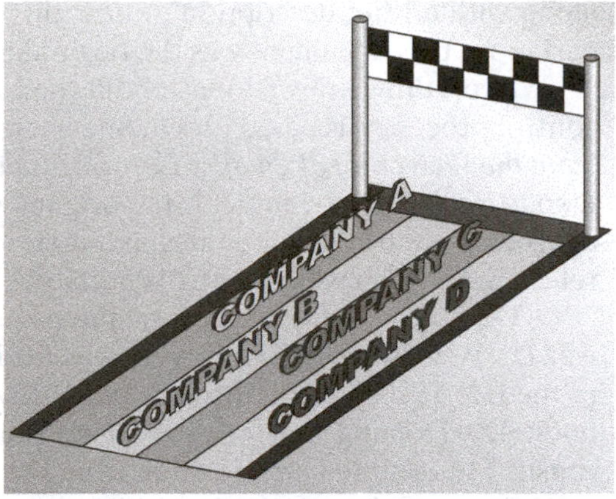

Let us examine some of these concepts further.

• Product quality versus total quality: Quality cannot be defined as just the product that you purchased—the definition of quality must include the service that came with the product. Most of us have had unpleasant experiences shopping for something. Often we decide where to shop based on our past service experiences or even based on the service experiences shared with us by friends and family.

• Performance: Quality must be about long-term performance as well as working in the present. Have you ever bought something that worked great on day one but failed within a matter of days or weeks of installation?

• Conformance: Quality must also be defined as more than just conformance to requirements. In the process industries the customer requirements are typically called product specifications. These product traits are guaranteed to the customer. Sometimes figuring out what the customers' expectations are and translating those expectations into specifications is part of the quality game.

• Meet or exceed: Meeting the expectations of the customer may keep you in business. Exceeding the expectations of the customer may in fact help grow your business. Giving the customer more than he or she asks for and more than is expected is often what it takes to ensure that the customer will in fact come back again and will speak positively about the experience.

• Relative quality: Like beauty, quality is in the eye of the beholder. Different customers have different requirements and different expectations.

• Consistency: Some people like fast food. Some people do not. In the process industries, the products must produce the same results every time. Your customers would no more tolerate inconsistency in their dealings with you than you would when you visit a restaurant. Because consistency is such an important concept when talking about quality, this book will spend much time teaching more about consistency, how to measure it, and how to improve it.

If you focus on only one aspect of quality, you lose. You may lose one customer. You may lose one sale. You may lose your business. Success requires that you view quality as a multifaceted concept seen from the customer's perspective. Ultimately you are in business to make money. The quality movement is not just an activity to perform when you have nothing else to do. It is not about creating busywork to make sure you are earning your pay. Quality is one of the key components in making your business successful. A successful business is one that makes money today and into the future.

History of the Quality Movement in the United States

In the 1920s Walter Shewhart showed his boss at Western Electric a one-page memorandum that pictured what would be called today a **control chart** (a graphic depiction of a process that is used to distinguish common cause from special cause variation) along with a brief description of how this chart could be used to characterize process variation. This document was the beginning of modern-day process quality control.

Everything known today about the use of control charts to understand and better control processes is based on Dr. Shewhart's work. His second book, *Statistical Method from the Viewpoint of Quality Control*, published in 1939 by the Graduate School of the Department of Agriculture, is the original work on the subject. As you will see in later chapters, when control charts are mentioned in technical training, they are often referred to as Shewhart control charts, paying homage to their originator.

One of the people who saw the beauty of and appreciated Dr. Shewhart's work was Dr. W. Edwards Deming, who was working for the U.S. Department of Agriculture in the late 1920s and all through the 1930s. In fact, Dr. Deming edited Dr. Shewhart's book. Dr. Deming was sent to Japan in the late 1940s to work on the 1951 Japanese census. He accepted an invitation from the Japanese Union of Scientists and Engineers (JUSE) to speak to them about statistical process control (SPC).

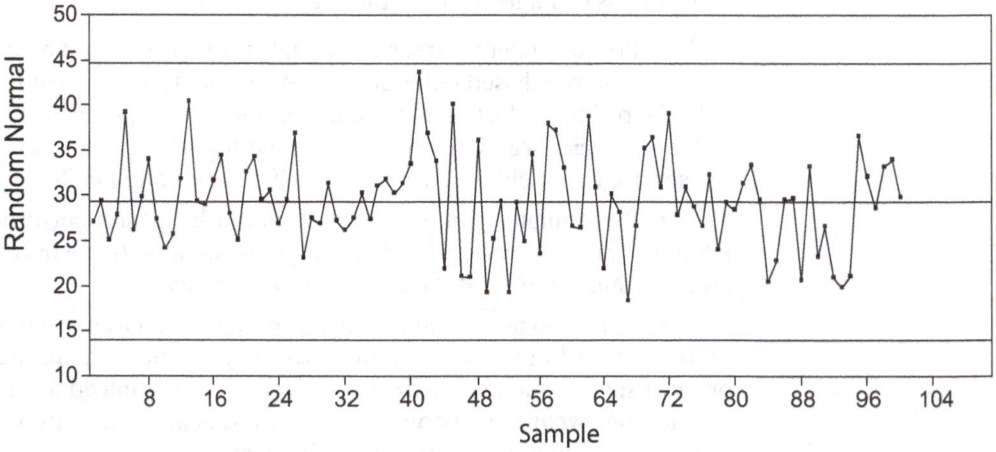

Over the coming years he continued to work with the Japanese as they rebuilt their nation following the destruction of World War II. The United States prospered during this time frame, having the only intact industrialized economy in the world. Anything that the United States made it could sell. Through the work of Deming and others, Japan began an intensive and society-wide effort to improve its quality and productivity.

In 1980 NBC aired a white-paper documentary entitled *If Japan Can . . . Why Can't We?* hosted by Lloyd Dobyns. This 70-minute video called attention to the reality that Japan was then an economic force to be reckoned with. Japan was busily modernizing its 20-year equipment while American companies were still running 50-year-old plants. Companies such as Westinghouse, Motorola, and Nashua were highlighted as they discussed their attempts to catch up to the Japanese economic threat.

American companies sought out the help of experts such as Dr. Deming, Joseph Juran, and Phillip Crosby. As each of these experts strove to make his brand of quality improvement the brand of choice, technical jargon in the quality field grew quickly. Companies would often align their quality improvement efforts with their chosen teachings. Some companies followed the teachings of Deming, others those of Juran. Which quality expert is the best? Which quality approach is the best? Let us take a look at the philosophies of the most recognized quality gurus and then maybe you can attempt to answer these questions.

Dr. W. Edwards Deming

The first quality guru is Dr. W. Edwards Deming, whose work in Japan is legendary. The seminars he taught to the Japanese Union of Scientists and Engineers (JUSE) were the beginnings of the Japanese post–World War II rebuilding efforts.

W. Edwards Deming
Courtesy of The W. Edwards Deming Institute®

The JUSE board of directors established the Deming prize to honor the man whom it credited with having more impact on Japanese manufacturing than any other person not of Japanese heritage. This coveted prize is still the highest honor a company can attain. In over 50 years only three American companies have ever achieved this honor.

In Deming's most recognized work, his 1986 book, *Out of the Crisis*, he strongly advocated the use of statistical expertise for those who would choose to practice it. He went so far as to state that anybody teaching control charts should have at least a master's-degree level of statistical knowledge plus experience working with a master. Entire chapters are devoted to various statistical topics.

As you will see shortly, this view was not necessarily shared by other quality gurus. He never really defined quality in a concise manner, but spent nearly an entire chapter talking about how hard it is to define quality and suggesting that quality can be defined only by the customer. Deming was vocal regarding his views on management. He advocated the elimination of the following concepts: managing by objectives, focusing on zero defects, and even performance appraisals.

Deming stated that short-term profits were not reliable indicators of management performance and suggested that such short-term focus defeats constancy of purpose and inhibits long-term growth. He even voiced his opinion that tying chief executive officer (CEO) compensation to short-term results would result in CEOs' finding it unrewarding to do what was right for the company long term. One could summarize these comments by saying Dr. Deming was seeking the best for society in general, not just the best for any individual. Perhaps this societal thinking is something he brought back with him from his work in Japan.

Did You Know?

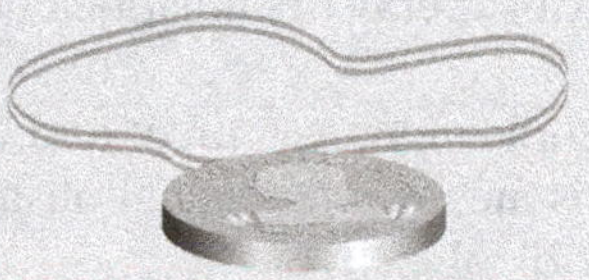

The only three Deming prizes given to American companies were to Florida Power & Light in 1989, AT&T Power Systems in 1994, and Sanden International in 2006.

Deming was not opposed to the use of advanced technology, but he made it clear that the use of computers, robots, and gadgets would not save American industry. Only hard work and application of his improvement methodologies could accomplish that great feat. He advocated doing things manually in order to get the worker closer to the product. As a proponent of training and continuing education, he was adamant that every process must be constantly and continually improved upon. He stated that putting out fires is not improvement; instead it only puts the process back to where it should have been in the first place.

He recognized that improvement would best be accomplished by true teamwork but declared that annual ratings instilled fear and defeated the teamwork. Deming recognized that the involvement of everyone in the company would be needed to accomplish the quality transition and advocated breaking down of barriers between the layers of the company. He felt that the barriers were put in place by misguided management who saw employees as the root of the problem.

Deming stated during an interview that 85% of the problems in a company are caused by management. In that same interview he also declared that inspecting a product cannot make it a quality product—you have to make it right in the first place. He saw the divide between management and workers as a key stumbling block to success.

During his later years, he was often sought after to consult in large American companies. If the CEO would not show up for the meeting, Deming would leave, believing that top-down action would be required for success.

His philosophy about quality is best summarized in his famous 14 points:

1. Create constancy of purpose toward improvement of product and service, with the aim to become competitive, stay in business, and provide jobs.
2. Adopt the new philosophy. We are in a new economic age, created by Japan. Western management must awaken to the challenge, must learn responsibilities, and take on leadership for change.
3. Cease dependence on inspection to achieve quality. Eliminate the need for inspection on a mass basis by building quality into the product in the first place.
4. End the practice of awarding business on the basis of price tag. Instead, minimize total cost. Move toward a single supplier for any one item on a long-term relationship of loyalty and trust.
5. Improve constantly and forever the system of production and service, to improve quality and productivity, and thus constantly decrease costs.
6. Institute training on the job.
7. Institute leadership. The aim of leadership should be to help people, machines, and gadgets to do a better job. Supervision of management is in need of overhaul, as well as supervision of production workers.
8. Drive out fear, so everyone may work effectively for the company.
9. Break down barriers between departments. People in research, design, sales, and production must work as a team to foresee problems of production and in use that may be encountered with the product or service.
10. Eliminate slogans, exhortations, and targets for the workforce that ask for zero defects and new levels of productivity.
11. **(a)** Eliminate work standards (quotas) on the factory floor. Substitute leadership.
    (b) Eliminate management by objective. Eliminate management by numbers, numerical goals. Substitute leadership.
12. **(a)** Remove barriers that rob the hourly worker of his right to pride of workmanship. The responsibility of supervisors must be changed from stressing sheer numbers to quality.
    (b) Remove barriers that rob people in management and engineering of their right to pride of workmanship. This means, *inter alia,* abolishment of the annual merit rating of management by objective.
13. Institute a vigorous program of education and self-improvement.
14. Put everybody in the organization to work to accomplish this transformation. The transformation is everybody's job. (Courtesy of *Out of the Crisis* by W. Edwards Deming, pages 23–24.)

IN A NUTSHELL

Deming
Advocated strong leadership, elimination of numerical targets and quotas, teamwork, and the reliance on statistics to drive decisions.

Joseph Juran

Joseph Juran, who defined quality as fitness for use, is the second guru. Like Shewhart, Juran worked at the Western Electric Hawthorne Works in the 1920s. The JUSE invited Juran, as they had invited Deming, to travel to Japan and assist in the rebuilding of its economy after World War II.

His works include *Juran's Quality Control Handbook* (1951), which focused on the statistical and technical aspects of quality improvement, and *Juran on Leadership for Quality* (1989), which was directed at managers in corporate America.

Whereas Deming was a huge proponent of the use of statistics, Juran focused more on managerial issues and the use of breakthrough teams to accomplish quality improvement. Juran was not against the use of statistics, but several times in his book on leadership he warned that managers should be careful that the statistics do not become an end in themselves. Juran recognized, as did Deming, that it is all too easy for managers to manage the numbers instead of managing the process and the people.

Juran was a strong proponent of the use of flowcharts, Pareto diagrams, and process capability indexes by project teams. In his leadership book, and his training videos from the 1980s, he declared that all improvements are accomplished only project by project.

Joseph Juran
Courtesy of the Juran Institute

Juran certainly did not invent the Pareto principle, also called the 80-20 rule, but he may have been the first to popularize its use in characterizing problems as either the 20% "vital few" issues that cause all the grief or the 80% "trivial many" issues that should be looked at, but with a lower priority.

Unlike Deming, who was adamant about eliminating all numerical goals, Juran taught that goals could be acceptable as long as they were legitimate, measurable, attainable, and equitable. He was an advocate for training at all levels and in all functions in a company, suggesting that quality training must be a prerequisite for advancement.

He was especially keen on ensuring that managers should be the first to acquire quality training in order to lead the quality revolution. He taught that quality improvement must be established as a continuing process that goes on year after year and that management's job is to create the proper atmosphere and environment to allow for continuous improvement.

A prime message of *Juran on Leadership for Quality* is that management must take the lead in implementing quality. Management must be the first to be trained in quality, must be the first to serve on quality improvement teams, and must personally and extensively participate in the entire quality improvement effort if there is to be any hope of success. Everyone must participate in the quality effort, from engineers to process technicians to the clerical staff, but managers must lead the quality effort.

Juran suggested that attempting to delegate the quality function to subordinates had been tried many times with disappointing results. He recognized the need for employee involvement and the value that the workers on the floor brought to the system but strongly asserted that without the leadership and participation of committed managers, the effort would ultimately be in vain.

He proposed a trilogy of managerial processes for managing quality:

1. Quality planning—Who are your customers? What do they need? What product features could the company provide to meet that need? What processes does it need to put in place to achieve those product features? How does it institutionalize these processes in the workplace?

2. Quality control—Evaluate actual quality performance. Compare it to desired quality performance. Act on the differences.

3. Quality improvement—Organize to achieve quality improvement. Identify problem areas. Establish project teams to address the issues. Provide the resources, training, and motivation needed to achieve success.

People's jobs are sometimes impacted by the planning and control phases, but most often they are called upon to improve an existing process or product. Juran taught quality improvement through the use of breakthrough teams—teams that seek not an incremental improvement in a process but a breakthrough to a new level.

Juran's philosophy regarding breakthrough teams was that they are quality practitioners on a series of journeys. During the first journey, called the diagnostic journey, the team seeks to understand the root cause(s) of the problem by following a flowchart like the one shown in Figure 1-1.

The diagnostic journey begins with gaining an understanding of the symptoms—exactly what is it that management or the customer is upset about? What is the complaint? Next theories about what might be causing these symptoms are listed, usually by employing techniques such as brainstorming. With a list of possible causes in hand, the team works on testing its theories. Teams may employ many tools to either validate or invalidate a potential cause.

Did You Know?

In companies in which different divisions followed different quality teachings, Deming followers were referred to as "Deming disciples" and Juran followers were called "Juran journeymen."

FIGURE 1-1 Example of a Diagnostic Journey

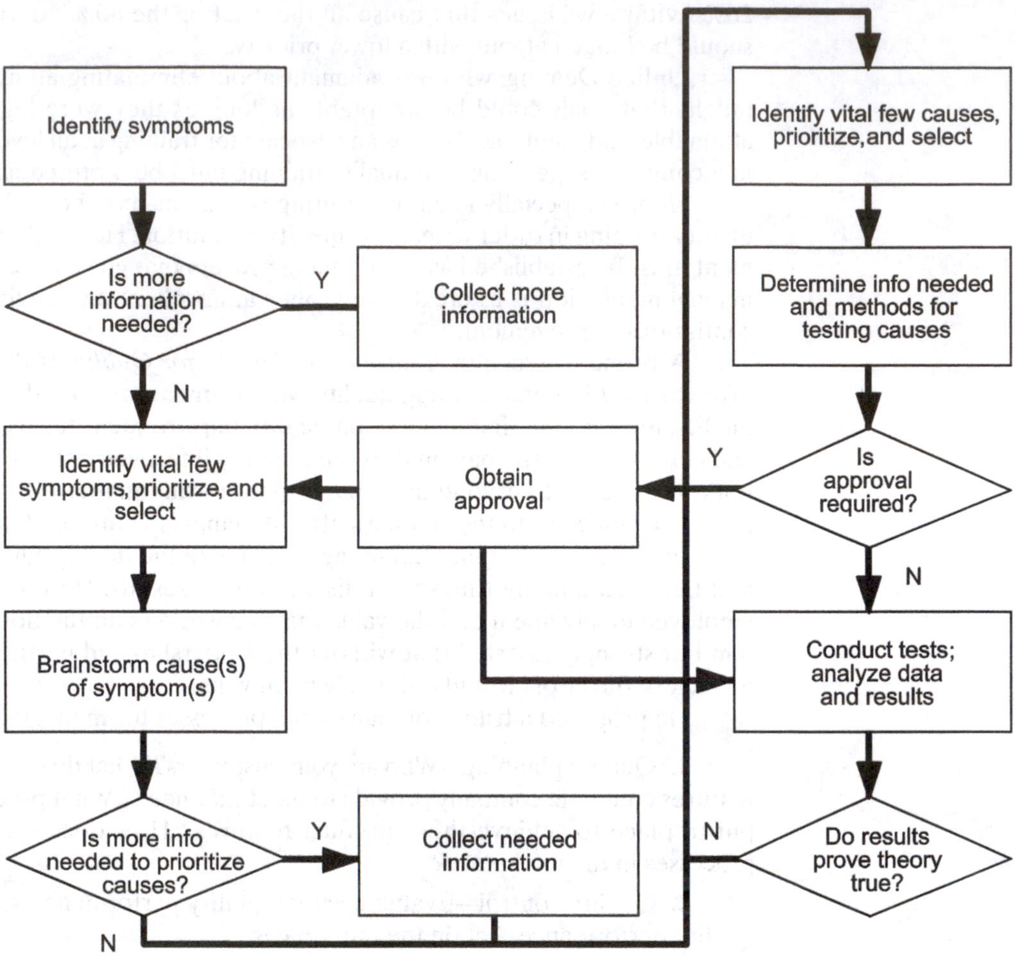

Once the team has a list of validated root causes, it has completed its diagnostic journey and can move to the remedial journey, which is where it remedies the situation. An example flowchart for the remedial journey is shown in Figure 1-2.

The basic flow of the remedial journey is rather straightforward: First the team solicits options for establishing a remedy. It might use brainstorming techniques, Internet searches, or past history to get started. Then it needs to find a way to evaluate the proposed solutions. Which one costs the most? Which one does a better job of addressing the root cause? Which one can be implemented the most quickly and most easily?

IN A NUTSHELL

Juran

Advocated strong leadership participation and the use of breakthrough teams to identify and eliminate problems.

Once the breakthrough team has narrowed down the list of possible solutions and prioritized them, it needs to test them. Given that it is not always possible to test every proposed solution, a small-scale test or experiment to validate that the solution does in fact work is always desirable. Otherwise a team might find out that it fixed a cause that was not the greatest contributor to the problem after all. Once its solution(s) have been validated, the team must take steps to establish control systems to maintain the gains. Too many times companies end up fixing the same problems over and over again.

Philip Crosby

The third guru in the world of quality is Philip Crosby, who defined quality as zero defects or, in other words, simple conformance to requirements. Two of his most noted works are *Quality Is Free* (1979) and *Quality without Tears* (1984). You will not find a big discussion about the use of statistics in Crosby's books because his mantra was zero defects, which is accomplished through prevention.

FIGURE 1-2 Example of a Remedial Journey

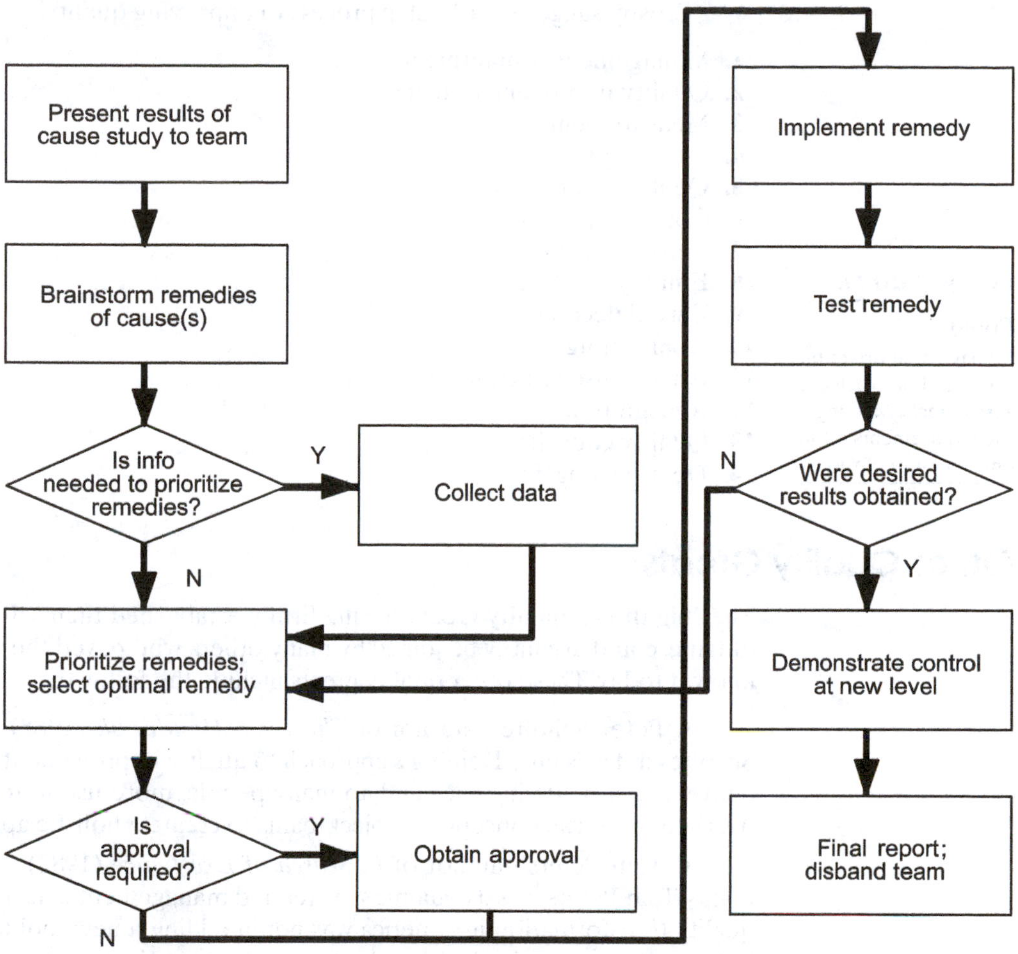

A common acronym attributed to his work is DIRTFT, which stands for "Do It Right The First Time." This was hardly a unique perspective, as Deming and others were saying the same thing. With quality defined as conforming to requirements, Crosby put a lot of stock in measurement of conformance, focusing on the results of the process.

As was the case with the first two quality gurus, Crosby held that management commitment was absolutely critical to success and that training of the employee base was a key component of achieving that success as quickly as possible. He also recommended the use of project teams as the mechanism to accomplish quality improvement.

Unlike the other two gurus, he did not offer much in the way of a tangible mechanism to accomplish the quality goals. His books and teachings were more directed toward convincing management that something must be done and helping them organize

Did You Know?

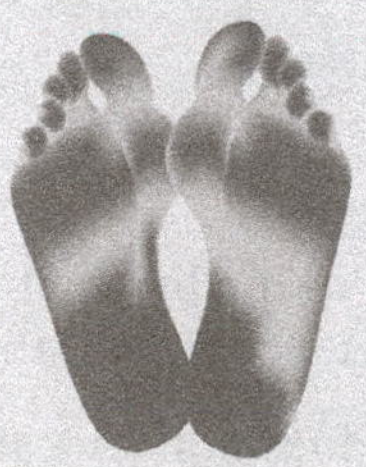

DIRTFT was often displayed on banners and posters as a mechanism to inspire the workforce in companies that followed Crosby's teachings. The slogan was pronounced "dirty feet."

to achieve quality improvement. His work in bringing the quality crisis to the forefront of management's attention earned him a significant following in industry.

Crosby suggested a 14-step process to improving quality:

1. Management commitment
2. Quality improvement team
3. Measurement
4. Cost of quality
5. Quality awareness
6. Corrective action
7. Zero defect planning
8. Employee education
9. Zero defect day
10. Goal setting
11. Error-cause removal
12. Recognition
13. Quality councils
14. Do it over again

IN A NUTSHELL

Crosby

Management drives the organization to achieve zero defects. Meeting the requirements is the only measure of success.

Other Quality Greats

The "big three" quality greats, having firmly established themselves in the quality hall of fame, could arguably be joined by many others who paved the way for quality as we know it today. These other quality greats include the following:

• Peter Scholtes—author of *The Team Handbook* (1988). In many ways what Scholtes did was take Deming's approach to quality improvement and add a structured delivery that made it easier and, to many people, more useful to implement. He also overlaid the Juran concept of project teams to create a holistic approach to quality.

• Tom Peters—author of *In Search of Excellence* (1982) as well as many other books. Tom Peters was a dynamic speaker and management consultant. His value to the quality transformation in America was not in adding a new tool to the quality toolbox but in getting everybody from the top of the organization to the shop floor worker excited about quality. His PBS specials and videos were widely viewed in the 1980s as a mechanism to infuse energy into the organization. He promoted the concept of a quality champion, stating that at whatever level your champion existed, that was the level of quality success you could expect. He also defined quality in terms of the wow factor, teaching that companies who could provide above and beyond what the customer expected were the companies that would ultimately succeed.

• Kaoru Ishikawa—author of *Guide to Quality Control* (1982). His work included implementing basic statistical techniques such as those covered in later chapters of this book. One of these techniques is called an Ishikawa diagram, or fishbone chart.

• Genichi Taguchi—a Shewhart-Deming disciple. Taguchi is widely recognized as an innovator in the field of designed experiments and made many contributions to the field of quality control. He popularized the concept of off-line quality control, a term that means you should build quality into the process during the design and construction phase so you do not experience quality problems during the manufacturing phase. His experimental design methods are often referred to as Taguchi methods, or robust design.

Key Quality Concepts

So what are some of the key concepts related to quality that are evident through this chapter's review of the quality gurus and their teachings? Let us make a list:

1. Management commitment. If quality does not start at the top, it probably will never get going at all.

Five Key Quality Concepts

1. Management commitment
2. Management systems
3. Statistics
4. Teams
5. Training

2. Management systems. Standardization of how things are done to ensure consistency of operation can only serve to drive consistency in the results. ISO 9000 and ISO 14000 are examples of internationally recognized management systems.

3. Statistics. Most of the works on quality are going to spend a considerable amount of effort showing you how to use statistical tools to run your process as efficiently as possible. Oftentimes people use the terms *SQC* and *SPC* when discussing the application of statistics to the field of quality. **Statistical quality control (SQC)** is the application of statistical techniques to the output (quality) of a process. **Statistical process control (SPC)** is the application of the same statistical techniques to the process that generates the product.

4. Teams. Nearly all the quality greats agree that two heads (or three or four) are better than one.

5. Training. If your teams are not properly equipped with the knowledge of management systems, statistics, and improvement methodologies, how can you expect them to succeed?

By virtue of the fact that you are studying quality, it is assumed that the company you hope to work for is demonstrating its commitment to quality in general and to your training in the field. Because the concepts of management systems, basic statistics, and teamwork will be covered in the remaining chapters of this book, it would appear that you are on your way to a broad-based understanding of how to apply quality improvement in the process industries. It is also necessary to mention **total quality management (TQM),** which is the pulling together of many different components of quality into a single program. TQM is not a separate topic but rather a melding together of many different quality concepts.

QUALITY TOOLS

Each of these concepts comes with a wide variety of quality tools to be employed in making improvements. Examples of some of these tools include control charts, process capability, designed experiments, management systems, Pareto charts, and root cause analysis.

All of these tools and many others will be discussed in more detail as you progress through this book. For some of these tools you will need to gain a detailed understanding. For others you will require only a basic understanding. In each case this text pursues the topic from the viewpoint of how this tool is used by, is impacted by, or affects process technicians.

Summary

Quality is a key component of the success of any business today. This applies to the quality of the product the business sells as well as the service provided during and after the sale. Defining quality requires understanding the needs and expectations of the customer. In the end, quality is making the customers happy so they continue to do business with you. Quality is a multifaceted issue with many pieces similar to a jigsaw puzzle. Each piece plays a role in helping to understand the total quality picture. Leaving out any one piece would result in an incomplete picture.

To understand how quality is used in the process industries requires a basic understanding of key quality concepts such as statistical thinking, customers, management systems, improvement strategies, root cause analysis, and quality costs. These pieces of the puzzle will be covered in more detail in the following chapters so that by the end of this book you will see the complete quality picture clearly.

The quality picture that we have today was developed as a response to economic threat from the Japanese as they struggled to rebuild their economy after World War II. Many of the quality experts who assisted Japan in their efforts were American scientists and mathematicians such as Dr. W. Edwards Deming and Joseph Juran. The different views of quality worked together to hasten the Japan economic success. These men and other quality experts such as Philip Crosby and Tom Peters set the foundation for the vast majority of the quality efforts applied in the United States in the last 20 to 30 years.

The detailed teachings of these experts filled your quality toolbox with many useful techniques such as control charts, Pareto charts, designed experiments, and capability analyses. Each of these tools has value to your quality management and quality improvement efforts. Each can be used to drive continuous improvement. No single concept or tool is sufficient in and of itself. What you should have discovered is that like any tool in a toolbox, each has its own purpose and value to the worker.

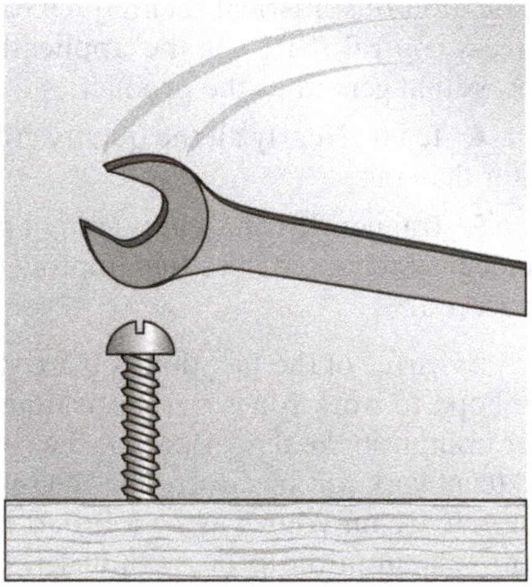

There is a time for teamwork and a time to just get the job done alone. There is a time for control charts, a time for designed experiments, and a time for standardized work processes. You could use a wrench to drive in a screw if you worked at it, but the process would be much more efficient if you used a screwdriver for screws, a hammer for nails, and a wrench for bolts. As with many other jobs, oftentimes quality is about selecting the right tool for the right job at the right time.

Checking Your Knowledge

1. Define the following key terms:
   a. Quality
   b. Statistical process control
   c. Statistical quality control
   d. Total quality management
2. (*True or False*) Quality is an easy term to define.
3. Which of the following concepts must be considered when defining quality?
   a. Performance
   b. Conformance
   c. Consistency
   d. All of the above
4. (*True or False*) A business has to consider quality of its products and services to ensure long-term success.
5. List the "big three" quality greats.
6. List the five key quality concepts.
7. (*True or False*) The growth of the quality movement in the United States in the last 20 years is largely attributable to the success of the Japanese in rebuilding their economy after World War II.

8. Match the following quality tools to their purpose:

1.	Control charts	A.	Used to judge the acceptability of the process variation seen against the needs of the customer
2.	Process capability	B.	Used to standardize the work process so that everybody does the same job the same way every time
3.	Pareto charts	C.	Used to generate a mathematical model of the process so you can better predict future performance
4.	Designed experiments	D.	Used to distinguish between normal process variation (common cause) and problem process variation (special cause)
5.	Management systems	E.	Used to determine the root cause of a problem prior to initiating any corrective action in order to ensure you fix the right thing
6.	Root cause analysis	F.	Used to identify the vital few problems from among the trivial many so you can prioritize where to work first

9. (*True or False*) Dr. Deming was adamant that only by the immediate implementation of computerized robots would the United States ever hope to catch up to the Japanese.
10. What is a diagnostic journey?
    a. An annual pilgrimage to Japan made by quality professionals
    b. The evaluation of data to determine (diagnose) the root cause of a problem
    c. A review of quality procedures to determine conformance to requirements
11. What is a remedial journey?
    a. The process of selecting the proper fix (remedy) for a problem
    b. A quality class for slow learners
    c. An annual pilgrimage to the United States by Japanese quality professionals
12. (*True or False*) Management commitment is an absolute MUST if the quality effort is to succeed.
13. Zero defects was a key component of the thinking of which quality guru?
    a. Deming c. Crosby
    b. Juran d. Peters
14. (*True or False*) All three of the quality greats agree that training of personnel was a "nice to have" but not a required part of the quality process.
15. (*True or False*) Management systems are computers used by large companies to control executive and mid-level managers.

Activities

1. Write out your own definition of quality. This could be from a personal perspective as a consumer or as a process technician.
2. Compare and contrast the major similarities and differences among the teachings of the "big three" quality greats.
3. Describe briefly the growth of the quality movement in the United States since the 1980s, and explain what has caused this growth.
4. What is the difference in how to approach statistical quality control (SQC) and statistical process control (SPC)?
5. Explain why teams are an important mechanism for implementing quality.

CHAPTER 2

Variability Concepts

"Commit yourself to quality from day one. Concentrate on each task, whether trivial or crucial, as if it's the only thing that matters (it usually is). It is better to do nothing at all than to do something badly."

~McCormack

Objectives

Upon completion of this chapter you will be able to:

- Explain the concept of variability and use a histogram to illustrate the distribution of data.
- Calculate measures of central tendency.
- Calculate measures of spread.
- Explain the differences between common cause and special cause.
- Recognize and explain the differences between a stable process and an unstable process.
- List and explain the types of data that are represented by normal, Poisson, and binomial distributions.

Key Terms

Binomial distribution—distribution of discrete data that are made up of data that have only two choices (e.g., pass/fail).
Central tendency—center or middle of a distribution of data; it is measured as the mean, median, or mode.
Common cause variation—variables in your process that cause the process to vary and are built in and inherent to the process. The variation of your process due to common causes is the variation that is there all the time when things are running normally.
Continuous data—numbers that can be of any value or fraction of a value.
Discrete data—numbers that are not continuous (e.g., whole numbers). Also known as variables data.
Histogram—bar chart that shows the distribution of a set of data.
Mean—arithmetic average of a set of data.
Median—geometric center of a data set.
Mode—number that occurs most frequently in a data set.
Normal distribution—distribution of continuous data that is bell shaped. The data in a normal distribution are distributed such that approximately 68% of the data fall within 1 standard deviation of the mean, approximately 95% of the data fall within 2 standard deviations of the mean, and approximately 99.7% of the data fall within 3 standard deviations of the mean. Also referred to as a Boltzman distribution.
Poisson distribution—distribution of discrete data that is made up of counting data. In this case only whole numbers are present.
Range (R)—difference between the high and low values in a group of numbers.
σ—lowercase Greek letter sigma; used in mathematics to denote the standard deviation.
Σ—uppercase Greek letter sigma; used in mathematics to indicate a summation.
Special cause variation—event-related items in the process that cause the process to vary outside of its normal operation. These causes are in addition to the common cause variation sources. Also referred to as assignable cause variation.
Spread—the broadness of the distribution of the data set; it is measured as the difference between the highest and lowest values in the data set or by using the standard deviation.
Stable process—state of your process when only common cause variation is present; also stated as the absence of special or assignable causes. A stable process is, by definition, predictable.
Standard deviation—statistical measure of the spread of a set of data. Often referred to as sigma (σ), which is always used to denote the standard deviation.
Unstable process—state of your process when both common cause and special cause variation are present. An unstable process is, by definition, unpredictable.
x-bar ($\bar{x}$)—used in mathematics to indicate the mean.

Introduction

Chapter 1 defined and discussed the definition of quality. One major component of the definition of quality is consistency.

Consistency is uniformity or agreement among things or parts. Customers like consistency. They want a product to look and act the same way every time they buy it. The opposite of consistency is variability.

Variability is observable differences or characteristics between seemingly similar products. Variations in product quality or variation in the service provided is not a good thing from a customer perspective.

Think about the stores you have been to where the service is inconsistent—where you sometimes get great service and other times there is no one there who can help you. Now think about those stores where you have consistently good service every time. Chances are you will be much more inclined to return to stores with consistent service instead of those where service is inconsistent or variable.

FIGURE 2-1 Sample Histogram

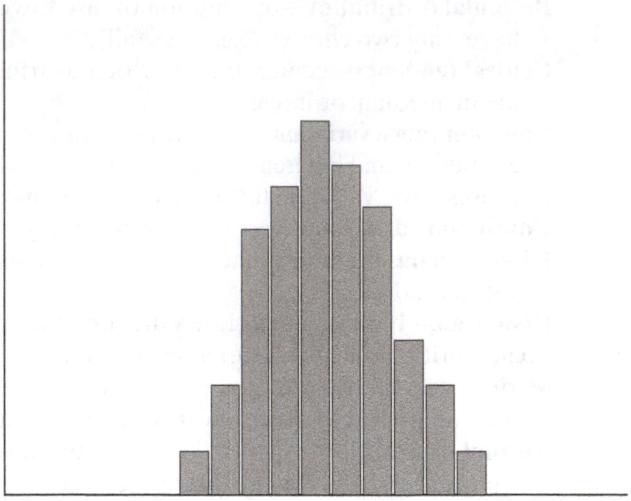

This chapter explores the concept of variability and proposes some ways to measure and define variation in terms that you can use to improve your processes. The graphic technique that will be employed to represent variability in the process is called a histogram. A **histogram** is a bar chart that shows the distribution of a set of data. Figure 2-1 shows an example of a histogram.

Data distributions can take on many different shapes depending on the type of data that you are collecting and the stability of the process that is generating the data. This chapter explores several different types of data that are important in the processing industries and the types of distributions that best represent those data types.

Regardless of what type of distribution you are dealing with, there are two fundamental ways of describing a distribution using mathematics: the central tendency and the spread.

You will learn how to measure each of these important concepts and how to use these measures to analyze your processes to determine when the process is stable, exhibiting only common cause variation, versus unstable, exhibiting special cause variation.

The specific branch of mathematics that helps in understanding variation and distributions is called statistics. Statistics can be confusing for some, but do not be alarmed. This chapter approaches statistics from the standpoint of making it as simple as possible to use for improving your processes, not from the standpoint of making statisticians out of you.

The concepts introduced in this chapter are the foundation for what will be explored in the next few chapters. It is important to take the time to make certain that you understand these basics.

This is the first of several chapters on statistics, which is one of the most important pieces of the quality puzzle.

Variability

Everything varies. This is so true that it might be considered a law of nature. If you find something that does not appear to vary, it probably means the variation is so small that you cannot measure it; but it does vary.

Common examples that everybody is familiar with include snowflakes and fingerprints.

The forces of nature that create snowflakes have never been found to create exactly the same crystalline structure. This can be easily understood when you consider the temperature and moisture in two different places. On the other hand, consider snowflakes created at the same place on the same day at the same time—they are still different.

There are billions of people in the world and yet no two fingerprints have ever been found to be the same. When we consider that these people have different parents, different genes, and different environments, we can accept that they will have different fingerprints.

Think about siblings that have the same parents, are raised in the same environment, and come from the same gene pool; they still have different fingerprints. Even though identical twins may be so similar that only those closest to them can tell them apart, they are different people with different personalities and different fingerprints.

The products that are sold also vary. Many things contribute to the variability of these products. The raw materials that are used vary. This should not come as a big surprise because our raw materials are just somebody else's product. The process used to convert raw materials into a new product varies also.

Think about your family car. You buy gas from the same station and burn that gas in the same car, but your gas mileage varies from tank to tank. Some of this variation could be due to different loads of gasoline in the station's storage tank, but it could also be due to the driving conditions experienced.

The most extreme example would be a full tank of only highway driving versus a full tank of only stop-and-go driving in the city. Even two tanks of stop-and-go driving will give you two different results depending on how much stopping and going you do.

Even if the discussion is limited to only highway driving and using the exact same stretch of road, the car itself has some inherent (built-in) variability. By the time you

FIGURE 2-2 Breaking of Paper Clip

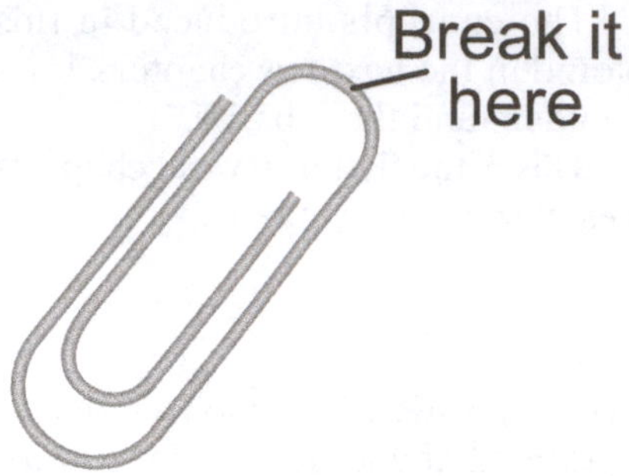

start your second trip to demonstrate that the car gets the same mileage every time, the tires are a little more worn and the spark plugs have more hours of use on them.

The process technician (the person behind the wheel) also plays a role in this variation. In this case, the process technician can easily affect the results by how he or she operates the vehicle.

Statistical quality control (SQC) and statistical process control (SPC) are built on the premise that all processes have inherent variation and that this variation does the following:

- Behaves in a predictable manner
- Is stable over time
- Is measurable
- Is random by nature

Once you accept these truths you can begin your work to improve your processes. Your job is to quantify the variation exhibited by your process. Let us create a process, run the process to collect some data, and then plot the data to demonstrate the journey of quality improvement.

In your plant, paper clips are purchased as raw materials. Your process is to break the paper clips in half in order to sell the parts to your customers.

Gather 50 paper clips. Notice that one side of the paper clip has two curved portions; the other side has one curved portion. Bend the paper clip back and forth and break the paper clip at the curve on the side that has just one curve. See Figure 2-2. Count the number of bends it takes to break the paper clip.

On the chart in Figure 2-3, color in the box along the bottom of the graph (X-axis) that corresponds to the number of bends required to break the first paper clip. In the case of the example, the first paper clip broke in 6 bends, so the box above the 6 is colored.

Break the second paper clip. Record the results on the chart in Figure 2-3. Repeat this process for the remaining paper clips.

After this same process was run for 100 paper clips, the chart seen in Figure 2-4 was generated.

Did You Know?

The word *data* is plural? The proper word for a single number is *datum*. That is why we use phrases such as "the data are distributed" instead of "the data is distributed."

FIGURE 2-3 Paper Clip Blank Histogram

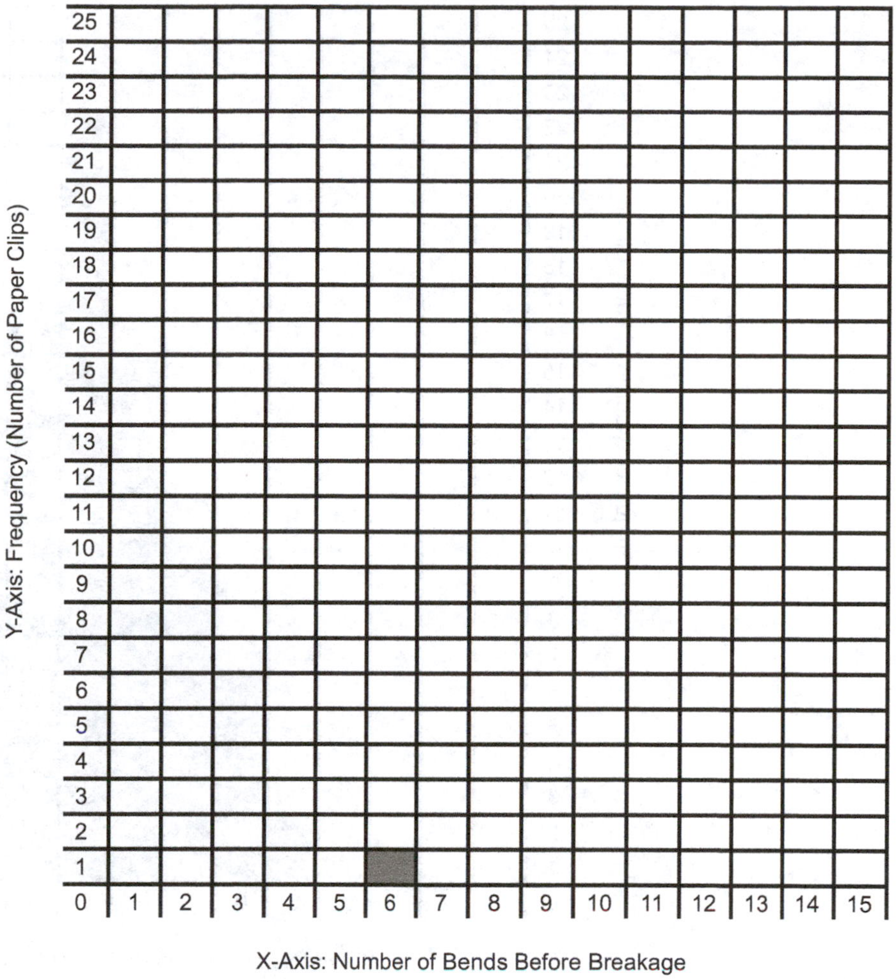

The charts that you made are called histograms. Histograms are just bar charts showing how frequently each result is obtained from a process. These histograms show how the data are distributed. These distributions are the foundation upon which SQC and SPC are founded.

If you take samples of data from your process periodically and plot the results in a histogram, assuming your process has not changed, the results would look similar—not exactly the same, but similar.

Your chart will not look exactly like the example in Figure 2-4, nor will it match exactly the charts of others in your class. Some examples of process variations that might contribute to the variability seen in your final product include the following:

- Speed of bending might heat up the metal and make the paper clips break faster.
- Degree of bending might vary.
- Bending straight versus bending with a twist would affect results.
- Different students would no doubt have different paper clips of different sizes and from different manufacturers.

Variability might be reduced if you were to implement a form of process training. Each person could work close to the same speed and bend his or her paper clips the same way. If there were only one bender (one process technician), that factor would be eliminated.

You could also make sure that all of your paper clips came from the same box to eliminate the size and composition variables. Of course, if you are going to train people

FIGURE 2-4 Example of Paper Clip Histogram

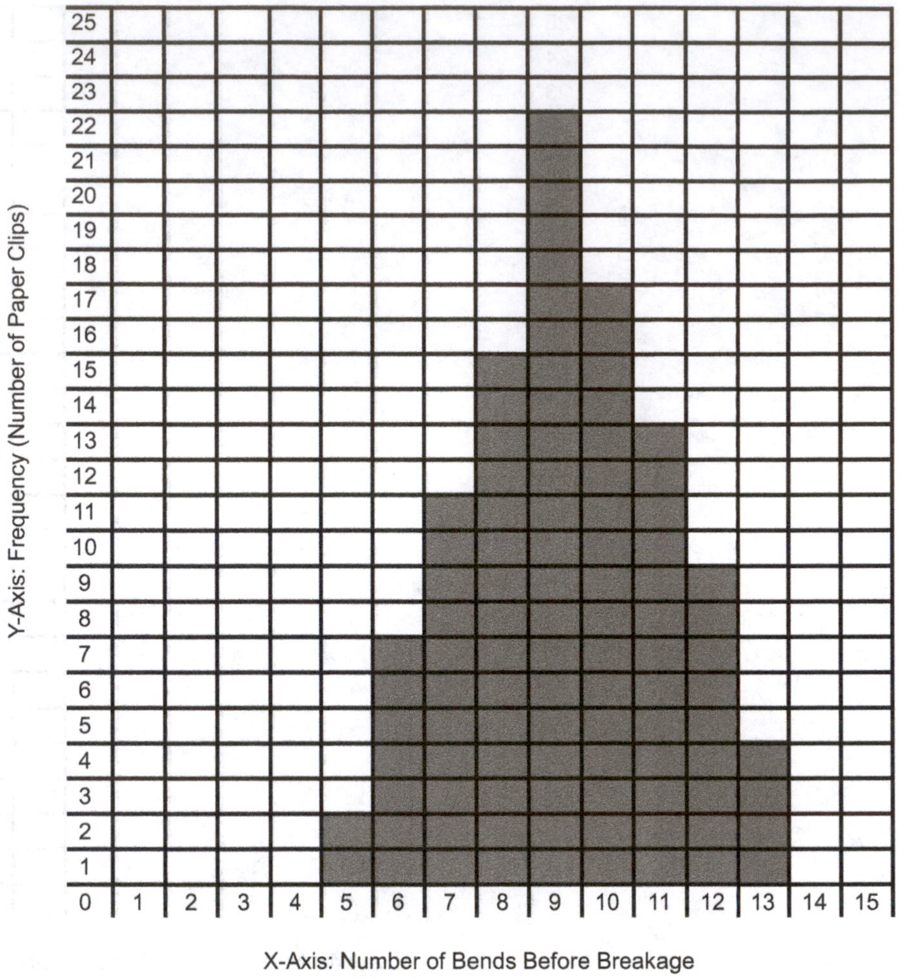

to operate the process the same way every time, you are going to need a procedure to train them against.

Customer specifications will determine the proper procedure. You will have to analyze your process to determine the best way to run the process in order to give the customer the desired product consistently.

In order to produce a consistent product, you need to run your processes in a consistent fashion. The purpose of documented work procedures is to define the best way of accomplishing a specific task so that everyone can do it the same way every time.

You might hear them called operating discipline, standard operating procedures (SOP), work instructions, or dozens of other names. They all have the same purpose, which is to document the best way to get the work done to ensure a consistent result.

Stories about process technicians who keep personal notebooks in their lockers in order to maintain a library of personal best practices are legendary. From their actions you would think that A shift is in competition with B shift, which is in competition with C shift, to see who can perform the best. These kinds of work practices are antiquality.

A shift, B shift, and C shift are not in competition with each other. Your company is in competition with the other companies that supply the same products to the marketplace.

The various shifts within your company have to work together to make the most salable product possible. If one shift comes up with a procedure that is better than the one currently documented, then by all means change the documentation so that everybody can employ the improved technique.

Consistent operation means a more consistent product, which means customer satisfaction, staying in business, and ultimately protecting our employment. We all have to work from the same book.

Central Tendency

There are two main measures or statistics used to describe a distribution: central tendency and spread (shown in Figure 2-5). Let us examine each in a little more detail.

The first measure is central tendency. **Central tendency** is the center or middle of a distribution of data; it is measured as the mean, median, or mode.

1. Mean–the arithmetic average of a set of data; calculated by adding up the individual numbers and dividing by the total number of data in a data set.

2. Median–the geometric center of a data set; determined by organizing data in numerical order and selecting the number that is in the center of the list. Literally it is

FIGURE 2-5 Spread and Central Tendency

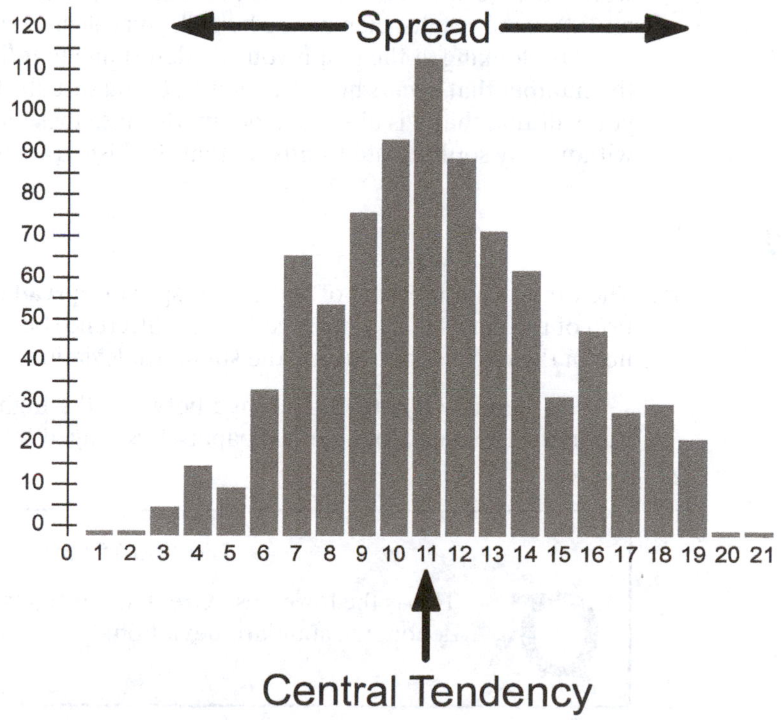

FIGURE 2-6 Mean Formula

$$\bar{x} = \frac{\Sigma x}{n}$$

the one in the middle. This measure is common in governmental statistics when referring to the median income or median home price.

3. Mode—the number that occurs most frequently in a data set. This measure of central tendency is not frequently used.

The measure that most of us are familiar with is the average or mean value. This is calculated by adding up the individual numbers and dividing by the total number of data in a data set. In order to write out the formula for this equation, you need to know a couple of terms:

- **x-bar**—the mean or average
- ***x***—the individual measurement
- ***n***—the number of measurements
- **Σ**—the uppercase Greek letter sigma; used in mathematics to indicate a summation

The mathematical formula for the mean is shown in Figure 2-6. It is read as x-bar equals the sum of the individual *x*s divided by the number (n) of *x*s.

Let us take the following five numbers: 2, 3, 9, 9, and 7

Add them up ($2 + 3 + 9 + 9 + 7 = 30$), then divide by 5 since there are 5 numbers. 30 divided by 5 equals 6, so our mean, or x-bar, is 6.

If you put these same 5 numbers in order and choose the middle number, you conclude that the median is 7. The mode is 9 because it is the number that occurs most frequently.

It is not important to know why, but when these three measures of the central tendency are close, it indicates that the data are normally distributed. When these three measures are not close, it indicates that the data are not normally distributed. The next section covers more about normal distributions.

Let us look again at the histogram of 100 broken paper clips in Figure 2-4. There were 2 results of 5 bends, 7 results of 6 bends, 11 results of 7 bends, and so on. If you add all 100 of the results, then divide by 100, you calculate a mean of 9.23.

By looking at the graph you can determine that the mode is 9 because 9 bends is the number that occurs most frequently. If you sort all 100 numbers in numerical order, you will find that 9 is also the median. Because these numbers are close, we can guess without any sophisticated statistics that the histogram is normally distributed.

Spread

The other main measure of statistics is spread. **Spread** is the broadness of the distribution of the data set; it is measured as the difference between the highest and lowest values in the data set or by using the standard deviation.

1. Range (R)—the difference between the high and low values in a group of numbers. The data from the 100 paper clips ranged from 5 to 13. Your histogram may

Did You Know?

σ This is the lowercase Greek letter sigma. It is used in mathematics to denote the standard deviation.

FIGURE 2-7 Standard Deviation Formula

$$\sigma = \sqrt{\frac{\sum (x - \bar{x})^2}{n - 1}}$$

have had a similar range or, if you used a different size or type of paper clip or a different technique for bending, could have been different.

2. Standard deviation—a statistical measure of the spread of a set of data. It is represented mathematically by the lowercase Greek letter sigma **(σ)**. It is commonly referred to as the sigma of your process.

The formula for the standard deviation is shown in Figure 2-7. An example calculation for this formula is shown in Chapter 4, where it is discussed in further detail. When in the field, you will not have to calculate this number by hand. There are certain tasks that are made for computers, and the calculation of standard deviations is one such task. You just need to know what it means and how to use it to improve your process.

The standard deviation is calculated using the formula shown in Figure 2-7 by doing the following:

1. Subtract the average (x-bar) from each individual measurement (x).
2. Square this value and then sum up all the squares.
3. Divide this sum by the number of measurements you have (n) minus 1.
4. Take the square root of the entire value.

Using statistical software, you can put in a formula to calculate the standard deviation of a set of data quickly and easily. Many inexpensive calculators have statistical functions built in to calculate standard deviation.

When your data fit under a bell-shaped curve, it means your data are normally distributed; this situation is commonly referred to as a **normal distribution** (or Boltzman distribution). An example of a normal distribution is shown in Figure 2-8.

FIGURE 2-8 Normal Distribution

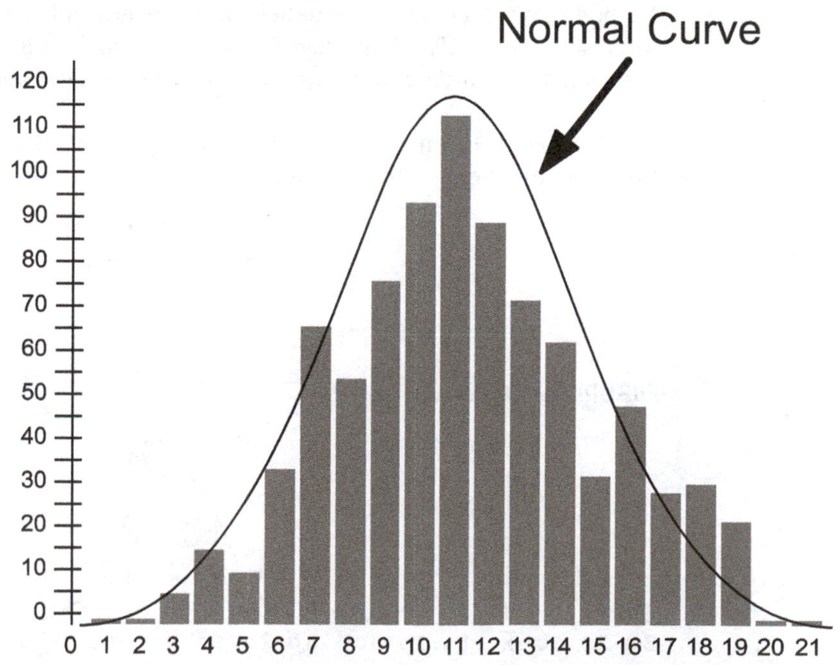

FIGURE 2-9 Standard Deviation Graph

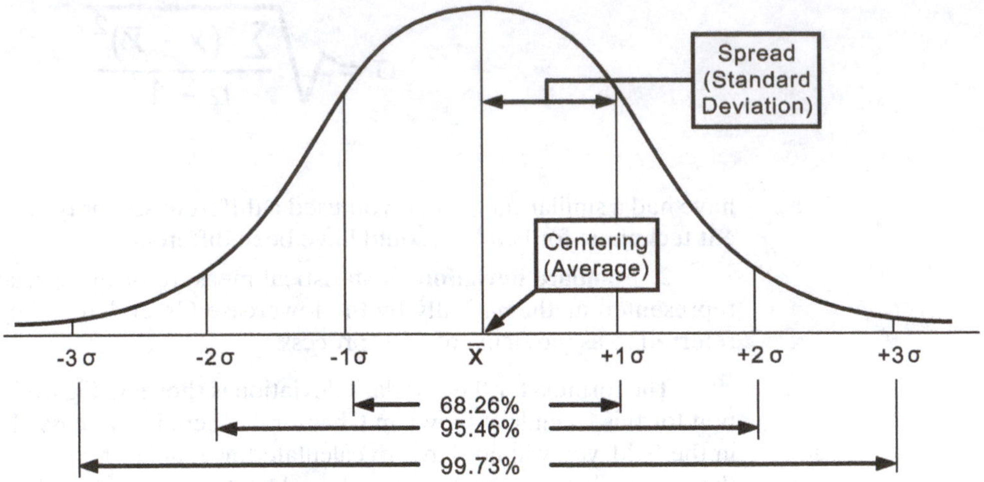

The standard deviation (Figure 2-9) can be used to ascertain how much of the data is gathered close to the center of the curve and how much is distributed out toward the edge of the curve. It can even indicate how far from the center the edges should be.

For a normal distribution, approximately 68% of the data fall within 1 standard deviation of the mean, approximately 95% of the data fall within 2 standard deviations of the mean, and approximately 99.7% of the data fall within 3 standard deviations of the mean, as shown in Figure 2-10.

Let us consider the case of our paper clip breaking example. We calculated a mean of about 9. It so happens that this distribution has a standard deviation of approximately 1.5. Using the information just provided, we could calculate that 68% of our data should fall within 1.5 units of 9.

$$9 + 1.5 = 10.5$$
$$9 - 1.5 = 7.5$$

We would conclude that 68% of the time our paper clips should break between 7.5 and 10.5 bends. We did not capture data in partial bends, but an examination of our histogram would show that 53% of our data fell between 8 and 10, and 77% fell between 7 and 11; it appears that our data fit the expected result of 68% within 1 standard deviation of the mean.

Let us go the next level. Ninety-five percent of our data should fall within 2 standard deviations of the average.

$$9 + (2 \times 1.5) = 12$$
$$9 - (2 \times 1.5) = 6$$

FIGURE 2-10 Deviation Graphs

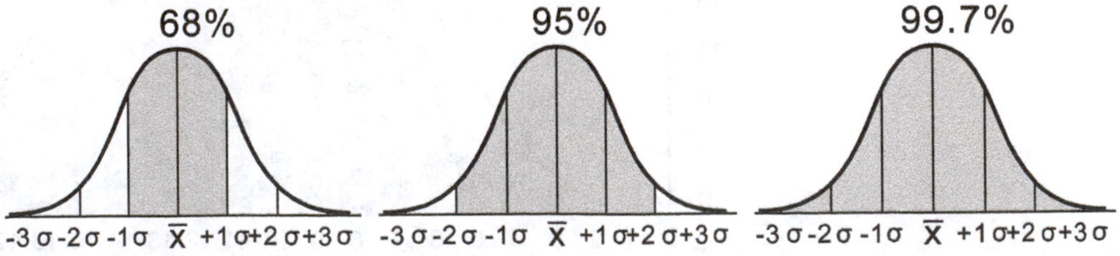

According to our actual data, 94% of the paper clips broke between 6 and 12 bends. Finally, we check to see how we did at 3 standard deviations from the mean:

$$9 + (3 \times 1.5) = 13.5$$
$$9 - (3 \times 1.5) = 4.5$$

Our data all ranged from 5 to 13, so 100% of our paper clips broke within 3 standard deviations of the mean. We expected 99.7% of them to break in that range, so it appears that our actual data fit what we would expect from a normal distribution. There are more accurate tools for determining whether our process is normally distributed, but as a quick check this works well.

Understanding the concept of how much of your data falls within a certain sigma range of the mean is critical to the quality improvement effort.

In checking the paper clip example against the expected norms, we found that our data fit. Knowing that your data fit a normal distribution and knowing how to describe that distribution using the mean and standard deviation mean you can predict the future.

If you are into gambling, you will like these odds. If you bet that the paper clip will break in 6 to 12 bends, you will win 95% of the time. If you bet that the paper clip will bend in 7.5 to 10.5 bends, you will win 68% of the time.

Assuming your process is normally distributed, and you will be shocked how many are, and as long as nothing strange is happening, like a pump breaking down or a line springing a leak, you can predict how your process will perform in the future.

COMMON CAUSE VERSUS SPECIAL CAUSE

When your process is performing normally, and the only variation evident is the random variation that is inherent in the process, you are seeing **common cause variation.**

When only common cause variation is present, expect to see 95% of your data within 2 standard deviations of the mean. The converse of that is to expect to see data as far away as 2 standard deviations of the mean only 5% of the time.

If you see more than 5% of the data 2 standard deviations away from the mean, it is a good indication that something is wrong. You may not know immediately what has changed, but you can be confident that something has changed.

The changes that cause your process to look different than you expect might be the shutdown of a pump, a leak in a pipeline, an instrument failure, or an electrical problem, just to name a few. These events are called special causes (or assignable causes) because they cause the process to vary outside of its normal operation. **Special cause variation,** or **assignable cause variation,** represents an abnormal variation that can be assigned to a specific event.

Later chapters will examine in more depth the kinds of indications that the data may provide to let you know when something is wrong with your process.

STABLE PROCESSES

The process is called stable when it looks exactly the same every time and is considered predictable. A **stable process** is the state of your process when only common cause variation is present; this is also stated as the absence of special or assignable causes.

When common cause variation is present, you know that whatever your process produces today is the same thing it will produce tomorrow and the day after that and the day after that. This is the desired state.

Given the representation of a process in Figure 2-11, it is possible to predict what the outcome from this process will be on day 4. This consistency is exactly what the customer desires. Customers want assurance that what they get tomorrow will look and act just like what they got yesterday and last month.

The process is called unstable when it looks different from time to time and is considered unpredictable. An **unstable process** is the state of your process when both common cause and special cause variation are present. When this situation occurs, it is uncertain whether what the process produces today will be the same thing it will produce tomorrow.

FIGURE 2-11 Stable Process

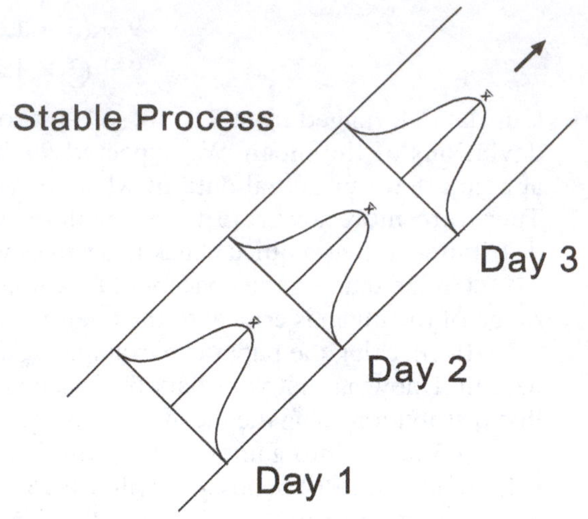

FIGURE 2-12 Unstable Process

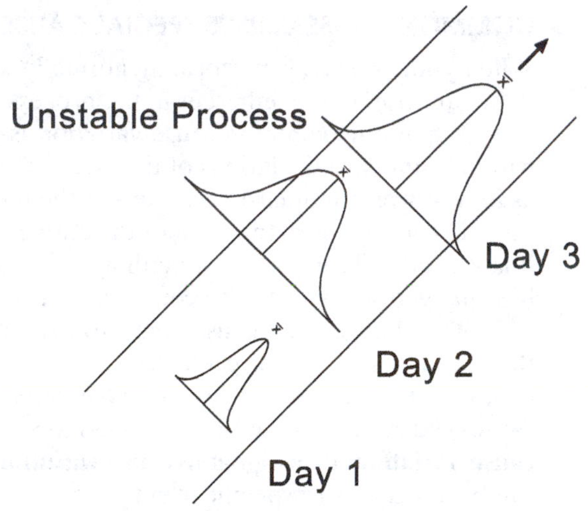

Given the representation of a process in Figure 2-12, it is almost impossible to predict what the outcome from this process will be on day 4. From the customer's perspective, your process is unstable, unpredictable, and undesirable.

Your customers will not know until they receive and try to use your product how it might behave in their process. This type of performance could cause you to lose business if your competitors are able to produce a stable product.

An unstable process may be evidence of special cause or assignable cause variation.

Types of Distributions

The discussion so far has been about normal distributions, but there are many other types of distributions. Each one has its own purpose and value. In the processing industries three types of distributions are encountered routinely.

The first is obviously the normal distribution (shown in Figure 2-13), which forms the foundation for much of the quality improvement work. Normal distributions are made up of **continuous data,** numbers that can be of any value or fraction of a value.

FIGURE 2-13 Normal Distribution

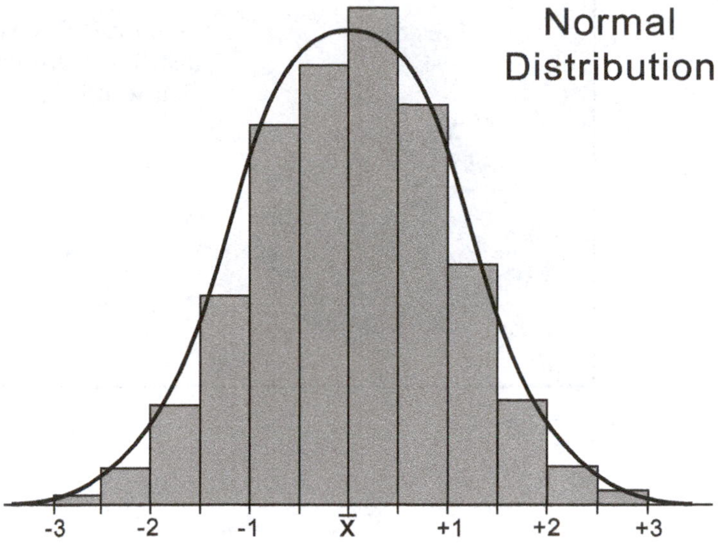

Given that description, you should have not used a normal distribution for your paper clip exercise because you did not capture fractional data. Chapter 4 will focus on how to use control charts based on the normal distribution to distinguish common cause variability from special cause variability. This will show when to investigate your process for problems.

Another common data type that process technicians encounter in the process industries is counting data. Counting data are **discrete data** (variables data), which are numbers that are not continuous—for example, whole numbers.

Examples of counting data include spills, injuries, illnesses, and anything that is a count of events. Let us consider injuries for just a minute. You cannot have 1.5 injuries. You either had 1 injury or you had 2 injuries.

Given enough data you might actually be able to apply a normal distribution to counting data, but the appropriate distribution would be the Poisson distribution (shown in Figure 2-14). A **Poisson distribution** is a distribution of discrete data that is made up of counting data.

When you look at a Poisson distribution, you will not see a smooth curve, because there are no data in between the whole numbers. In Chapter 5 you will learn how to use control charts based on a Poisson distribution.

FIGURE 2-14 Poisson Distribution

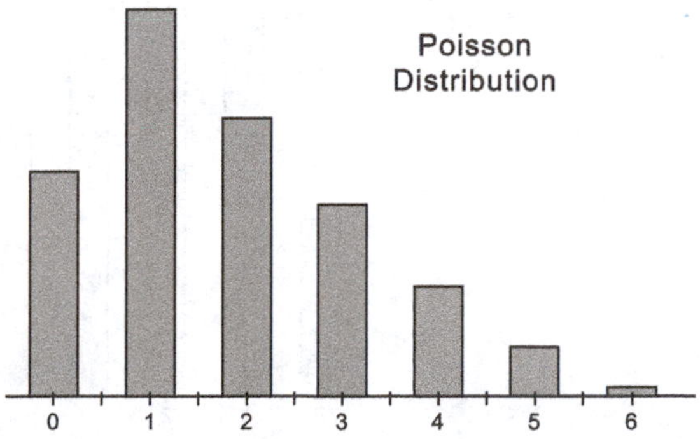

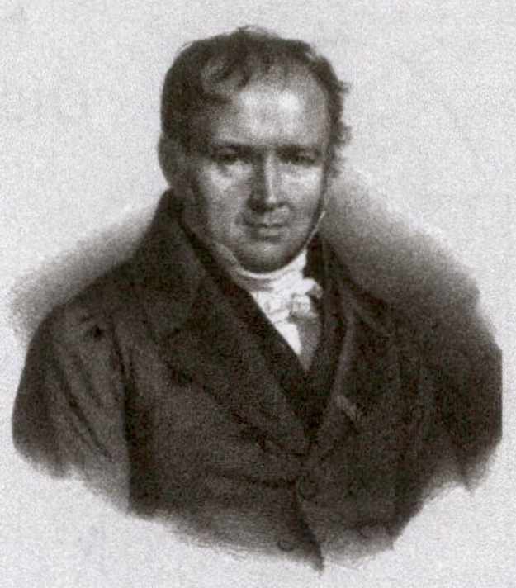

Did You Know?

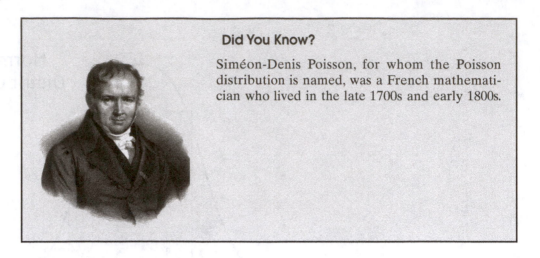

Siméon-Denis Poisson, for whom the Poisson distribution is named, was a French mathematician who lived in the late 1700s and early 1800s.

The last commonly used distribution discussed at any length here is the binomial distribution (shown in Figure 2-15). A **binomial distribution** is a distribution of discrete data that are made up of data that have only two choices. Binomial literally means "numbered or numbered in twos." Like Poisson data, binomial data are discrete, not continuous.

What makes it look a little confusing is that the binomial distribution counts the number of failures, so the binomial distribution ends up looking a lot like the Poisson distribution. Further discussion about how to use control charts based on the binomial distribution is in Chapter 5.

Poisson and binomial distributed data are both referred to as attribute data. Rather than measuring some variable such as length or weight of the product, these distributions attempt to measure a more generic attribute about the product such as number of defects on the product or the number of defective products produced.

As mentioned earlier, there are many other types of distributions, a few of which are shown in Figure 2-16, but these will not be explored any further in this book.

FIGURE 2-15 Binomial Distribution

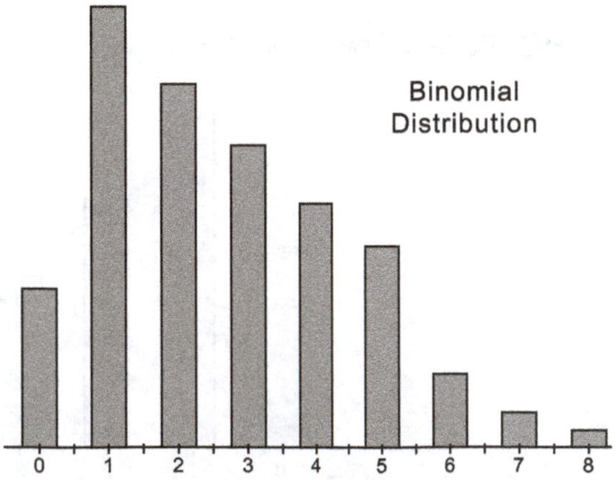

FIGURE 2-16 Other Distributions

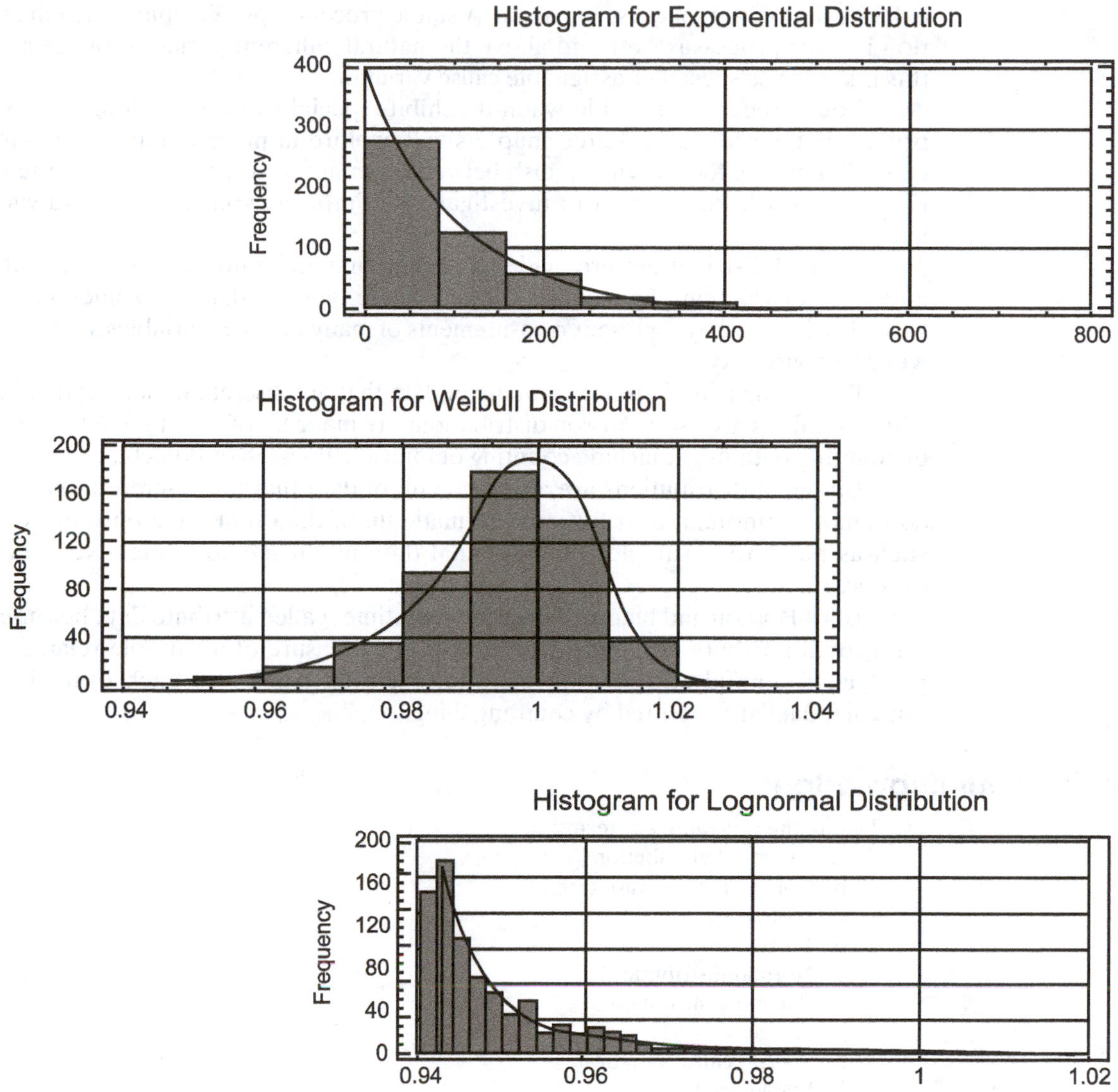

Summary

Consistency is a major part of the definition of quality. The opposite of consistency is variability. Because everything varies, variation is a natural part of all processes. By understanding the variability of your process and how that variation is distributed, you can use past data to predict future performance and use the data to be alerted when something unusual is happening. Histograms are an excellent graphic tool to help in understanding the distribution of your data.

The distribution of your data can be described using two key measures:

a. measures of central tendency–the most common being the mean or average
b. measures of spread–the most common being the standard deviation or sigma (σ)

The mean indicates where on the scale your process is centered, and the standard deviation indicates how much of the scale your process encompasses. For a normally distributed process, you can expect to find the following:

a. Approximately 68% of the data within 1 sigma of the mean
b. Approximately 95% of the data within 2 sigma of the mean and
c. Approximately 99.7% of the data within 3 sigma of the mean

When the only variation seen in your process is that variation that is inherent in the process, this is known as common cause variation. Your process is stable when it exhibits only common cause variation. A stable process is predictable. When the variation in your process is over and above the natural, inherent variation of the process, this is known as special or assignable cause variation.

Your process is unstable when it exhibits special cause variation. An unstable process is unpredictable. Later chapters will explore in more detail how to employ control charts as tools to distinguish between common cause and special cause variation so you will know when to investigate and improve your process and when to leave it alone.

Normal distributions are made up of data that are continuous in nature; that is, any value or fraction of a value is allowed. These types of data are sometimes called variables data as they represent measurements of many kinds of variables such as purity, weight, or viscosity.

Poisson distributions are made up of data that are discrete in nature; that is, only whole numbers are used. Poisson distributions are made up of counting data. Examples of counting data might include counting of injuries, illnesses, or defects.

Binomial distributions are also made up of data that are discrete in nature, not continuous. Binomial distributions are made up of data that have only two options such as pass/fail. Examples of binominal data might include defective data on a production line.

Both Poisson and binomial data are sometimes called attribute data because they refer to an attribute of the product instead of a measure of a variable related to the product. Binomial and Poisson distributions look similar to each other because binomial data are captured by counting things.

Checking Your Knowledge

1. Define the following key terms:
    a. Binomial distribution
    b. Common cause variation
    c. Histogram
    d. Mean
    e. Normal distribution
    f. Poisson distribution
    g. Range
    h. Special cause variation
    i. Stable process
    j. Standard deviation
    k. Unstable process
2. (*True or False*) Variability is the opposite of consistency.
3. Histograms are graphical tools used to:
    a. Provide justification for capital projects
    b. Display the distribution of data in a tabular format
    c. Display the distribution of data in a bar chart format
    d. Provide management with information about shift performance
4. Inherent variation:
    a. Behaves in a predictable manner
    b. Is stable over time
    c. Is measurable
    d. Is random by nature
    e. All of the above
5. (*True or False*) Consistent operation is required if consistent product quality is desired.
6. Choose the options below that can help ensure consistent operation (select all that apply):
    a. Standardized procedures
    b. Training
    c. Cheat sheets posted on the walls
    d. A degree in process technology
    e. Application of statistical tools to recognize special cause variation

7. (*True or False*) There are three key measures to understanding your histograms.
8. List three common measures of central tendency and circle the one most commonly applied:
   a. ________________
   b. ________________
   c. ________________
9. What is the most commonly used measure of spread?
   a. Range
   b. Standard deviation
   c. Variance
10. Which of these Greek letters is used to denote the standard deviation?
    a. Σ c. σ
    b. μ d. Ω
11. Which of these Greek letters is used to denote the summation?
    a. Σ c. σ
    b. μ d. Ω
12. In a normal distribution, how much of the data is encompassed within 1 standard deviation of the mean?
    a. 33% c. 95%
    b. 68% d. 99.7%
13. In a normal distribution, how much of the data is encompassed within 2 standard deviations of the mean?
    a. 33% c. 95%
    b. 68% d. 99.7%
14. In a normal distribution, how much of the data is encompassed within 3 standard deviations of the mean?
    a. 33% c. 95%
    b. 68% d. 99.7%
15. A distribution of counting data is called:
    a. Boltzman c. Poisson
    b. Normal d. Binomial
16. A distribution of pass/fail data is called:
    a. Normal c. Binomial
    b. Poisson d. Boltzman
17. (*True or False*) A process that exhibits only common cause variation is said to be stable.
18. (*True or False*) A process that exhibits common cause and special cause variation is said to be stable.

Activity

1. Obtain 50 paper clips and complete the exercise outlined in the chapter, including the charting of the histogram. Be prepared to share your results in classroom discussion.

CHAPTER 3

Process Capability

"Quality is not the exclusive province of engineering, manufacturing, or, for that matter, services, marketing, or administration. Quality is truly everyone's job."

~JOHN R. OPEL (IBM)

Objectives

Upon completion of this chapter you will be able to:

- Understand how process data relate to the needs of the customer.
- Calculate key process capability indexes (Cp, Cpk, and Ppk), and explain what each index is telling you.
- List and explain potential sources of process variability.
- Explain the concept of nested designs and how nested designs can be used to segregate the total variation into separate components (e.g., sampling and measuring variability).
- Explain the concept of measurement system variation and the importance of using the right tool for the job.
- Explain the importance of a representative sample.
- Explain the concept of overcontrol and its impact on process capability.

Key Terms

Analysis of variance (ANOVA)—a statistical tool that helps you understand the nature of variation in your process.

Lower specification limit (LSL)—the lowest level allowed by the customer for a specific quality measure.

Measurement capability index (Cm)—the ratio of your specification band to your measurement process variability.

Nested design—a statistical tool used to segregate total process variation into separate components. Also referred to as a component analysis of variance.

Process capability index (Cpk)—the ratio of the customer's specification range to your process variability taking the average (process center) into account; this index assumes the data have been "cleansed" of special causes.

Process performance index (Ppk)—the same calculation as the Cpk but without the assumption that the data have been cleansed of special causes.

Process potential index (Cp)—the simple ratio of the customer's specification range to your normal process variability without taking the average (process center) into account.

Upper specification limit (USL)—the highest level allowed by the customer for a specific quality measure.

Introduction

Now you understand that your processes vary. You know how to display the variability of your process graphically with a histogram. You know how to measure that variation mathematically and describe the distribution of your data by using the mean to define the center of the distribution and the standard deviation or sigma to define the spread. Your customers tell you that variation is not a good thing, and they desire consistency in the products they buy.

In this chapter, you will learn how to relate the distribution of your data to the needs of the customer. Completely eliminating all variation from your processes and thus your product is not possible. The question is not, How do I get rid of all variation? The question instead is, How much variation can the customer accept?

There is a practical limit to your ability to measure variation. There is also a certain amount of variation that your customers can tolerate without negatively impacting their process. You will calculate several indexes that will offer the ability to judge the adequacy of your process variability compared to what the customer needs. These include the process potential index (Cp), process capability index (Cpk), and process performance index (Ppk).

Once you can make a judgment about the acceptability of your process's variation, you can begin examining potential sources of variation. Sometimes different people on different shifts have different ways of accomplishing the same task. This can lead only to an increase in variation. If you accept that all processes vary, then you can extend your concept of a process to include the process of taking a sample, the process of analyzing your product, and even the process of storing your product.

Measurement variation is such an important topic that it has its own index called the measurement capability index (Cm). This chapter looks at each of these potential sources of variability in more detail and explains the importance of understanding the contribution each one makes to the total variability that the customer sees. One additional potential source of variability that will be explored is the concept of overcontrol.

Finally, this chapter discusses a statistical tool called a nested design. This tool is specifically designed to analyze data that separate the total variation into components so you know where to apply your improvement efforts. The data have to be collected in a specific designed fashion in order to use this tool.

Learning how to design these experiments and analyze the data will have to wait for a more advanced course, but knowing that the tool is out there, that it is being used, and that process technicians are the mechanism for gathering the data makes it important for you to understand what the tool does and how it helps make your processes better.

FIGURE 3-1 Example of the Relationship of Process Spread to Specification Limits

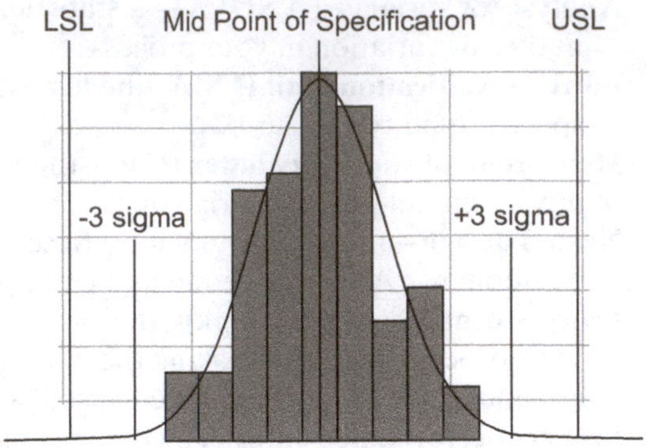

Process Capability Discussion

In the previous chapter you learned that for a normal distribution 99.7% of all the data are expected to be within 3 standard deviations (sigma, or σ) of the mean. With a 3 sigma range above the mean and a 3 sigma range below the mean, the total spread of your distribution is 6 standard deviations wide. This tells how much variability you have in your process.

Now you want to relate that variability to what is important to the customer. Most variables that are important to the customer will have a range of acceptability defined, typically called the tolerance, specification (spec), or specification range. The upper end of the specification range is referred to as the **upper specification limit (USL),** which is the highest level allowed by the customer for a specific quality measure. The lower end of the specification range is referred to as the **lower specification limit (LSL),** which is the lowest level allowed by the customer for a specific quality measure. These numbers define how much variability the customer is willing to accept.

Keep in mind that just because the customer accepts this much variability does not mean the customer wants this much variability. One way of comparing your process variability to the customer's needs would be to add the customer's specification limits to your histogram (Figure 3-1). This quick visual will tell you and the customer whether you can expect your process to meet the customer's needs. In this figure the +3 sigma and −3 sigma limits represent what is normal for your process. The USL and LSL limits indicate what the customer is willing to tolerate. Because it appears that your process falls well within the customer's tolerance limits, this process seems to be acceptable. What you need now is a simpler way of determining whether your process is acceptable.

Process Potential Index—Cp

The process potential index (Cp) was developed as a mechanism to provide a single number that would indicate whether your process is capable of meeting the customers' needs. The **process potential index (Cp)** is a simple ratio of the customer's specification range to your normal process variability range, with the normal process variability being defined as ±3 sigma, or 6 times sigma. The formula is as follows:

$$Cp = \frac{USL - LSL}{6 \times \sigma}$$

Logically, you want the process variability (the denominator) to be small compared to the tolerances (the numerator). If your process variability is greater than the customer's needs, it will result in a Cp index of less than 1.0; this indicates that some percentage of the time you are going to make product that will not meet the specifications.

FIGURE 3-2 Example of a Centered Process That Is Incapable of Meeting Customer Specifications

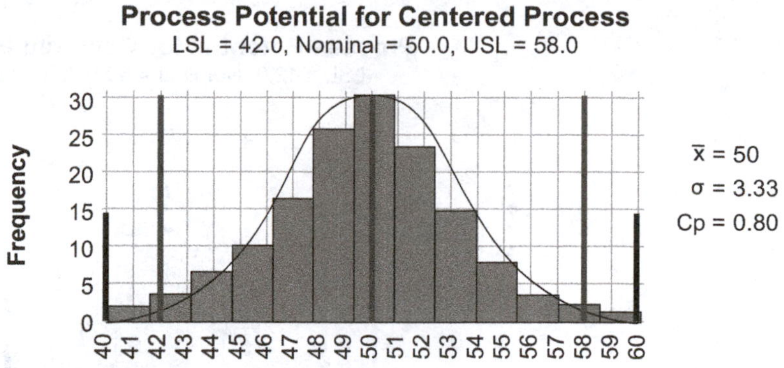

Figure 3-2 shows an example of a process. The specifications are set at 42 to 58 with a target, or nominal, value of 50. This process is centered around the nominal value but has a calculated standard deviation (σ) of 3.33, which means your process varies ± 10 ($3 \times \sigma$), but the customer desires a range of ± 8.

By replacing the terms in the formula (below) with the numbers provided, a Cp index of 0.8 is obtained. This index, being less than 1.0, tells you that your process will produce product a percentage of the time that does not meet the needs of the customer.

$$Cp = \frac{USL - LSL}{6 \times \sigma} = \frac{58 - 42}{6 \times 3.33} = \frac{16}{19.98} = 0.8$$

If the top and bottom of the equation were exactly the same, it would give you a Cp index of 1.0. This would mean that your product would normally vary within the boundaries of the customers' needs 99.7% of the time—which is pretty good for a lot of processes.

Although a Cp index of 1.0 is good (Figure 3-3), it does not leave you with any safety net or room for your process to wiggle. What you want is a process that fits well inside of the customer requirements.

If your process variability is less than the customers' needs, it will result in a Cp index of greater than 1.0 (Figure 3-4). This indicates that your process will virtually never make anything that will fail the specifications, which is exactly what you want.

From the customers' perspective, the higher the Cp the better they like it. The lower the Cp, the more variability your process exhibits in relation to their tolerances. By inverting the Cp and multiplying by 100, you can even calculate what percentage of the specification range your process takes up at different Cp indexes (shown below).

FIGURE 3-3 Example of a Centered Process That Is Marginally Capable of Meeting Customer Specifications but Has No Safety Net

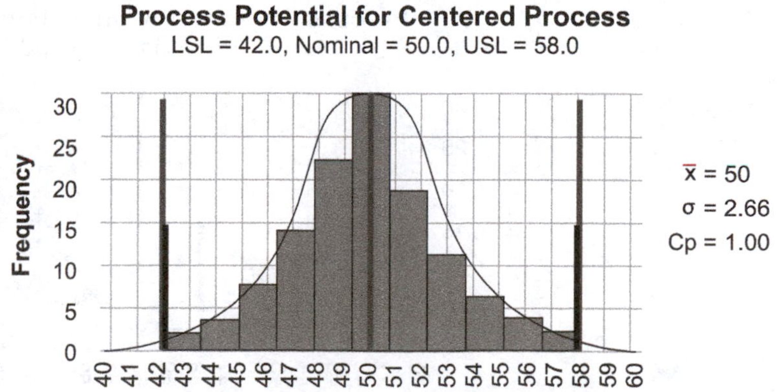

FIGURE 3-4 Example of a Centered Process That Is Capable of Meeting Customer Specifications with a Safety Distance on Both Sides

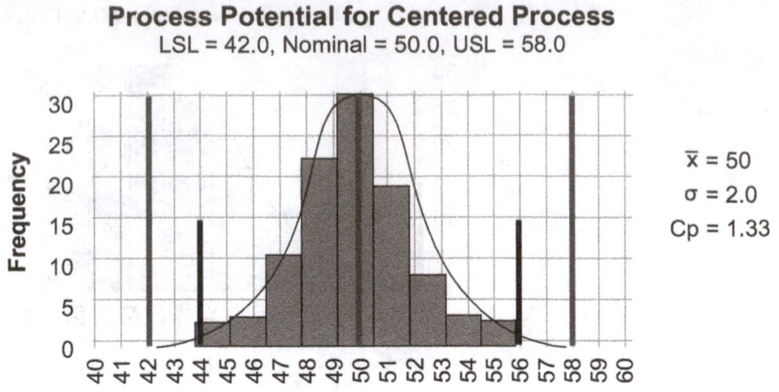

The customers want your product to use the minimum amount of their available specification range.

- A Cp of 2.00 (6 sigma) means 50% of the specification range.
- A Cp of 1.66 (5 sigma) means 60% of the specification range.
- A Cp of 1.33 (4 sigma) means 75% of the specification range.
- A Cp of 1.00 (3 sigma) means 100% of the specification range.
- A Cp of 0.66 (2 sigma) means 150% of the specification range.
- A Cp of 0.33 (1 sigma) means 300% of the specification range.

The problem with using this index is that it tells only what your process has the potential of doing if the process is perfectly centered within the specifications. Figure 3-4 is a process with a total specification range of 16 (USL − LSL = 58 − 42 = 16). The process variability is 6 times the sigma, which gives you a process spread of 12 ($6 \times \sigma = 6 \times 2 = 12$).

Using the formula for the Cp, you calculate the Cp index as $16 \div 12 = 1.33$. Your normal process variability is much less than the specifications. In this particular case there is a safety net of 1 standard deviation on each side of your process between the process and the specifications. A good Cp index tells you that your process is capable of meeting specifications as long it is running in the center of the range—in other words, that your process variability is within acceptable limits.

In Figure 3-5, there are the same numbers for your specification limits and the same process sigma, so you get the same Cp of 1.33. What is the problem? In this case your process is not centered at the middle or nominal value of the specification, which

FIGURE 3-5 Example of an Uncentered Process That Is Incapable of Meeting Customer Specifications

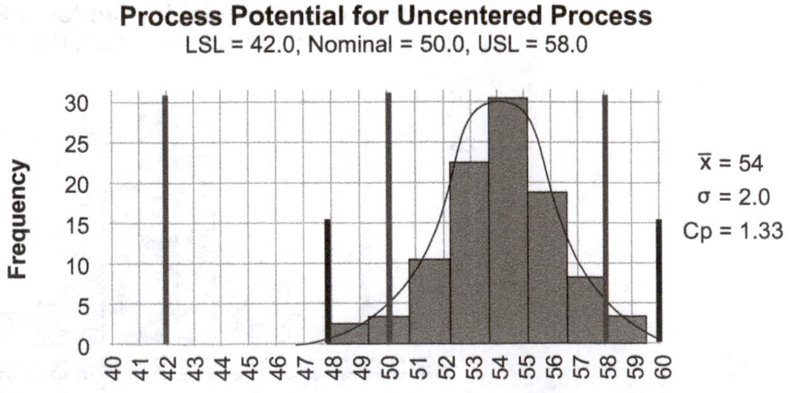

is 50. Instead your process is running at an average of 54. Because of this, some of your product will fail to meet the customer requirements.

The reason the Cp index is called the process potential index is that it tells how capable your process is if the process is perfectly centered within the specification range. Recall the two key measures of distributions of data—mean and sigma. The Cp index uses only the sigma in its judgment of acceptability, so it can only tell whether the spread of your data is good enough; it cannot tell whether your process is actually on target.

Process Capability Index—Cpk

The **process capability index (Cpk)** is the ratio of the customer's specification range to your process variability taking into account where your process is running in relation to the specification range. If your process is not centered, then it stands to reason that the process is running either to the left or to the right of the middle. The Cpk index is calculated by comparing the process to both the upper and the lower side of the specification, then reporting the worst-case scenario. Here are the formulas:

$$\text{Cpk(upper)} = \frac{\text{USL} - \bar{x}}{3 \times \sigma} \quad \text{or} \quad \text{Cpk(lower)} = \frac{\bar{x} - \text{LSL}}{3 \times \sigma}$$

The Cpk that you report is the minimum value of these two calculations. Let us try it out for Figure 3-5:

$$\text{Cpk(upper)} = \frac{58 - 54}{3 \times 2} = \frac{4}{6} = 0.67 \qquad \text{Cpk(lower)} = \frac{54 - 42}{3 \times 2} = \frac{12}{6} = 2.0$$

These values indicate that you have a process capability of 0.67 against the upper specification limit and a process capability of 2.0 against the lower specification limit. The interpretation of the Cpk index is the same as for the Cp index. A value of 1.0 is good. Values above 1.0 are better. Values below 1.0 are not good. You would then conclude that this process is capable of meeting the lower specification limit but not capable of meeting the upper specification limit. Because the value reported is the worse of the two, you would report your Cpk as 0.67 (Figure 3-6).

If the Cpk and the Cp calculations give exactly the same index, it indicates that your process is centered within the specifications. The Cp will always be equal to or greater than the Cpk because the Cp represents the best the Cpk can be if the process is perfectly centered. In this case, your Cpk is considerably lower than your Cp. You know that your process is not centered, and by looking at the Cpk calculations you know it is because your process is off center to the high side ($\bar{x} = 54$, but nominal or desired = 50). You also know that the total variation of your process is okay (because your Cp is greater than 1.0); if you could just shift your process down from 54 to 50, you would have a process that is capable of meeting all the customers' specifications all the time.

FIGURE 3-6 Example of an Uncentered Process That Is Incapable of Meeting Customer Specifications

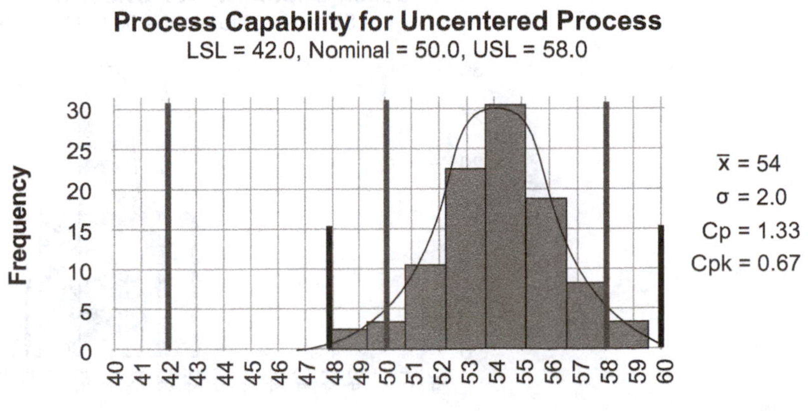

FIGURE 3-7 Example of an Uncentered Process Marginally Capable of Meeting Customer Specifications

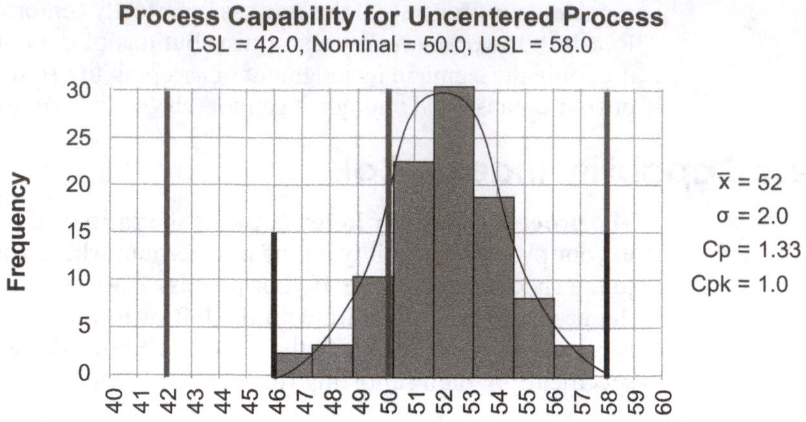

A Cpk of 1.0 (Figure 3-7) would indicate that your normal process variability, defined as ±3 σ from the mean, was bumping right up to one of the specification limits. If the Cp is greater than 1.0 for this process, then you know you just have to shift the process to the middle of the target range to achieve a capable process.

What your customers want to see is a process wherein the Cpk is well above 1.0, as shown in Figure 3-8. A Cpk index of 1.33 or greater is a common target in the processing industries. Mathematically this means that your specification limits are equal to 4 standard deviations of your process, but the math is less important right now than just knowing that above 1.0 is required and above 1.33 is desired.

Process Performance Index—Ppk

The Cpk index assumes that your process is stable and that there were no assignable causes in the data. Many times process technicians are compelled to use the data they have even though they know that there were "issues" with the process during the collection of the data. For this purpose, there is the process performance index (Ppk).

Many software packages provide an option to generate a Ppk index instead of a Cpk index. The **process performance index (Ppk)** calculations are identical to those of the Cpk index but without the assumption that the data have been cleansed of special causes. When you report the Ppk index, you are admitting that you just used the data

FIGURE 3-8 Example of an Uncentered Process That Is Capable of Meeting Customer Specifications

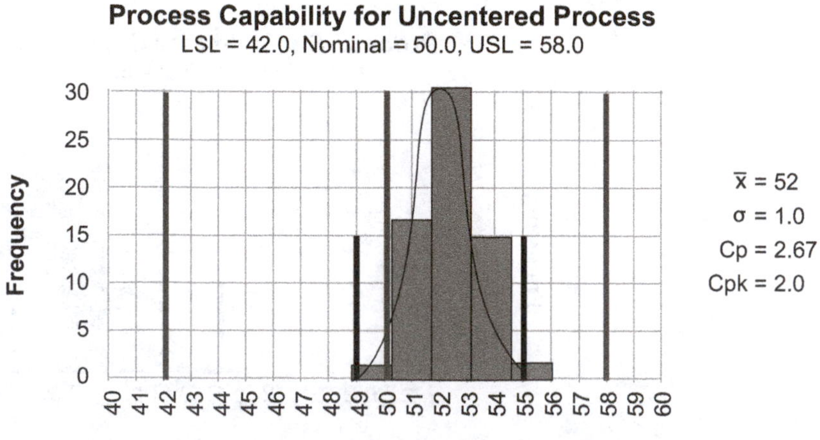

available and are acknowledging that the data may not be completely trustworthy. These data do not meet the rules of a process that is in control; for example, data are above or below the control limits.

When process technicians report the Cpk index, they are saying that the data have been cleansed (the process is in a statistical state of control) of all assignable (special) causes and that this index is truly reflective of what the process is capable of under the best of circumstances. Again there is no difference in calculations and no difference in interpretation of the results. There is only a difference in the assumptions about the integrity of the data.

CAPABILITY INDEX SUMMARY

1. The Cp index is the process potential index. It is the maximum your Cpk index can achieve. It does not take into account where your process is running (centered versus off center); consequently, it can be used only to judge how good the spread of your process is relative to the spread of the customers' specifications.

2. The Cpk index is the process capability index. It takes into account where your process is running in addition to how much spread there is in your process compared to the customers' specifications. The Cpk index assumes that your data have been cleansed of assignable causes and represent the capability of your process.

3. The Ppk index is the process performance index. It is calculated exactly the same as the Cpk but assumes that your data have not been cleansed of assignable causes and that the index is reflective of how your process is currently performing, not what it is capable of performing.

4. For all three of these indexes, values of 1.0 or greater are desired, with 1.33 being the industry standard target. Values below 1.0 are not desired, as this indicates that your processes are not capable of meeting the customers' needs.

5. For a perfectly centered process, the Cp and the Cpk are the same number.

6. For a process that is not centered within the specification range, the Cp is the maximum number that the Cpk could be if the process were to be shifted toward the center.

POTENTIAL SOURCES OF VARIATION

Now that you know how to tell whether your process is capable of meeting the needs of your customers, you need to put controls in place to monitor your process to ensure it stays capable and in control. You need to turn your attention to more pressing matters. Control charts are tools employed to monitor your processes for changes. The following chapters will talk more about the various types of control charts at your disposal.

For right now, let us talk about what to do if your process is not capable of meeting the needs of the customer. Let us assume that your process is stable, which means there are no special or assignable causes (just random variation), but your process variability either is not centered within the specification range or is too wide compared to the specification range. A process that is not centered within the range drives you to an easy conclusion: just make an adjustment and shift the process to the center of the range. But what if the variation is too wide? What can you do about a process that varies more than the customer can tolerate? The obvious answer is that you must attempt to reduce the variability, but how? How do you know where to apply your efforts? Some potential sources of variability may include the following:

- Shift-to-shift differences in operation
- Changes in raw material lots
- Seasonal variation (cooling water temperature can vary greatly from summer to winter, yielding different cooling capacities in the unit)
- Top of the storage tank to bottom of the storage tank differences
- Sampling variability
- Measurement system variability

How many more examples can you name?

Think of the total process variation that the customer sees as a block of variation. There are many components to this block of variation, and they all add up to a number that the customer cannot tolerate. Your job is to dissect this block into the individual components. Then you can see where to spend your time making improvements. Process technicians can choose to work on the easiest things to fix or on the things that add the most variation to the big block.

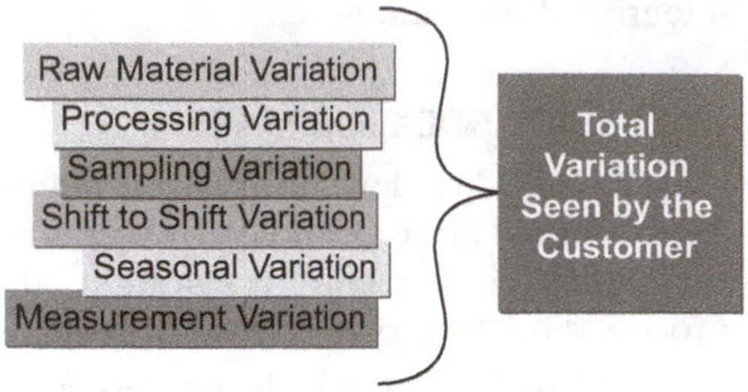

It all depends on how much the improvement will cost to make and how much of an improvement can be made. Something else to notice is that sometimes the variation overlaps. In other words, you cannot just add up the standard deviations of the components and get the standard deviation for the total. Conceptually, this is exactly what we do, but mathematically it takes a little more work. A commonly applied tool for segregating the total variation into separate components is called a nested design.

Measurement Capability Index—Cm

Before discussing nested designs, let us talk briefly about measurement system variation. The process of analyzing a sample has variation just like the process for making the chemical in the first place. Many of the same components of variation can exist within the measurement process.

For example, different lab analysts running the samples on different days can get different results. Many chemical analyses use chemical reagents. When one changes bottles of reagent, one gets different results. The lab instrumentation can "drift" off calibration just like the process equipment instrumentation.

From that standpoint, the analytical process must undergo the same kind of scrutiny that the main process undergoes. Your analytical process might be capable of meeting the needs of your process. A thorough understanding of analytical variability is a major necessity in understanding your process variability. The capability of your analytical process is such an important topic that it has its own capability index, the Cm. The **measurement capability index (Cm)** is the ratio of your specification range to your measurement process variability.

The good news is that the calculations for the Cm look familiar. Here is the equation:

$$\text{Cm} = \frac{\text{USL} - \text{LSL}}{6 \times \sigma}$$

It is the same formula as for the process potential index, or Cp. The only difference is that instead of using the sigma (σ) for the entire process, you use the sigma for the measurement process only. This is accomplished by taking the same sample of material and analyzing it 15 to 20 times. When the same analyst measures the same sample using the same equipment and the same reagents and gets different numbers, the standard deviation for the analytical method can be calculated. This takes out all the variation due to anything related to the process of manufacturing the product.

The interpretation of the Cm differs from that of the Cp. If you think back to the earlier discussion, a Cp of 1.0 means that the variation of your process is equal to what the customer needs or the specifications. Applying that same thinking here would lead you to conclude that a Cm of 1.0 means that the variation of the measurement process is equal to the specifications. Because the measurement process is just one of many components of the total variability, this cannot be acceptable.

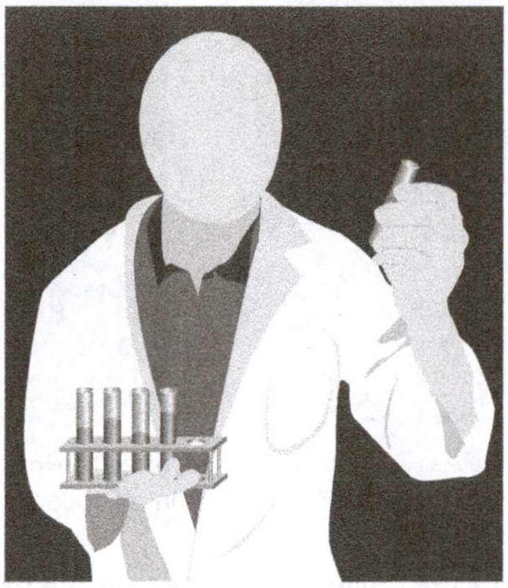

From a practical standpoint, if your measurement system varies within the entire range of the specifications, no variation is allowed in the rest of the process. A Cm of 1.0 or less means that when you get a result back from the lab, it could be on the high side of the specification or the low side of the specification or anywhere in between, and yet nothing has changed in your process.

Many companies in the processing industries have accepted a minimum Cm target of 5.0. This is equivalent to the measurement process's consuming 20% of the total specification range. Some Six Sigma work targets set the goal as high as 10.0.

The higher the Cm, the less variability that your measurement process has compared to the specification range, and the greater the chances that your total variability will be acceptable. The lower the Cm, the more variability that your measurement process has compared to the specification range.

It may not be the job of the process technician to analyze samples or determine the method capability; but without knowing how capable your analytical methods are, you will never know when to react to the lab results. If your analytical process is capable, that is, it has a Cm of at least 5.0, then you move on to other sources of variability; but what if it is not?

Sometimes the issue is that you are asked to use a ruler to measure the length of a stick to the nearest thousandth of an inch. The measurement tool may not be up to the task. In this case you really need a micrometer. The converse could also be true. If you need to know the pH of your chemical to the second decimal place, you will probably need a sophisticated analytical technique. If you need to know that pH only to the closest whole number, then litmus paper may be good enough.

As is the case with quality in general, the answer is to know what you need and use the right tool for the right job at the right time. Without knowing how variable your measurement process is or how much variability you can tolerate, you cannot select the right tools.

Nested Designs

Let us say you are running a process that produces product into a batch make tank (just a generic name). The batch make tank fills up twice a day. When it fills up, the process technician has to pump the product out into 10 individual drums. Customers tell you that different batches of your product behave differently in their process. They even tell you that different drums of the same batch behave differently when charged to their process.

When you try to understand the issue by taking more samples, you get even more confused because multiple samples taken at the same time during the drumming process give different results. How can this be? You know from the previous section that the analytical process has variation just like any other process, but how can you have different material in each of the 10 drums when all 10 came from the same tank?

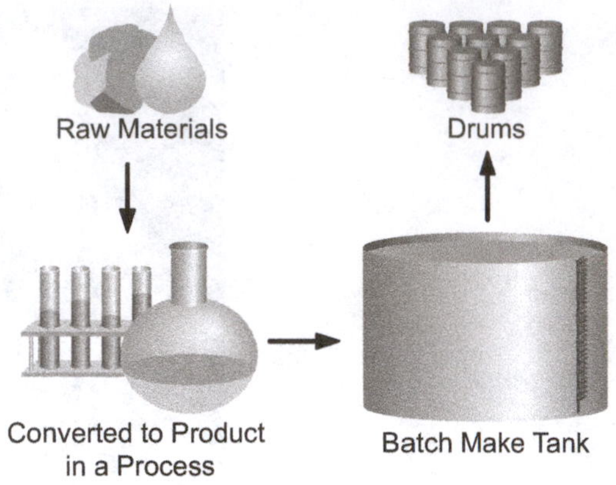

One answer might be that the tank is not well mixed. What if your process is drifting over time and the product is layering inside the batch make tank? As different layers are pumped out into drums, the contents of the drums will vary. Another answer might be that the samples are not representative of what is in the drums, but we will save that discussion for the next section.

After all the discussion about process variability, you are convinced that there is variation in your ability to manufacture the product. The issue you have now is to determine what to fix in order to satisfy the customer—which is, after all, the reason you are in business. You cannot afford to work on improving every process simultaneously. You need to prioritize your workload and spend your resources working on the biggest contributor to variation. Do you work on the process variability, the mixing of the batch make tank, or the analytical method? Everything you have learned so far has been about measuring variability in a single source. Now you need to split that variation into separate components.

If you take your samples in a certain structured manner, you can use a statistical tool called a nested design. **Nested designs** are used to segregate total process variation into separate components. In this case you want to separate the batch-to-batch variation from the drum-to-drum variation from the analytical variation. You would set up a three-level nested design like the one shown in Figure 3-9. For each batch produced, you will capture a sample of Drum 1 and Drum 10. If there is any difference between drums, this should show it.

For each drum sample that you take, you will have the laboratory run the analysis in duplicate—labeled A and B on the chart. Label these samples carefully to make sure that you know what to do with the results.

In this case you might label the samples B1-D1-A, B1-D1-B, B1-D10-A, and so on, to denote the sample came from Batch 1, Drum1, Test A; then Batch 1, Drum 1, Test B; and so forth. Setting up a nested design and analyzing the results is probably not something that will be expected from a typical process technician. Taking the samples at

FIGURE 3-9 Example of a Three-Level Nested Design

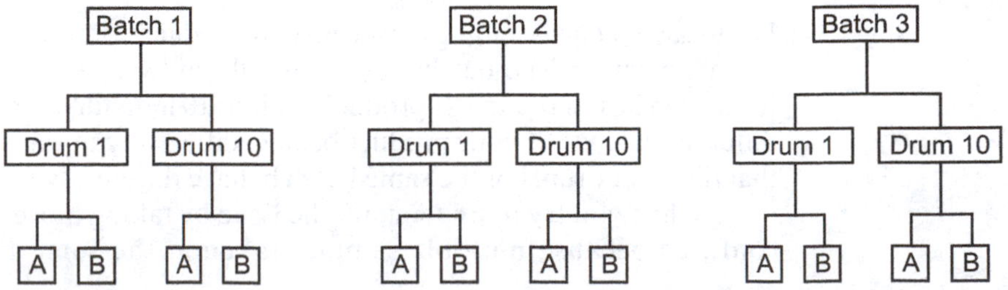

FIGURE 3-10 Example of an ANOVA Table (the Results of a Nested Design)

Analysis of Variance

Source	Sum of Squares	Df	Mean Square	Var. Comp.	Percent
TOTAL (Corrected)	0.00275187	119			
Batch	0.000999538	5	0.000199908	0.00000902655	36.64
Drum	0.00108533	54	0.0000200987	0.00000449101	18.23
Analysis	0.000667	60	0.0000111167	0.0000111167	45.13

the appropriate time, labeling them accurately, and collecting data to make the nested design are part of the process technician's role.

Many times, a designed experiment has been wasted because the technical staff did not think the technicians needed to know. Our philosophy is that the technicians on the floor can make or break you. If we are all in this together, then everybody should understand not only what is going on but also why it is going on so we can work together for success.

When collected properly and analyzed with the right statistical software, data from a nested design will literally calculate a standard deviation for each level of the design, showing you exactly where you should concentrate your improvement efforts. Many statistics packages even calculate the percentage of the total variation due to each level in the design for you.

Analysis of variance (ANOVA) is a statistical tool that helps you understand the nature of variation in your process. Figure 3-10 shows an example analysis of variance table. The only information you really need to understand from this is that 36.64% of the variation is from the process (the batch level); 18.23% is from the mixing/drumming operation (the drum level); and 45.13% is from the analysis (the test level). With this information in hand, you know that nearly one-half of the total variation is due to the test method, so you would concentrate your initial improvement efforts on getting better analytical data. Because less than 20% of the total variation is coming from the mixing/drumming operation, do not pursue any improvement efforts at that stage of the process.

Nested designs are not limited to three levels and do not have to be as symmetrical as the one used in this example. They may take many different forms. The keys to success are having the following:

- A well-thought-out design that separates the total variation into components that you are interested in investigating
- A well-carried-out design that ensures samples are taken and labeled appropriately to ensure accurate analysis of the data

Representative Samples

A sample that is not representative is worse than worthless. Poor samples can result in drawing the wrong conclusions from your data and potentially making matters worse.

FIGURE 3-11 Tank with a Recirculating Line

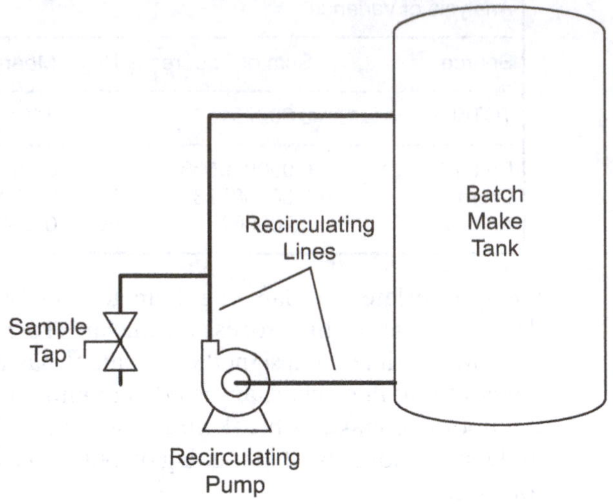

If you are trying to understand drum-to-drum variability, then it is absolutely critical that the samples be taken from the specified drum.

Samples taken from the wrong location because of convenience can sabotage an improvement effort. Samples taken from a tank that is not well mixed will tell you nothing.

If the nested design example had concluded that the drum-to-drum variation was too large, the results from any single sample from the batch make tank would tell the quality only of what was in the sample bottle. The results would not assess the quality of the product in the batch make tank. In order to understand the quality of the product in the batch make tank, you have to know that the contents in the tank are homogeneous or well mixed.

Finally, there has to be a sampling methodology in place that ensures your samples are meaningful. In Figure 3-11 there is a tank with a recirculating line. There is a small sample tap on the discharge of the pump, from where you obtain your samples. The tubing on this sample tap is over 2 feet long and can hold nearly a pint of liquid. When you open the valve to take your sample, are you really getting a sample that is representative of what is in the recirculating line? No, in this case you need to ensure that the sample tap is adequately flushed prior to taking the sample.

Overcontrol

This text has covered variation and has established that the total variation can come from a wide variety of sources. Sometimes the sources of the variation are inherent in the process (normal variation), and sometimes they are episodic (abnormal variation).

In the next couple of chapters the topic of control charts will be covered. The purpose of control charts is to distinguish between common cause and special cause variation so you know when to react and when not to react. Without the appropriate knowledge of what is happening in your process, people react to numbers that they should not react to. When you react to common cause variation, you actually add more variation into the process. If you are measuring the process capability of a process that is being overcontrolled, you will naturally find that the process is less capable.

Dr. Deming would to walk into a manufacturing plant and tell all the workers to take their hands off of the equipment so he could see what the equipment could do on its own. He documented an average 40% reduction in variation just by eliminating the overcontrol that is so common in industry. He also played an interesting game with managers from the companies for which he consulted called the red bead

FIGURE 3-12 Dr. Deming's Red Bead Experiment

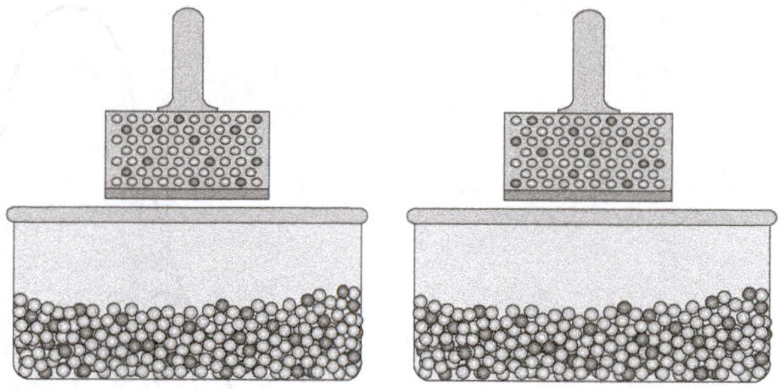

experiment (shown in Figure 3-12). Deming used a sample bowl with red and white beads to demonstrate the futility of reacting to common cause variation. The concept of overcontrol can be demonstrated in the classroom using a quincunx (shown in Figure 3-13).

Let us say your process variability has been well documented, and you know that your process varies ± 3 from the target value when all is running well. This ± 3 variation includes process and analytical variation. When you get any value within this range, you should just leave the equipment alone because it is doing what it is capable of doing. If, on the other hand, you get a value that is 2 units lower than target, and you adjust your process up by 2 units to compensate, now your entire process spread of 6 units (± 3) will no longer be centered; and the probability of getting a higher number increases.

When you get that higher number, let us say you adjust the process down to compensate again. In the end what has happened is that you have created a grand total process made up of several smaller ones that you generated by moving your process around manually. The total process capability is greatly reduced, and the variability of your product is greatly increased—much to the dissatisfaction of your customers. Figure 3-14 shows how one big histogram may in fact be made up of several smaller histograms that process technicians created themselves by not understanding the basic concepts of variation.

FIGURE 3-13 Quincunx

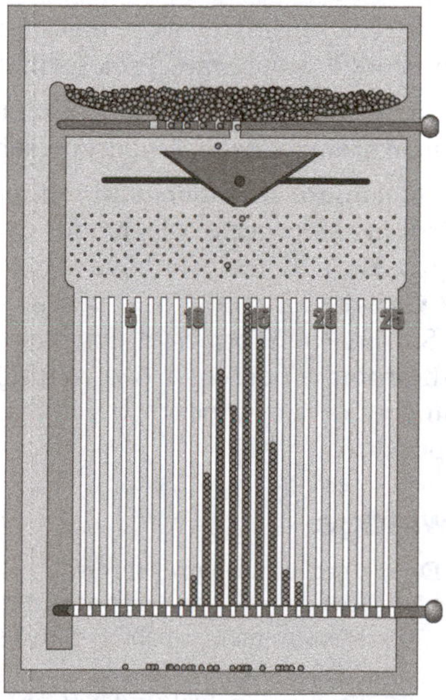

FIGURE 3-14 Multiple Subprocesses Make Up the Grand Process

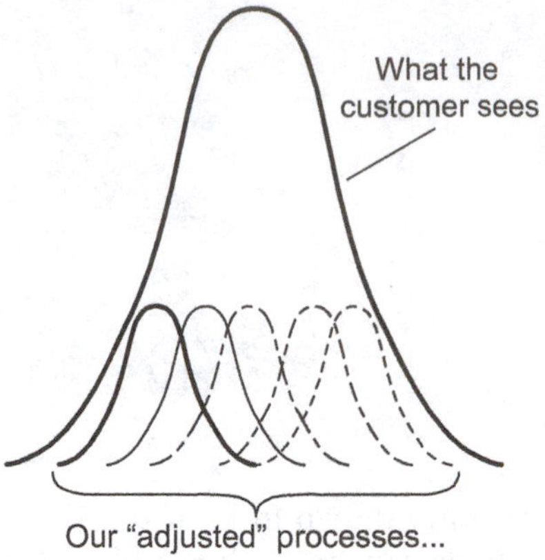

Summary

Knowing that your processes vary is a good start. Relating that variation to the needs of the customer is an important next step. The process potential index (Cp), process capability index (Cpk), and process performance index (Ppk) provide simple numerical indexes to help you understand the relationship between the variability of your processes and the needs of the customer.

You must understand several concepts in order to make proper use of your knowledge of the process capability:

1. Variation can come from a variety of sources. Some examples include seasonal variation, shift-to-shift variation, sampling variation, and measurement system variation.

2. All capability indexes are calculated using measures from your process or products; therefore, understanding the amount of variation that comes from the measurement process is critical to your improvement efforts. The measurement capability index (Cm) provides a simple numeric evaluation of the adequacy of your measurement system.

3. Statistical tools such as the nested design are employed in the processing industry to segregate the total variability into separate components so you can know where to apply your improvement efforts.

4. The sensitivity of the measurement device must be adequate for the measurement needed. Sometimes a ruler will do; other times you need a micrometer.

5. In the processing industries, the "goodness" of the sample is as important as the measurement itself. A nonrepresentative sample is of no value.

6. Failure to understand the measurement system variation can lead to overcontrol of the process, which adds even more variability to that which is inherent in the process. Did you ever have someone set the thermostat to 65 because he was too hot? Can the air conditioner run any cooler? No. Either it blows cold air or it does not. Setting the thermostat down a degree or two will cause it to come on and cool off the room a degree or two. Setting the thermostat down lower will not make the room cool off any quicker. It will just get too cold and cause you to overcontrol the temperature.

Checking Your Knowledge

1. Define the following key terms:
   a. Lower specification limit (LSL)
   b. Measurement capability index (Cm)
   c. Nested design
   d. Process capability index (Cpk)

e. Process performance index (Ppk)
f. Process potential index (Cp)
g. Upper specification limit (USL)

2. You have been working on understanding your process variability and how this variation might be perceived by your customers. Given the following information, what is your Cp?

$$\bar{x} = 102$$
$$\sigma = 2$$
$$USL = 106$$
$$LSL = 94$$

a. 0.8 c. 1.0
b. 0.9 d. 1.1

3. Using the same information from question 2, what is your Cpk?
a. .67 c. 1.0
b. .76 d. .5
4. Using the same information from question 2, what is your Ppk?
a. .67 c. 1.0
b. .76 d. .5
5. Given a specification range of 95 to 110 and a measurement system standard deviation of 0.5, what is your Cm?
a. 1.0 c. 5.0
b. 3.0 d. 7.0
6. The minimum Cm to ensure that your measurement system is adequate is what?
a. 1.0 c. 5.0
b. 3.0 d. 7.0
7. (*True or False*) The process potential index and the measurement capability index use exactly the same calculations.
8. (*True or False*) Nested designs are always balanced or symmetrical in nature.
9. List some potential sources of variability:
a. ________________ c. ________________
b. ________________ d. ________________
10. To ensure that your process is capable of meeting the customers' needs, they desire a capability index of:
a. Less than 1.0
b. Exactly 1.0
c. Greater than 1.0
11. (*True or False*) Measurement system variation does not impact the total variation in the process.
12. (*True or False*) A representative sample is needed to ensure that the measurement is useful.
13. What happens to your process variability when you adjust your process in response to common cause variation?
a. The variability gets better.
b. The variability gets worse.
c. There is no change in the process variability.
14. (*True or False*) Any measurement device will do as long as it gives us a number.
15. A Cpk index of __________ or greater is a common target in the processing industries.
a. 0.67 c. 1.33
b. 1.00 d. 3.00
16. A statistical tool used to segregate total process variation into separate components is called:
a. Design flaw c. Nested design
b. Design for Six Sigma d. Nested variance

Activities

1. List the strengths and weaknesses of a Cpk index versus a Cp indes.
2. Compare and contrast the differences in the calculation of a Cpk index, a Cp index, and a Ppk index.
3. Why would you use a Cpk index over a Cp index? Explain.

CHAPTER 4

Variables Control Charts

"The goods come back, but not the customer."
~ROBERT W. PEACH (SEARS, ROEBUCK & CO.)

Objectives

Upon completion of this chapter you will be able to:

- Explain the purpose of variables control charts and how they relate to the distribution of the data.
- Given a data set, construct an x-bar and R control chart.
- Given a data set, construct an individuals control chart.
- Explain the differences between individuals and x-bar and R control charts, and explain when each one is used.
- Explain the purpose of a zone chart.
- Explain the purpose of a cumulative summation (CUSUM) chart.
- Explain the purpose of an exponentially weighted moving average (EWMA) chart.

Key Terms

Cumulative summation (CUSUM) chart—a special type of control chart used to detect small shifts in the process mean.
Exponentially weighted moving average (EWMA) chart—a special type of control chart that takes previous data into account when evaluating each point; also capable of detecting small shifts in the mean.
Lower control limit (LCL)—typically 3 standard deviations (σ) below the mean.
Moving range (MR)—difference between two consecutive points.
Upper control limit (UCL)—typically 3 standard deviations (σ) above the mean.
x-bar and R ($\bar{X}R$)—a control chart used when data are grouped together.

Introduction

This is the third chapter that focuses on the statistical piece of the quality puzzle. Control charts are some of the most basic quality improvement tools. Walter Shewhart initially proposed their benefits in the 1920s and published the first work on the subject in 1939. Control charts are the main tools Dr. Deming employed in his work to help the Japanese rebuild their industrial infrastructure in the 1950s.

Control charts have been around a long time and have proven their worth over and over again. The two preceding chapters developed a thorough understanding of process variability and how that variability relates to the needs of your customers. With that understanding as your foundation, start digging deeper into the various tools available to improve your processes. Recall that Chapter 2 discussed different kinds of distributions used to describe different types of data. For continuous data the normal distribution is used. The tools covered in this chapter are some of the different kinds of variables control charts including the following:

1. $\bar{x}$ and R charts
2. Individuals charts
3. Zone charts
4. CUSUM charts
5. EWMA charts

As has been stated, there are many different tools in the quality toolbox, and each serves a different purpose. You need to select the right tool for the job in order to accomplish your goals as efficiently as possible. Selecting the right tool for the job requires an understanding of the purpose of the tools and a basic knowledge of when each should be applied, so this chapter discusses the value of each of these types of control charts. You will also learn how to construct several of them so you can apply your knowledge to everyday situations.

Purpose of Control Charts

Control charts serve a specific purpose in your quality improvement efforts: they help distinguish between common cause and special cause variation. If you know the variation that you are seeing is inherent in the process, then just accept it and keep moving forward. If you can see that the variation is not inherent in the process, you know that something special has occurred; and your job becomes to figure out what happened or assign a cause to the event.

Process technicians naturally assume that a special cause is a bad thing, but that may not be the case. If the special cause makes the process behave worse than normal, it is important to understand this cause in order to prevent its recurrence. If the special cause makes the process exhibit less variation than normal, there is just as much need to investigate the issue. In this case process technicians need to understand the cause so they can incorporate the change into their standard operating procedures so their process can operate at this improved level all the time.

Let us say that you are installing a new procedure or a new piece of equipment in order to effect an improvement in your process. The control chart can tell you pretty

FIGURE 4-1 Normal Distribution

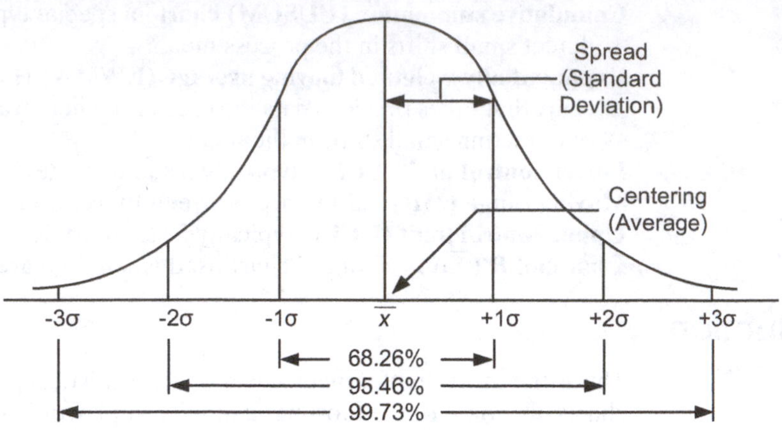

quickly whether the improvement worked or not by letting you know whether the variation seen is the same as you have always seen or whether it has changed.

Relating Control Charts to Distributions of Data

Recall the normal distribution (shown in Figure 4-1). It is shaped like a bell, high in the center and tapering off to both sides. It is symmetrical, with the tapering pretty even to the left and to the right.

The variation about the mean ($\bar{x}$) is distributed such that 68% of the numbers fall within 1 standard deviation (σ) of the mean, 95% fall within 2 standard deviations of the mean, and 99.7% fall within 3 standard deviations of the mean. For all intents and purposes, you would expect all of your process data to fall inside the 3 sigma limits. Only 3 times out of 1,000 would a point outside of the 3 sigma limits be normal.

The limitation in using this distribution is that you have to collect a lot of data before you can draw the curve. It takes at least 20 to 30 data points before you can calculate the mean and standard deviation. If your process makes only one batch a day, it may take you a month to collect enough data to describe your process with a distribution. If you make a change to your process, do you want to wait a month to find out whether the change was good or bad?

Plotting a histogram gives a snapshot of what your process looked like during the time frame when you were collecting the data. It does not tell anything about what the process was doing during that time frame.

Table 4-1 shows a single list of 100 numbers generated to be normally distributed with a mean of 30 and a standard deviation of 5. The list is presented twice. The set on

TABLE 4-1 List of 100 Numbers in Random and Sorted Order

Random										*Sorted*									
27	29	25	28	39	26	30	34	27	24	18	19	19	20	20	21	21	21	21	21
26	32	40	29	29	32	34	28	25	33	22	22	23	23	23	24	24	24	25	25
34	30	31	27	29	37	23	27	27	31	25	25	26	26	26	26	27	27	27	27
27	26	28	30	27	31	32	30	31	33	27	27	27	27	27	27	27	28	28	28
44	37	34	22	40	21	21	36	19	25	28	28	28	28	29	29	29	29	29	29
29	19	29	25	35	24	38	37	33	27	29	29	29	29	30	30	30	30	30	30
26	39	31	22	30	28	18	27	35	36	30	31	31	31	31	31	31	31	31	32
31	39	28	31	29	27	32	24	29	28	32	32	32	32	33	33	33	33	33	33
31	33	29	20	23	29	30	21	33	23	34	34	34	34	34	35	35	36	36	36
27	21	20	21	36	32	28	33	34	30	37	37	37	38	39	39	39	40	40	44

FIGURE 4-2 Histogram of Data in Table 4-1

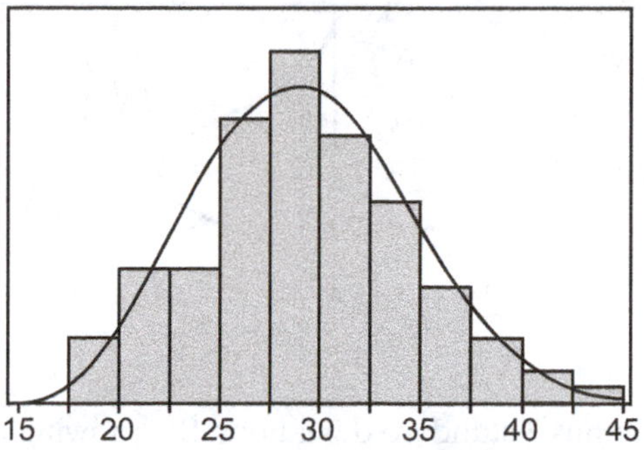

the left is presented in random order. The set on the right has been sorted from lowest to highest.

Both of these data sets would produce the exact same histogram (Figure 4-2) because the data are identical. The order in which the data were collected has no bearing at all on the distribution of the data. As you can see, the data are pretty much centered around 30 and appear to vary from roughly 15 to 45, which is what you would expect given a standard deviation of 5.

Notice that the histogram is not absolutely perfect. Even though these data were generated with computer software to force the mean and standard deviation, the random variability in the software functionality yielded a calculated mean of 29.3 and a calculated standard deviation of 5.1.

To provide the element of time to your understanding of the distribution of these data, you would use a run chart or control chart. A run chart is just a sequential charting of the data. The random data from this example have been plotted (Figure 4-3). As you can see, the numbers move up and move down without anybody doing anything to this process. Nothing but common cause variation is seen here.

If you moved all these numbers to the left side of the graph, they would stack up and build the histogram that you have become so familiar with. When you do this, you see that your linear graph of numbers can easily be related to the underlying distribution (Figure 4-4).

Because you know how to calculate the mean and standard deviation of these data, you do not have to present the histogram to show the distribution of the data. You can simply draw a line to represent the mean (average) of the data and draw lines to show the upper and lower limits of where you expect the data to be located (Figure 4-5). These upper and lower limits are called the **upper control limit (UCL)** and **lower control limit (LCL).** They are traditionally set at the mean plus 3 standard deviations and the mean

FIGURE 4-3 Run Chart of Data in Table 4-1

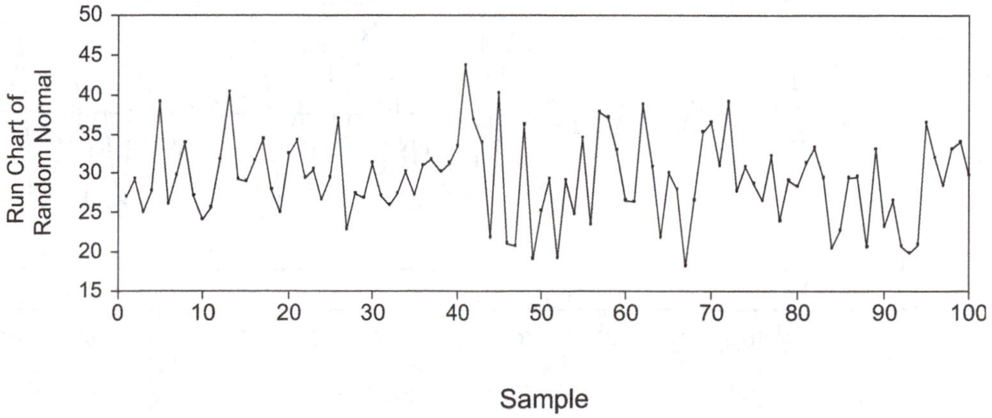

FIGURE 4-4 Histogram Derived from Figure 4-3

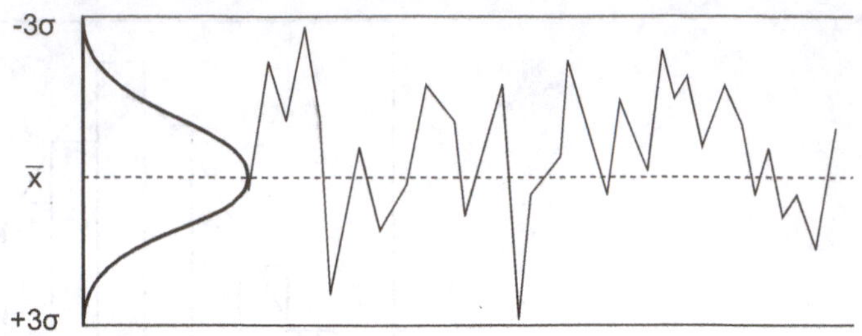

minus 3 standard deviations. By knowing this, you also know that 99.7% of the time your numbers should fall inside these limits.

For this set of data, the calculations would look like this:

$$UCL = \overline{x} + 3\sigma = 29.3 + (3 \times 5.1) = 44.6$$
$$LCL = \overline{x} - 3\sigma = 29.3 - (3 \times 5.1) = 14.0$$

With a control chart in hand, you no longer have to wait until you collect a minimum of 20 to 30 new data points to check on your process. Once you have the control limits calculated, you can plot each new point as it is collected and check to see whether this point is "in control" or "out of control."

All of the control charts that will be discussed are based on the same basic principles. First, calculate control limits based on the underlying distribution of the data and then use these control limits to provide signals as to when your process is starting to behave differently than expected.

As mentioned before, different types of data have different underlying distributions and will thus require different kinds of control charts. Even data that come from the same underlying distribution may require different kinds of control charts. Next is an examination of a couple of the most commonly used variables control charts.

X-Bar and R ($\overline{X}R$) Control Charts

X-bar and R control charts are used when data are grouped together. An x-bar and R control chart consists of two charts, one that depicts the average (x-bar) and one that depicts the range (R). In order to determine the control limits, you must also calculate the standard deviation.

FIGURE 4-5 Run Chart Showing Upper and Lower Control Limits

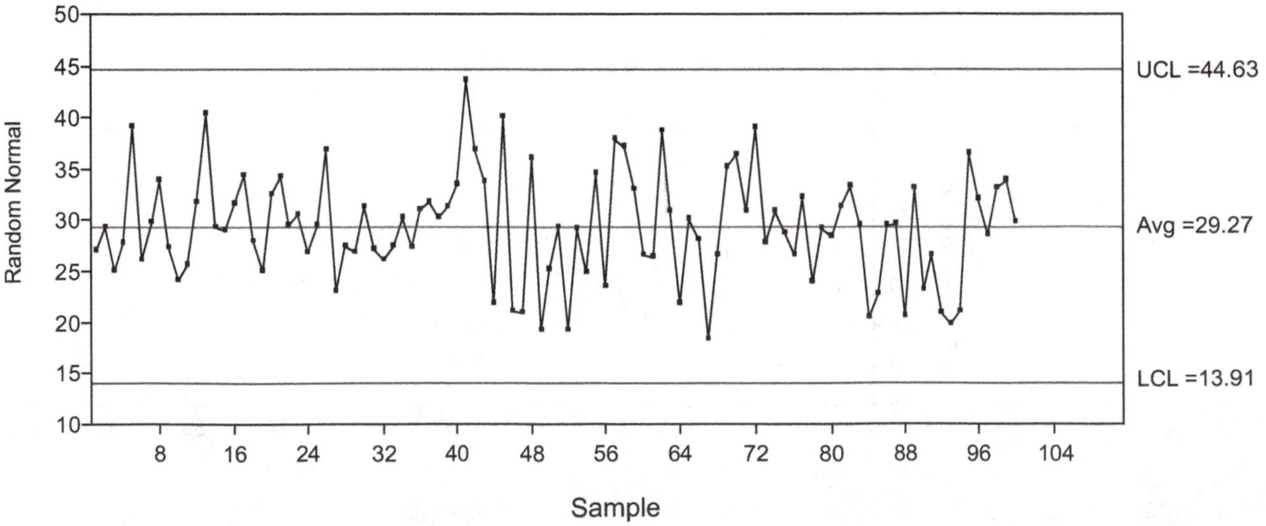

Today you can use a handheld calculator or easily available software packages to calculate the standard deviation of a set of data. You are not expected to be able to calculate standard deviations by hand, but just so you appreciate what comes next, here is a quick example of a set of five data points; these are our individual *xs*: 26, 24, 33, 27, 40. As a reminder, here is the formula for the standard deviation presented in Chapter 2:

$$\sigma = \sqrt{\frac{\Sigma(x - \bar{x})^2}{n - 1}}$$

In order to calculate the standard deviation, follow these steps:

1. Add up these 5 numbers and then divide the total by 5 to get $\bar{x} = 30$. Now to accomplish the part of the equation inside the parentheses, subtract this average from each one of the individual values:

x	$x - \bar{x}$	Difference
26	26 − 30 =	−4.00
24	24 − 30 =	−6.00
33	33 − 30 =	3.00
27	27 − 30 =	−3.00
40	40 − 30 =	10.00

2. Square (multiply by itself) each one of these differences:

Difference	Square That Difference
−4.00	16.00
−6.00	36.00
3.00	9.00
−3.00	9.00
10.00	100.00

3. Sum (Σ) these squared differences:

$$16 + 36 + 9 + 9 + 100 = 170$$

4. Divide this sum by $(n - 1)$, where n is the number of data in the data set. In this example, there are 5 numbers in the data set, so $n - 1 = 4$.

$$170 \div 4 = 42.5$$

5. The final step is to take the square root of 42.5. The $\sigma = 6.5$.

Can you imagine doing this for a data set of hundreds of numbers? Can you imagine doing it by hand without a calculator like Shewhart and Deming had to in their early days? Fortunately for them they found a way to estimate the standard deviation without having to crank through all these numbers. They came up with a series of factors that would estimate the control limits based on the range (the difference between the high and low values) of data in a subset.

Did You Know?

$(-) \times (-) = +$

Two negative numbers multiplied together yield a positive number as a result.

The earliest application of control charts was in what is sometimes called the widget industries, which make discrete things like nuts and bolts and tires that can be individually measured. Process technicians typically make drums or tanks of materials that have to be sampled and analyzed.

The application of control charts when you make bins of hundreds of bolts each is to sample a few bolts out of each bin and measure them. You capture the average measurement and also the range of the measurements. After you have captured 20 such samples, you can calculate the average of the ranges ($\overline{R}$).

Shewhart came up with a set of factors (A_2) that could estimate the control limits using this average range value. It is a set of factors instead of a single factor because the factor changes if you have 2 bolts in your sample versus 3 or 4 or 5 bolts in your sample. The formulas he used for the upper and lower control limits looked like this:

$$\text{UCL} = \overline{x} + A_2\overline{R}$$
$$\text{LCL} = \overline{x} - A_2\overline{R}$$

Instead of adding and subtracting 3 standard deviations from the mean ($\overline{x}$) to get the control limits, you multiply the average range ($\overline{R}$) by some factor (A_2) based on the number of units in your sample, and this is your estimate for the control limits. Tables of these factors can easily be found in many books on control charting (shown in Table 4-2).

Let us create a scenario in which you might use this technique in the processing industries, build an example data set, and create an $\overline{XR}$ chart. Let us say you make gallon bottles of antifreeze. The chemical comes from a large tank that has already been analyzed for chemical composition, but you need to know how well you are filling the bottles.

Your target is to fill each bottle to 128 ounces plus or minus 2 ounces. Each tank of chemical that you produce is enough to fill 500 bottles. There are 4 shifts per day, and each shift is expected to capture a new point on the control chart using whatever batch of product happens to be in the tank when it comes time to capture that shift's control chart data.

You decide to sample 3 bottles at random from each tank of chemical. Each bottle is to be weighed to the nearest tenth of an ounce. Each shift will capture the net weight of each of the 3 bottles chosen at random for its sample. That shift will record the descriptive information such as date, shift, and batch number as well as the results from its 3 bottle weights. Then it will calculate the average ($\overline{x}$) of the 3 bottle weights as well as the range between the lowest and the highest bottle weights.

Your control chart (Figure 4-6) begins on April 1. A shift pulls 3 gallon bottles of antifreeze off the filling line while batch 339 is being bottled. The 3 bottles are weighed to the nearest 1/10th of an ounce. In this case, the bottles weighed 129.6 ounces, 129.2 ounces, and 127.9 ounces. The shift averages these three points and determines that the $\overline{x}$ is 128.9. The shift also subtracts the lowest number (127.9) from the highest number (129.6) and determines that the range, R, is 1.7.

Later that day B shift pulls its 3 bottles off the line while batch 348 is being bottled and records the weights as 129.7 ounces, 128.9 ounces, and 127.1 ounces. B shift calculates its $\overline{x}$ as 128.6 and the range as 2.6. Shifts C and D perform the same function

TABLE 4-2 Factors for Estimating Control Limits

Number of Observations	*Factor A_2*
2	1.88
3	1.023
4	0.729
5	0.577

FIGURE 4-6 Sample Control Chart

ANY Corp							
Product	Antifreeze						
Date	4/01	4/01	4/01	4/01	4/02	4/02	4/02
Shift	A	B	C	D	A	B	C
Batch No.	339	348	355	362	370	378	385
Bottle-1	129.6	129.7	131.1	127.2	131.0	128.6	129.9
Bottle-2	129.2	128.9	127.6	130.1	128.8	128.9	127.7
Bottle-3	127.9	127.1	130.2	127.2	132.0	129.3	131.0
X-Bar	128.9	128.6	129.6	128.2	130.6	128.9	129.5
R	1.7	2.6	3.5	2.9	3.2	0.7	3.3
Observation	1	2	3	4	5	6	7

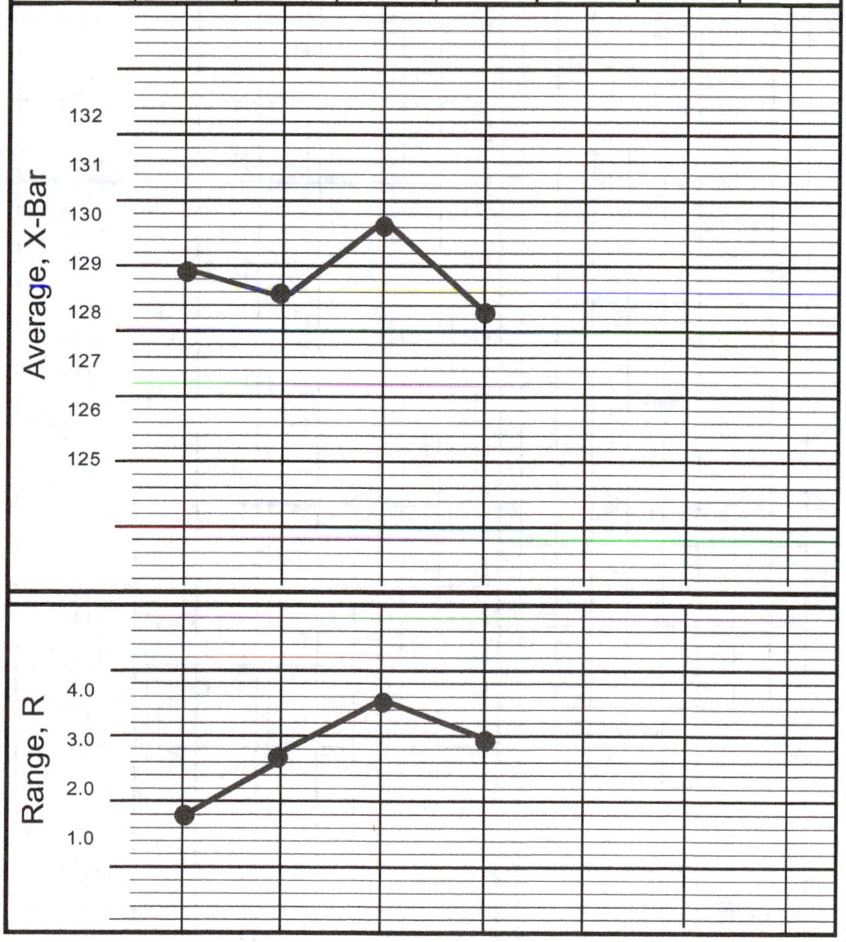

on their respective shifts. Filling in the data that have been collected, you can then plot the $\bar{x}$ and the range on the control chart as shown in Figure 4-6 on this small portion of your full-page control chart.

Figure 4-7 provides the rest of the data. You should be able to fill in the $\bar{x}$ and R for each of the remaining observations and plot the data on the control chart.

Once you have calculated the range for each of your 20 sets of 3 bottles, you can then calculate the average of your subgroups as well as the average range ($\overline{R}$) within the subgroups. From Table 4-2, we see that the A_2 factor for a sample size of 3 is 1.023. Let us assume that you calculated an $\bar{x}$ of 129.3 and an $\overline{R}$ of 2.2. With these numbers in hand, you can calculate control limits for your data as follows:

FIGURE 4-7 Extended Sample Control Chart

ANY Corp

Product: Antifreeze — Property: Net Weight — Sample Point: Bottle Filler

Date	4/01	4/01	4/01	4/01	4/02	4/02	4/02													
Shift	A	B	C	D	A	B	C	D	A	B	C	D	A	B	C	D	A	B	C	D
Batch No.	339	348	355	362	370	378	385	393	400	408	416	423	431	438	446	454	461	469	476	484
Bottle-1	129.6	129.7	131.1	127.2	131.0	128.6	129.9	129.1	130.7	127.2	128.6	128.9	127.3	128.1	131.8	130.7	128.1	127.6	130.8	127.1
Bottle-2	129.2	128.9	127.6	130.1	128.8	128.9	127.7	127.8	131.7	129.3	128.0	127.0	128.9	127.9	129.2	132.0	131.4	129.5	131.3	129.2
Bottle-3	127.9	127.1	130.2	127.2	132.0	129.3	131.0	129.6	130.1	127.0	130.0	129.7	130.1	129.3	131.0	131.7	130.6	128.3	131.2	127.3
X-Bar	128.9	128.6	129.6	128.2																
R	1.7	2.6	3.5	2.9																
Observation	1	2	3	4	5	6	7	8	9	10	11	12	13	14	15	16	17	18	19	20

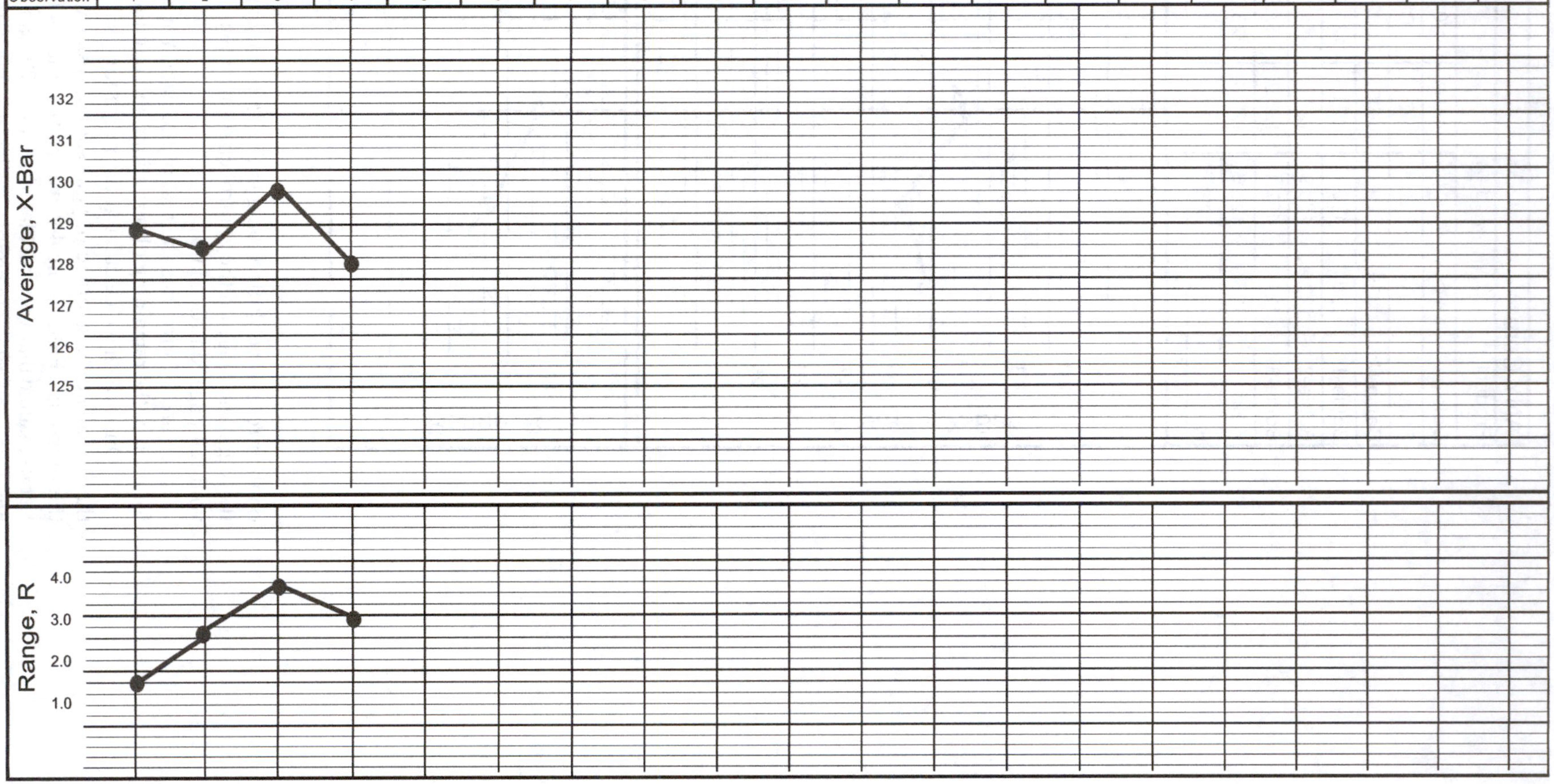

$$UCL = \overline{x} + A_2\overline{R} = 129.3 + 1.023\ (2.2) = 131.5$$

$$LCL = \overline{x} - A_2\overline{R} = 129.3 - 1.023\ (2.2) = 127.0$$

IN A NUTSHELL

$\overline{X}R$ charts are made from averages of rational subgroups of individual measurements. Use them when your data can be grouped rationally, for example, with SPC measurements where you can collect multiple measurements per batch.

Always complete the multiplication and division first, then the addition and subtraction.

Lines can then be drawn on the chart to show the limits within which your data should fall. The final product is shown in Figure 4-8.

It appears that your data "fit" well within the control limits that you calculated. Because this was your first control chart for this process, you had to wait until you had collected enough data in order to establish control limits. Now that you have control limits, you can draw these limits onto a blank chart.

Each time a subgroup of 3 bottles is weighed, you can then draw the point on the graph and see instantly whether your new data point falls inside or outside the control limits. In addition to seeing whether the data fall within the control limits, these charts can provide other signals to let you know something is amiss. Because these signals or interpretation rules are fairly common for all control charts, they will be covered in more detail at the end of Chapter 5.

One more comment about $\overline{X}R$ charts: the creation of subgroups needs to be based on meaningful rationale. Just grouping things together for convenience may prevent you from learning what you need to learn. In the preceding example, you chose bottles all from the same lot number so that each bottle would have some meaningful relationship to the other bottles filled during that same lot. This concept, routinely used in the industry, is called rational subgrouping.

Individuals Control Charts

Individuals control charts are plots of individual measurements instead of averages of subgroups. One perfect example of an appropriate application for an individuals control chart is your vehicle's gas mileage. You can collect data on your gas mileage and plot it on a control chart, but grouping tanks of gas together would not make any sense. What would you expect to see if you did this? Perhaps higher values would result when the entire tank of gas was expended on highway driving. Perhaps a string of low values would indicate that your vehicle is in need of a tune-up.

Oftentimes in the process industries there is no need to subgroup the data. For example, a chemical is made in a continuous process that fills up a tank every 12 hours. When that tank is full, a sample is taken for compositional analysis. When a passing analysis is obtained, the contents are transferred to a bulk storage tank.

If you are applying SQC techniques to the production facility, you have a single set of analytical numbers to look at. You do not really have a grouping of numbers to average. For this reason you may expect to find individuals control charts applied more frequently in the process industries than $\overline{X}R$ charts.

The basic methodology is the same:

1. First, collect 20 to 30 data points so you can set the control limits.

2. Then, apply these control limits to new data that are collected, and monitor the process for out-of-control signals.

3. The control limits can be calculated using the standard deviation, or they can be estimated using the range. In the case of individual measurements, there is not a subgroup of data to obtain a range, so the absolute value (always a positive number) of the moving range is used. The **moving range (MR)** is the difference between two consecutive points.

In your process you are making a chemical called dibutyl futile and are tracking the purity of this chemical. You make two tanks a day, so you have two purity analyses a day. If there is some rational reason to group the data from these two tanks together, you might use an $\overline{X}R$ chart; but if you just happen to make two tanks a day and there is no real expectation that the data from one day are capable of being grouped

FIGURE 4-8 Completed Sample Control Chart

ANY Corp — Variable Control Chart

Product: Antifreeze | Property: NetWeight | Sample Point: Bottle Filler

Date	4/01	4/01	4/01	4/01	4/01	4/01	4/01	4/01	4/01	4/01	4/01	4/01	4/01	4/01	4/01	4/01	4/01	4/01	4/01	4/01
Shift	A	B	C	D	A	B	C	D	A	B	C	D	A	B	C	D	A	B	C	D
Batch No.	339	348	355	362	370	378	385	393	400	408	416	423	431	438	446	454	461	469	476	484
Bottle-1	129.6	129.7	131.1	127.2	131.0	128.6	129.9	129.1	130.7	127.2	128.6	128.9	127.3	128.1	131.8	130.7	128.1	127.6	130.8	127.4
Bottle-2	129.2	128.9	127.6	130.1	128.8	128.9	127.7	127.8	131.7	129.3	128.0	127.0	128.9	127.9	129.2	132.0	131.4	129.5	131.3	129.6
Bottle-3	127.9	127.1	130.2	127.2	132.0	129.3	131.0	129.6	130.1	127.0	130.0	129.7	130.1	129.3	131.0	131.7	130.6	128.3	131.2	127.6
X-Bar	128.9	128.6	129.6	128.2	130.6	128.9	129.5	128.8	130.8	127.8	128.9	128.5	128.7	128.4	130.7	131.4	130.0	128.4	131.1	127.4
R	1.7	2.6	3.5	2.9	3.2	0.7	3.3	1.8	1.6	2.3	2.0	2.7	2.8	1.4	2.6	1.3	3.4	1.9	0.5	2.1
Observation	1	2	3	4	5	6	7	8	9	10	11	12	13	14	15	16	17	18	19	20

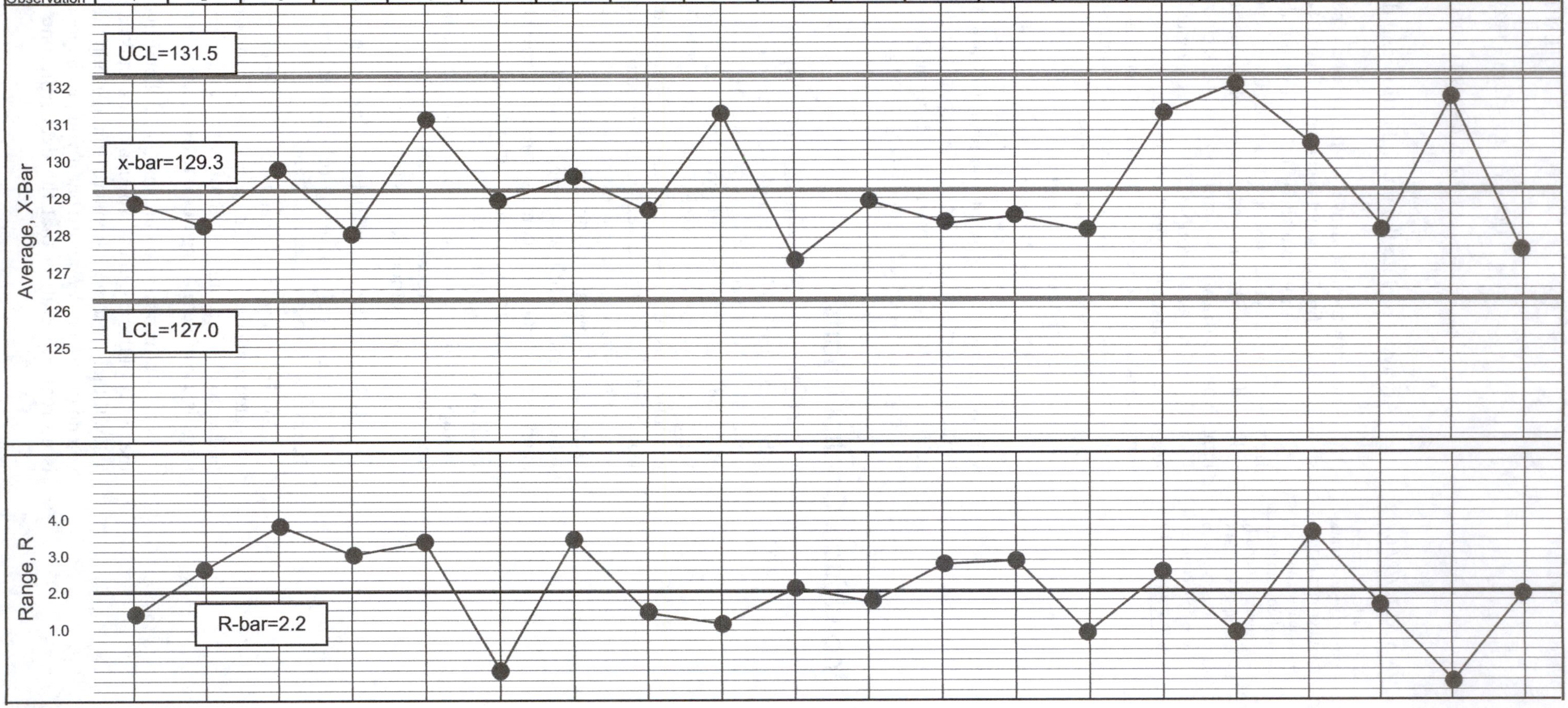

TABLE 4-3 Dibutyl Futile Data Set

Date	5/1	5/1	5/2	5/2	5/3	5/3	5/4	5/4
Batch No.	331	332	333	334	335	336	337	338
Result	98.2	98.8	99.4	99.5	98.7	99.4	98.3	99.8
MR		0.6	0.6	0.1	0.8	0.7	1.1	1.5

together, you might apply an individuals control chart instead. The beginning of your data set is shown in Table 4-3.

You have captured the purity values from each make tank and have calculated the moving range (MR) between the pairs. There cannot be a moving range value for the first data point, and all of your moving range values are positive because you use the absolute value of the difference. Some experts would suggest that plotting the moving range as well as the $\overline{x}$ is the right way to go because a significant change in your ranges is also an indication of abnormality. Others would suggest that focusing on the variable of interest will tell you what you need to know and is much simpler to manage. For our use here, we will adopt the simpler option, but be aware that in industry you may see individuals control charts with a moving range section at the bottom much like your $\overline{X}R$ charts showed.

The formulas for estimating the control limits for an individuals control chart based on the moving range are the same as for $\overline{X}R$ charts. The A_2 factor for individuals control charts is always 2.66.

$$\text{UCL} = \overline{x} + 2.66 \times \overline{MR}$$
$$\text{LCL} = \overline{x} - 2.66 \times \overline{MR}$$

Following the preceding process, you would collect 20 to 30 data points and then calculate your average, $\overline{x}$, and your average moving range, $\overline{MR}$. The following 25 values are the data collected from this example:

99.0 98.5 99.4 99.5 99.8 99.8 99.2 99.7 99.0 99.6 99.1 99.4 99.7
98.9 99.5 99.0 98.6 99.5 99.2 99.4 99.6 99.3 99.5 98.9 99.6

Average these data to obtain an $\overline{x}$ of 99.3. Between each of the pairs of data, you calculate a moving range:

0.5 0.8 0.1 0.3 0.0 0.7 0.5 0.7 0.6 0.5 0.3 0.3
0.8 0.6 0.5 0.5 1.0 0.3 0.2 0.1 0.2 0.2 0.4 0.7

You average these moving ranges to obtain an $\overline{MR}$ of 0.5. Applying the control limit formulas above, you calculate the UCL to be 100.6 and the LCL to be 98.0:

$$\text{UCL} = \overline{x} + 2.66 \times \overline{MR} = 99.3 + 2.66 \times 0.5 = 99.3 + 1.3 = 100.6$$
$$\text{LCL} = \overline{x} - 2.66 \times \overline{MR} = 99.3 - 2.66 \times 0.5 = 99.3 - 1.3 = 98.0$$

Notice that you have an UCL of 100.6. Should you plot an upper control of 100.6 when you are plotting the purity of a chemical? No, statistics is sometimes a funny business. The computers and calculators will provide numbers that are not always meaningful. In this case it is obvious that something cannot be more than 100% pure. Just because you calculated a control limit above 100% does not make it rational. In this case you would set your UCL at 100 because this is the practical limit for the measurement.

You can expect to see the same phenomenon when you are plotting impurities that show up at low levels. In those cases the LCL will sometimes be calculated as a number less than zero, but again, to have less than zero percent impurities is not possible, so you would set your lower control limit at zero. The result of your first control chart for this parameter (plotting the data and moving ranges above) is shown in Figure 4-9.

Take a look at this chart to see whether it looks "normal." The data seem to bounce around the average in a random pattern. Nothing appears abnormal, so you

FIGURE 4-9 Sample Individuals Control Chart

ANY Corp

Individuals Control Chart

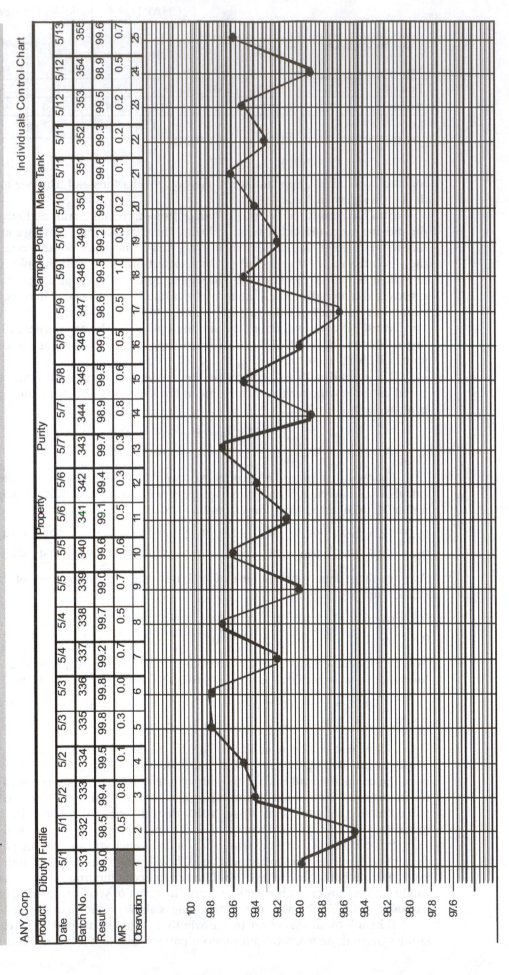

Product	Dibutyl Futile										Property		Purity						Sample Point		Make Tank				
Date	5/1	5/1	5/2	5/2	5/3	5/3	5/4	5/4	5/5	5/5	5/6	5/6	5/7	5/7	5/8	5/8	5/9	5/9	5/10	5/10	5/11	5/11	5/12	5/12	5/13
Batch No.	331	332	333	334	335	336	337	338	339	340	341	342	343	344	345	346	347	348	349	350	351	352	353	354	355
Result	99.0	98.5	99.4	99.5	99.8	99.8	99.2	99.7	99.0	99.6	99.1	99.4	99.7	98.9	99.5	99.0	98.6	99.5	99.2	99.4	99.6	99.3	99.5	98.9	99.6
MR		0.5	0.8	0.1	0.3	0.0	0.7	0.5	0.7	0.6	0.5	0.3	0.3	0.8	0.6	0.5	0.5	1.0	0.3	0.2	0.1	0.2	0.2	0.5	0.7
Observation	1	2	3	4	5	6	7	8	9	10	11	12	13	14	15	16	17	18	19	20	21	22	23	24	25

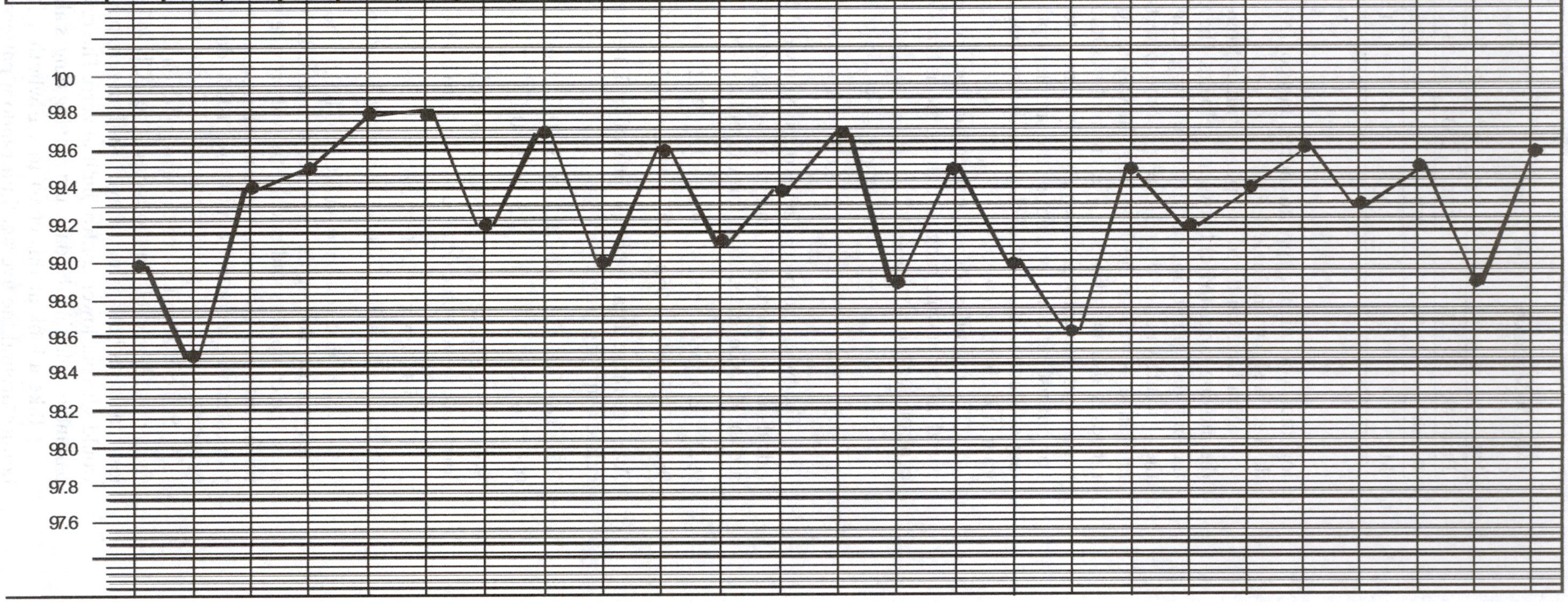

IN A NUTSHELL

Individuals charts are made from individual measurements. Use them when your data have no rational subgrouping, for example, SQC applications where you have a single measurement per batch.

accept these control limits as normal for your process. Your next step would be to create a new control chart that already has these control limits drawn on it. Then you can continue your data collection and begin examining each new batch as it is analyzed to monitor for process shifts.

Which control chart should you use—$\overline{X}R$ or individuals? Use the one that makes the most sense for your application. They are both valid tools and both serve a useful purpose. In fact, it is much more important to apply the tools you learn here to learn something about your process so you can improve it than it is to worry about exactly which tool may be the most technically correct tool for the job.

It is important to use the right tool, but sometimes there is more than one right answer. For example, you could use an open-ended wrench, a box wrench, or a socket to tighten a nut on a bolt. They are different tools but they are all perfectly valid options. Just do not try to use a screwdriver. Results are what count. These quality technologies are just tools to help you get results.

Zone Charts

A technique that has been developed a little more recently (at least compared to the 1920s) is that of the zone chart (Figure 4-10). It is basically an individuals control chart that has been made even easier to plot and manage.

Think back to the discussion about variability concepts: 68% of your data fall within 1σ of your mean, 95% fall within 2σ, and 99.7% fall within 3σ. Zone charts are designed on the premise that you can then predict how frequently data should fall within each of these sigma zones. You can set a numerical value to data that fall within each zone.

There are many scoring systems in use, depending on how reactive the user wants the control chart to be, but the most commonly used system is a score of zero for data that fall within 1σ, a score of 1 for data that fall within 2σ, a score of 2 for data that fall within 3σ, and a score of 4 for any data that fall outside the 3σ control limits. A zone chart template might look like the example shown in Figure 4-11.

Setting up the control chart is exactly the same as setting up an $\overline{X}R$ or individuals control chart. First you have to collect 20 to 30 data points. Then you calculate the zone limits or estimate them based on the range or moving range. With this information in hand, you can construct the control chart for your future monitoring. Here is where the zone chart gets easier to use.

When plotting data on a zone chart, you do not have to worry about the exact scale—just plot the point inside the appropriate zone based on the standard deviation calculation or estimate. Every time you collect a new data point, ask yourself the question, "Did my plotted line just cross the mean?" If so, simply record the score for the new data point. If not, add the score of the new point to the existing score and record the sum. Whenever your score exceeds 4, your control chart is giving you a signal that something is wrong and you should investigate to determine the special cause.

FIGURE 4-10 Sample Zone Chart

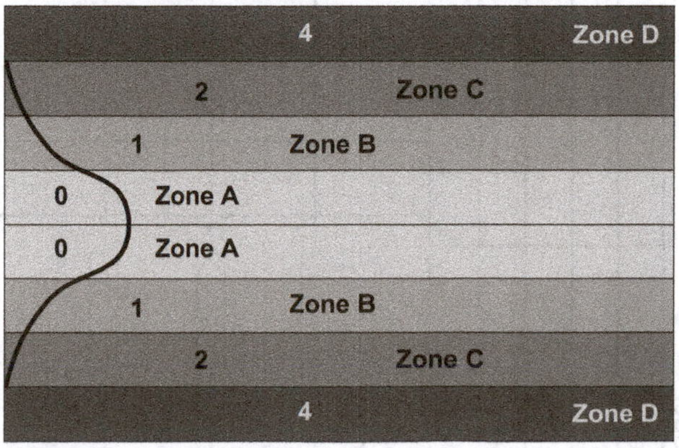

FIGURE 4-11 Example of a Zone Chart Template

ANY Corp — Zone Chart

Product: Property: Sample Point:

Date																										
Batch No.																										
Results																										
Score																										
Observation	1	2	3	4	5	6	7	8	9	10	11	12	13	14	15	16	17	18	19	20	21	22	23	24	25	
Out of Control																										4
Zone C																										2
Zone B																										1
Zone A																										0
Zone A																										0
Zone B																										1
Zone C																										2
Out of Control																										4

FIGURE 4-12 Zone Chart Limits of Dibutyl Futile

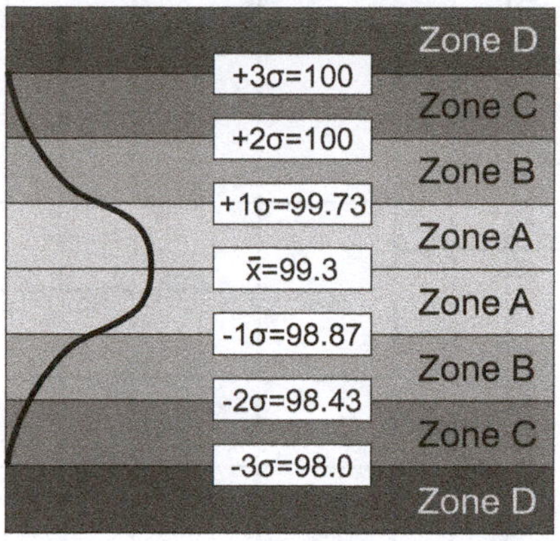

Let us construct an example using the data from the individuals control chart on the preceding pages. You will estimate the sigma zones using the moving range information already provided.

Recall that your $\overline{x}$ was 99.3 and your $\overline{MR}$ was 0.5. You used the following formula to set your upper control limit at 100.6: UCL $= \overline{x} + 2.66 \times \overline{MR}$. You then used this formula to set your lower control limit at 98.0: LCL $= \overline{x} - 2.66 \times \overline{MR}$.

Because you know that these control limits represent the $+3\sigma$ and -3σ limits, you can easily calculate the single standard deviation by taking the difference between the UCL and the $\overline{x}$ and dividing by 3. You can then add this single standard deviation to the mean and subtract it from the mean three times to set the rest of the zone limits.

For this case you already calculated that the UCL was 100.6. Subtracting your mean (99.3) from this and dividing by 3 tells you that a single standard deviation is estimated to be $(100.6 - 99.3)/3 = 1.3/3 = 0.433$. You fill in the zone chart template with the $\overline{x}$ of 99.3 and with your zone limits of $99.3 \pm 1 \times 0.433$, $99.3 \pm 2 \times 0.433$, and $99.3 \pm 3 \times 0.433$. Try this on your own and see whether you get zone limits like the ones shown in Figure 4-12.

If these limits had already been put in place, and then you collected the same data that were previously shown in your individuals control chart example, your zone chart would have looked like the one shown in Figure 4-13.

Notice that because your process is running so close to the practical limit of 100%, you rarely achieved a high score. You keep adding the new score to the old score until your line crosses the mean, then you start over with your score.

Zone charts exist to make setting up, plotting, monitoring, and interpretation of control charts as simple as possible. They do this by having you plot your data within sigma zones instead of on an exact numerical scale and by replacing the more complex interpretation rules (which will be discussed in Chapter 5) with a simpler scoring mechanism.

The types of charts explored so far are expected to be constructed with data that are completely independent of each other and for a process that is stable over time. This is not always the case, so you may see a couple of specialty control charts in the workplace. Because they are much more difficult to set up and operate, the only goal in this book is to explain in general terms what they are so you will recognize them when you see them again.

FIGURE 4-13 Zone Chart of Dibutyl Futile

ANY Corp — Zone Chart

Product: Dibutyl Futile | Property: Purity | Sample Point: Make Tank

Date	5/1	5/1	5/2	5/2	5/3	5/3	5/4	5/4	5/5	5/5	5/6	5/6	5/7	5/7	5/8	5/8	5/9	5/9	5/10	5/10	5/11	5/11	5/12	5/12	5/13
Batch No.	331	332	333	334	335	336	337	338	339	340	341	342	343	344	345	346	347	348	349	350	351	352	353	354	355
Results	99.0	98.5	99.4	99.5	99.8	99.8	99.2	99.7	99.0	99.6	99.1	99.4	99.7	99.9	99.5	99.0	99.6	99.5	99.2	99.4	99.6	99.8	99.5	99.9	99.6
Score	0	1	0	0	1	2	0	0	0	0	0	0	0	0	0	0	1	0	0	0	0	0	0	0	0
Observation	1	2	3	4	5	6	7	8	9	10	11	12	13	14	15	16	17	18	19	20	21	22	23	24	25

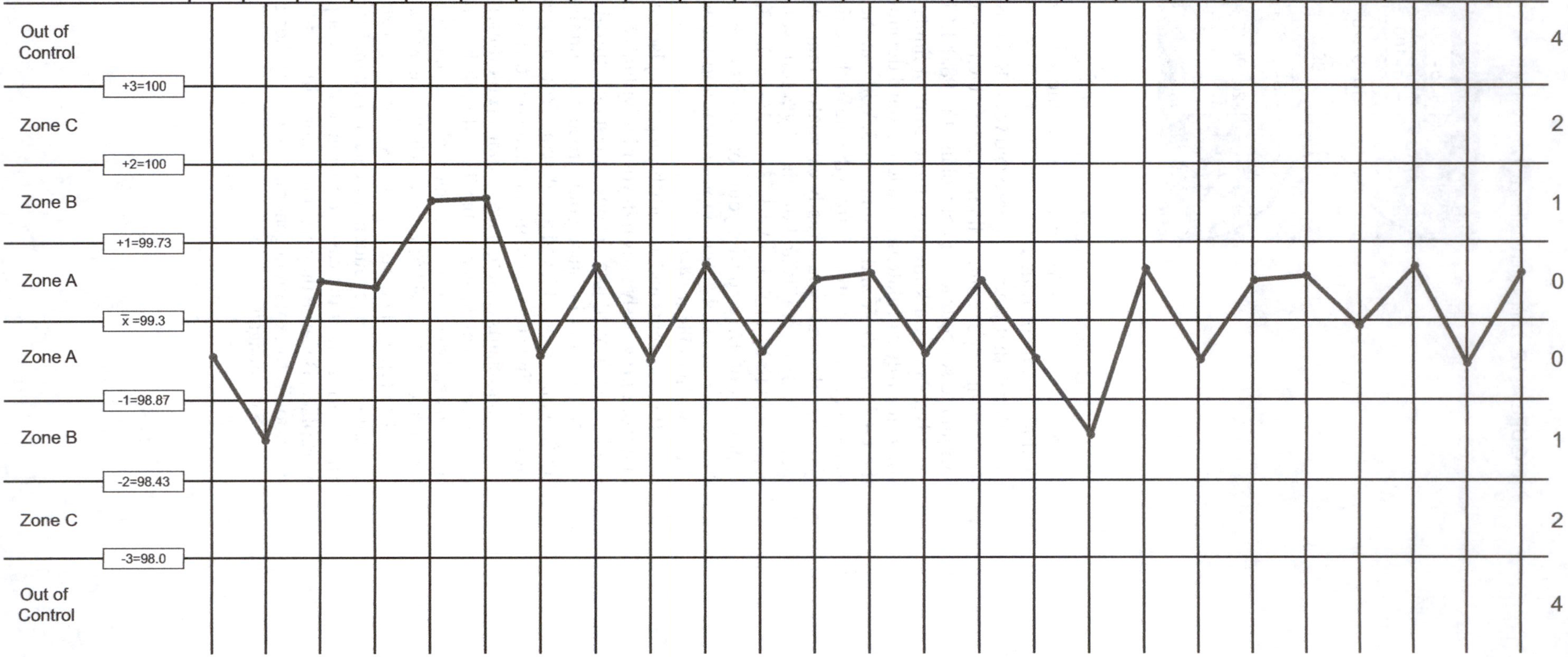

FIGURE 4-14 Example of a CUSUM Chart

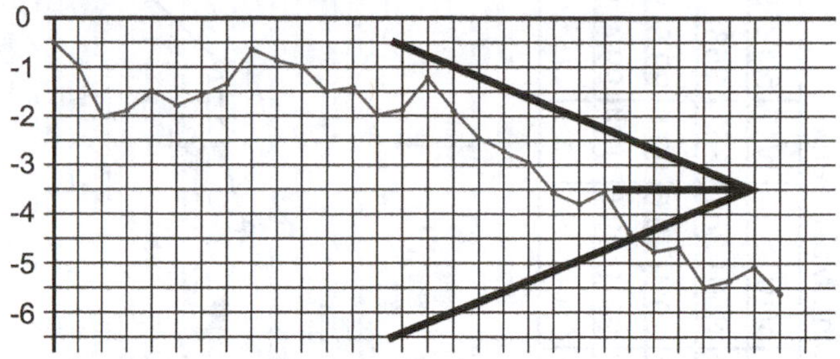

Cumulative Summation (CUSUM) Charts

Cumulative summation (CUSUM) charts are a special type of control chart designed to detect small shifts in the process mean. A CUSUM chart (shown in Figure 4-14) is a cumulative sum of the deviations away from the target, which accumulate over time.

A CUSUM chart is more similar to an individuals control chart than to an $\overline{X}R$ chart because only the individual data are plotted. The difference is that you do not calculate straight line control limits for the CUSUM as you do for the other control charts discussed.

A "V" mask is used with the chart to track the process. The mask is cut to tolerate only the allowable variation in the process as defined by the specifications, customers, and so on. One point outside the mask is an indication that the process needs immediate attention. In this way CUSUM charts achieve their sensitivity by taking into account the history of the process, not just the last result. They are easy to recognize because of the V mask, which is used in place of the control limits, as you can see in Figure 4-14.

Exponentially Weighted Moving Average (EWMA) Charts

Traditional Shewhart control charts use the last data point plotted for the decision making. The CUSUM chart just reviewed takes into account all of the previous data within the V mask. An **exponentially weighted moving average (EWMA) chart** is another way to take previous data into account when evaluating each point and making "in-control" or "out-of-control" decisions.

In many continuous processes raw materials are flowing into a reactor at the same time as final product is flowing out the other side. You will learn more about residence time in other classes, but for now just consider that if you are putting 10 gallons per minute into a reactor that holds 3,600 gallons, it will take 6 hours to fill the reactor. Once the reactor is full, you do not change out the entire volume of the reactor every 6 hours because some of what you are putting in is mixing around with what was already there and not flowing straight through (turbulent flow versus plug flow).

The engineering rule of thumb is that it takes 2½ residence times to completely change out the volume of the vessel. For this example, that means you need 15 hours to make a complete change in the reactor contents. This particular reactor is sampled every 6 hours for compositional analysis.

If a process technician is not careful, he or she could make a change to the process conditions, then take a sample 4 hours later, obtain a result that is not consistent with the new conditions, and so be fooled into making another change in the process conditions. This would be a horrible mistake because the change has not been given time to take effect yet.

In this case, the data that you are collecting every 6 hours are not totally independent of each other. The sample taken 6 hours after a process change will show only a little effect of the change. The 12-hour sample will show most of the change. The 18-hour sample will be completely representative of the change being implemented. Each data point is related to the last couple of data points, but more heavily related to the most recent point.

For cases such as these, the best choice is an EWMA chart where you plot the individual points as well as the exponentially weighted moving average, but you set your

FIGURE 4-15 Example of an Exponentially Weighted Moving Average (EWMA) Chart

ANY Corp — Zone Chart

Product: Dibutyl Futile — Property: Purity — Sample Point: Make Tank

Date	5/1	5/1	5/2	5/2	5/3	5/3	5/4	5/4	5/5	5/5	5/6	5/6	5/7	5/7	5/8	5/8	5/9	5/9	5/10	5/10	5/11	5/11	5/12	5/12	5/13
Batch No.	331	332	333	334	335	336	337	338	339	340	341	342	343	344	345	346	347	348	349	350	351	352	353	354	355
Results	99.0	98.5	99.4	99.5	99.8	99.8	99.2	99.7	99.0	99.6	99.1	99.4	99.7	99.9	99.5	99.0	98.6	99.5	99.2	99.4	99.6	99.8	99.5	99.9	99.6
EWMA			99.1	99.4	99.7	99.8	99.4	99.6	99.2	99.4	99.2	99.3	99.8	99.5	99.3	99.2	98.8	99.2	99.2	99.4	99.5	99.4	99.4	99.1	99.4
Observation	1	2	3	4	5	6	7	8	9	10	11	12	13	14	15	16	17	18	19	20	21	22	23	24	25

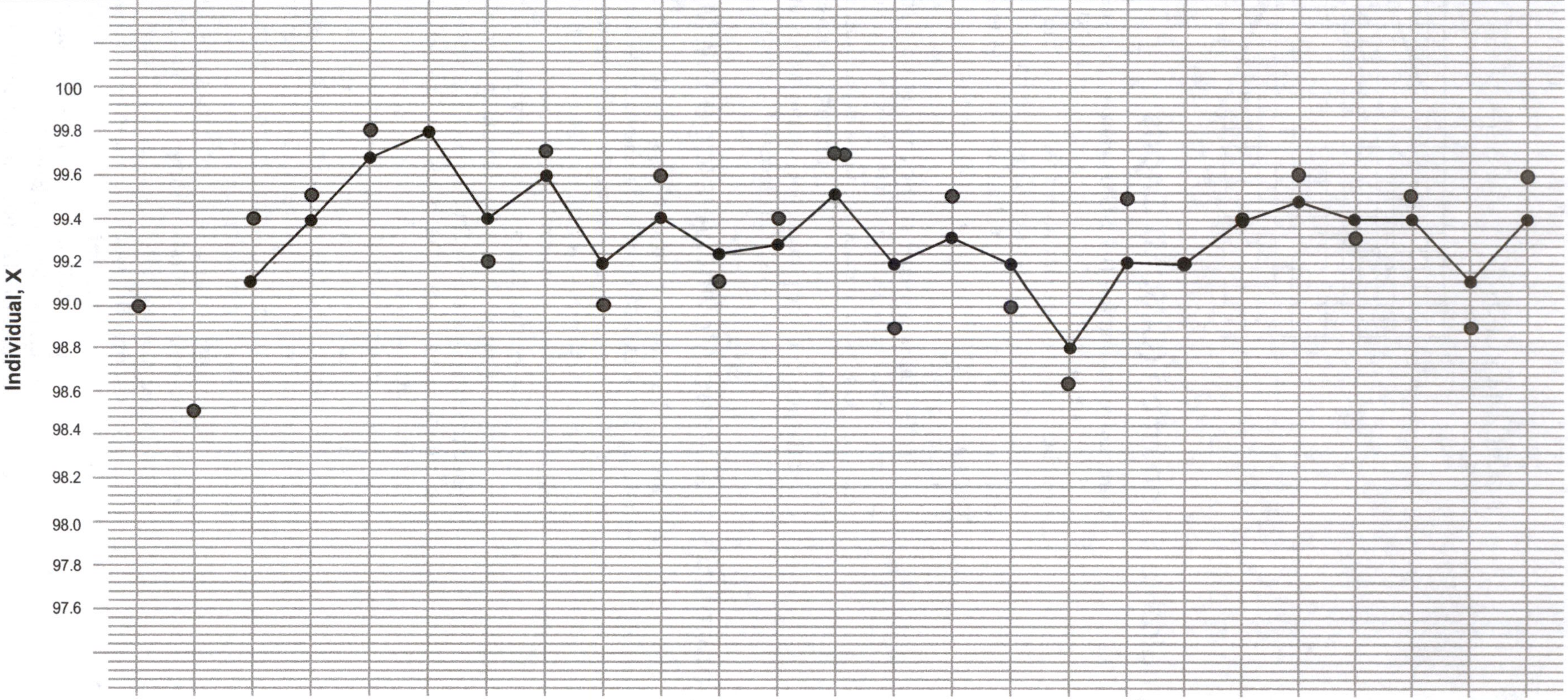

control limits around the EWMA and use those numbers to make your decisions. EWMA charts are also designed to be sensitive to small changes in the mean. Figure 4-15 shows an example of an EWMA chart. Remember, for an EWMA chart you are not concerned when an individual point is out of control, but only when the EWMA is out of control.

Summary

Control charts are some of the most basic of the quality improvement tools. Some were developed many decades ago, whereas others are much more contemporary. Control charts are a mechanism to look at your data in real time, as opposed to a snapshot in time, such as is provided by a histogram, and to use the data to monitor your process. All of the control charts discussed in this chapter are based on variables or continuous data and the normal distribution.

If your process data can be rationally subgrouped, then you can apply an $\overline{X}R$ chart. If your data are structured such that they would better be examined individually, then you should apply an individuals control chart.

If you would like to implement the simplest possible control chart, then you could substitute a zone chart for an individuals control chart.

If you have a specific need to detect small shifts in the process mean, you could apply CUSUM or EWMA charts. If your data are not independent of each other, then by all means you should be applying the EWMA chart.

As has been stated before, there are many different tools in the quality toolbox and each serves a different purpose. You need to select the right tool for the job in order to accomplish your goals as efficiently as possible. Selecting the right tool for the job requires an understanding of the purpose of the tools and a basic knowledge of when each should be applied.

By now you should know which of these control charts applies to which situation and should be able to construct and use both an $\overline{X}R$ chart and an individuals chart.

Checking Your Knowledge

1. Define the following key terms:
   a. LCL
   b. MR
   c. UCL
   d. $\overline{X}R$
2. What is the purpose of variables control charts?
   a. They help distinguish between common cause and special cause variation.
   b. They help distinguish between uncommon cause and special cause variation.
   c. They help distinguish between uncommon cause and assignable cause variation.
3. (*True or False*) Special causes are always bad.
4. Control charts add the element of __________ to your understanding of the distribution of data.
   a. Histograms
   b. Variation
   c. Time
   d. Money
5. What is the difference between individuals control charts and $\overline{X}R$ charts?
   a. There is no difference—they are interchangeable.
   b. Individuals control charts use individual measurements, whereas $\overline{X}R$ charts use averages of subgrouped measurements.
   c. $\overline{X}R$ are easier to spell.
   d. Individuals control charts cannot be used by teams.
6. Use $\overline{X}R$ charts:
   a. When the boss tells you to
   b. When your data are grouped into rational subgroups
   c. When your data cannot be grouped into rational subgroups
   d. As the default when you do not know what else to do
7. Use individuals control charts:
   a. When the boss tells you to
   b. When your data are grouped into rational subgroups
   c. When your data cannot be grouped into rational subgroups
   d. As the default when you do not know what else to do
8. What is the purpose of zone charts?
   a. They make setting up, plotting, monitoring, and interpretation of control charts as simple as possible.
   b. They are the technically correct charting technique when you are in the quality zone.

c. They can detect small shifts in the mean.
d. They take all previous data into account when determining how good your process is.

9. To detect a small shift in the mean, which control chart might be useful?
   a. $\overline{X}R$
   b. Zone
   c. CUSUM
   d. Individuals
10. A control chart that takes previous data into account but gives more weight to the most recent data is the __________ chart.
    a. EWMA
    b. CUSUM
    c. FMEA
    d. MAIC
11. CUSUM charts use a __________ instead of traditional control limits.
    a. Sea mask
    b. Mask aira
    c. C mask
    d. V mask
12. Which is the best control chart to use in the processing industries?
    a. $\overline{X}R$
    b. Zone
    c. CUSUM
    d. Individuals
    e. Whichever one is the best fit for the particular need
13. CUSUM, EWMA, zone, individuals, and $\overline{X}R$ charts are all based on the __________ distribution.
    a. Poisson
    b. Binomial
    c. Normal
    d. Rectangular
14. (*True or False*) You need to understand the purpose of the various quality tools available in order to choose the right tool for the right job at the right time.

Activities

1. You are operating a process that makes approximately 150 gallons of dibutyl futile. Due to the changes in the viscosity in this product as it is drummed, you must sample each drum as it is filled. Because each drum within a batch is logically linked to the other drums, you decide to implement an $\overline{X}R$ chart for this activity. Given the initial data below,

Observation	Drum 1	Drum 2	Drum 3
1	55.8	56.8	55.8
2	54.3	53.3	53.4
3	52.7	52.4	54.8
4	53.7	54.9	59.3
5	55.8	54.7	55.0
6	57.8	57.2	55.5
7	54.2	53.5	54.8
8	57.0	57.6	57.8
9	57.4	57.3	59.5
10	54.1	55.6	56.6
11	56.1	55.4	57.6
12	57.6	59.1	59.9
13	48.3	62.0	58.1
14	56.1	56.6	57.2
15	54.0	55.6	56.5
16	57.5	56.8	54.1
17	54.1	54.6	58.2
18	57.4	56.4	53.9
19	57.3	56.5	55.4
20	53.2	54.2	55.3
21	55.2	53.3	52.3
22	55.4	56.9	58.1
23	55.8	56.6	58.2
24	53.7	55.4	52.1
25	55.5	55.3	57.3

calculate the $\bar{x}$ and range for each batch, the mean and $\overline{R}$ for the entire data set, and the control limits that you would expect to set for this activity. Using the template at the end of this chapter, plot your data with the control limits. What can you conclude from your initial chart?

2. In the process of making dibutyl futile, you also check the modulus (measurement of product quality) of the chemical in the reactor. You get only one measurement per batch, so you implement an individuals control chart. Your initial set of data is provided below. Calculate the $\bar{x}$ and moving range for these data, the mean and $\overline{MR}$ for the entire data set, and the control limits that you would expect to set for this activity. Using the template at the end of this chapter, plot your data with the control limits. What can you conclude from your initial chart?

Observation	Modulus
1	152.7
2	152.4
3	152.6
4	151.6
5	152.8
6	152.6
7	152.9
8	151.5
9	152.6
10	154.1
11	153.7
12	151.7
13	152.2
14	153.9
15	151.6
16	152.0
17	151.0
18	151.5
19	153.1
20	152.1
21	151.1
22	153.7
23	152.8
24	152.6
25	152.7

3. After completing your individuals control chart above, you implement this chart and use it to monitor your process for the next 2 months. At the end of that time you rerun all the calculations. You determine that the process mean is 153 and the single standard deviation is 0.8. Using these numbers, calculate the zone limits for your process, and plot the following data on the zone chart template provided at the end of this chapter. What can you learn from this chart? How is your process doing?

Observation	Modulus
1	152.7
2	152.4
3	152.6
4	151.6
5	152.8
6	152.6
7	152.9
8	151.5
9	152.6
10	154.1
11	153.7
12	151.7
13	152.2
14	153.9
15	151.6
16	152.0
17	151.0
18	151.5
19	153.1
20	152.1
21	151.1
22	153.7
23	154.8
24	155.2
25	154.7

Control Chart Templates

Product | Property | Sample Point

Date																				
Shift																				
Batch No.																				
Drum-1																				
Drum-2																				
Drum-3																				
X-Bar																				
R																				
Observation	1	2	3	4	5	6	7	8	9	10	11	12	13	14	15	16	17	18	19	20

Average, X-Bar

Range, R

Individuals Control Chart

Product											Property	Purity						Sample Point							
Date																									
Batch No.																									
Result																									
MR																									
Observation	1	2	3	4	5	6	7	8	9	10	11	12	13	14	15	16	17	18	19	20	21	22	23	24	25

Zone Chart

Product: Property: Sample Point:

Date																									
Batch No.																									
Results																									
Score																									
Observation	1	2	3	4	5	6	7	8	9	10	11	12	13	14	15	16	17	18	19	20	21	22	23	24	25

Out of Control																										4
Zone C																										2
Zone B																										1
Zone A																										0
Zone A																										0
Zone B																										1
Zone C																										2
Out of Control																										4

CHAPTER 5

Attributes Control Charts

"Unfortunately, the company is usually the last to know of client dissatisfaction."
~ERNST & YOUNG

Objectives

Upon completion of this chapter you will be able to:

- Explain the purpose of attributes control charts and how they relate to their specific distributions of data.
- Explain the differences between Poisson (C and U) control charts versus binomial (P and NP) control charts, and explain when each one is used.
- Given a data set, construct a C chart.
- List the three main rules for interpreting control charts.
- Apply the rules for interpreting control charts to determine the presence of special causes.

Key Terms

Attributes data—data from a discrete distribution wherein only whole numbers (counts) are possible.
C chart—control chart for counts of defects (Poisson distribution) with a constant sample size.
NP chart—control chart for counts of defectives (binomial distribution) with a constant sample size.
Out of control—points outside the 3 sigma (σ) control limits.
P chart—control chart for counts of defectives (binomial distribution) with a variable sample size.
Shifts—runs of 8 points above or below the average.
Trends—runs of 8 points going up or going down.
U chart—control chart for counts of defects (Poisson distribution) with a variable sample size.
Variables data—data from a distribution wherein any number or fraction of a number is possible.

Introduction

The majority of the discussion thus far has been in regard to data that are based on the normal distribution as well as the proper application of quality tools such as control charts to those data. This type of data is called variables data. **Variables data** are data from a distribution wherein any number or fraction of a number is possible.

In your role as a process technician, you will frequently encounter other types of data. For example, every process industries plant tracks safety statistics such as injuries, illnesses, and spills. It also tracks quality data such as rejected batches and customer complaints.

These types of non–normally distributed data, as mentioned briefly in Chapter 2, are collectively referred to as **attributes data,** which are data from a discrete distribution wherein only whole numbers (counts) are possible. Although attributes data are typically counted as whole numbers, you will see later in this chapter that you may count whole numbers and then report fractions.

The purpose of attributes control charts is exactly the same as the purpose for variables control charts: to distinguish common cause and special cause variation. The only difference is that because the data have a different underlying distribution, you have to calculate your control limits differently.

This chapter explores several control charts that are based on various attribute distributions. Finally, some basic rules for interpreting both variables and attributes control charts are defined. This is the last chapter on basic statistics. After this chapter, the text will examine some of the other pieces of the quality puzzle.

Attribute Distributions

When it comes to health, safety, environmental, and quality data, process technicians are typically counting things or events. They count injuries, customer complaints, spills, and rejects. These counted data generally fall into one of two distributions, depending on how the data are collected.

These two distributions are the Poisson and the binomial. The Poisson distribution is the underlying distribution for the counting of defects. The binomial distribution is the underlying distribution for the counting of defectives. Until you have heard them a few times and seen them in practice, these descriptions sound much too similar to distinguish, but they are different. Let us look at each one separately to make the distinction clearer.

A Poisson distribution (shown in Figure 5-1) is a discrete distribution, made up of whole numbers only. It represents the counts of defects.

FIGURE 5-1 Example of a Poisson Distribution

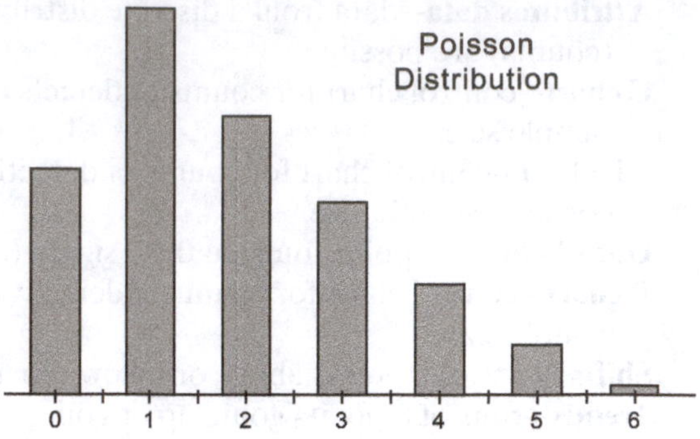

If you define your refinery population as the target of your study, then every injury or illness is a defect. If you define a single drum of chemical as the target of your study, then every dent, paint chip, or crooked label could be defined as a defect. In each of these cases what matters is that you define what it is you are studying so you can apply the appropriate improvement tool.

In each of these cases you can have multiple defects counted and still have an acceptable working process. You can experience one or two injuries per year and still operate your plants. You might receive a couple of customer complaints about dented drums or scratched labels, but you can continue to sell product to these customers. These counts indicate your process is not perfect and perhaps provide information about how to improve your process, but they do not invalidate the entire process.

A binomial distribution (shown in Figure 5-2) is a discrete distribution, made up of whole numbers only. It represents the counts of defectives.

If you define each person in your refinery population as the target of your study, then any person who is unable to fulfill his or her job duties due to an injury is defective.

FIGURE 5-2 Example of a Binomial Distribution

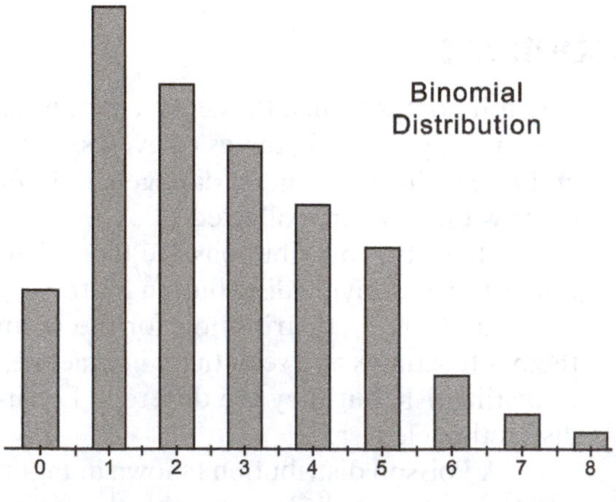

If you define a single drum of chemical as the target of your study, then you must examine it to determine whether it is defective. If you can ship it, then it may contain defects but it is not defective. If you cannot ship the drum, then you must define it as defective. If you define your process for delivering quality goods to the customer, then any batch of product that cannot be sold is defined as defective. You reject this product and do not ship it to the customer.

In each of these cases, what matters is that you define what it is you are studying so you can apply the appropriate improvement tool. As indicated in the case of the Poisson distribution, the object of your study may not be perfect and you counted the defects. In the case of the binomial distribution, the object of your study is unacceptable.

An easy way to remember the options for a binomial distribution is that there are only two options: pass or fail; accept or reject. A binomial distribution has two answers, just as a bicycle has two wheels.

How often do process technicians need to collect their counting data? The answer is, It depends on what you are trying to accomplish. If you are charged with improving the annual performance of a manufacturing facility that has a 20-year history of quality problems, you might start with an analysis of the annual customer complaint data.

Let us say you are assigned to a newer plant that has been running for only 5 years. You will probably want to start with a quarterly or monthly data analysis because you will not have enough annual data to be of value to you. Counting customer complaints per month or per quarter is a fairly common practice. Counting customer complaints per day or per week would result in too little information included in each count.

In answering the question about how to collect data, sometimes your sample size is constant and sometimes your sample size is variable. For example, let us count the number of customer complaints per month. This sounds like a constant sample size; but if you made 50 shipments last month and only 25 this month, a count of 2 complaints in each month carries a different meaning. Two complaints out of 25 shipments is much worse performance than 2 complaints out of 50 shipments. For each of the two distributions just discussed, there are two different control charts: one for a variable sample size and one for a constant sample size. Let us take at look at them.

Control Charts for Poisson Distributed Data

Having determined that your data are Poisson distributed, examine the source of your data to determine whether your sample size is constant or whether it varies. If your sample size is constant, then the appropriate control chart is a **C chart.** If your sample size varies, then the appropriate control chart is a U chart. Because a C chart is much simpler than a U chart, let us start there.

The overall methodology for control charting is exactly the same.

1. Collect a reasonable amount of data to estimate the control limits.
2. Plot the data versus time to see what you might see.
3. Impose the control limits on the plotted data.
4. Evaluate the data to determine whether there are special causes present.
5. When satisfied with the initial data collection, set up blank charts using these control limits to evaluate future data as they are collected.

Here is where this particular control chart gets really easy. The standard deviation for a Poisson distribution is the square root of the average. The formula looks like this:

$$\text{Sigma}_C = \sqrt{\overline{C}}$$

The control limits for a C chart are the same as for variables charts. The control limits are set at 3 standard deviations above the average ($\overline{C}$) and 3 standard deviations below the average. The difference is that the calculation for a standard deviation for a Poisson distribution is different from the calculation for a standard deviation for a normal distribution. Putting this formula for a standard deviation together with our 3 standard deviation control limits results in these formulas for the control limits for a C chart:

$$UCL = \overline{C} + 3\sqrt{\overline{C}}$$
$$LCL = \overline{C} - 3\sqrt{\overline{C}}$$

Let us suppose you have a product called isopropyl soapyl, and your upper-level managers are wondering how your business is doing commercially. You gather the monthly data for the last couple of years in order to have at least 20 data points. You count the number of customer complaints received every month and decide to include both major as well as minor complaints for your study. Table 5-1 shows the result.

With these data in hand, you pull up your attributes chart template and fill in the numbers. Then, you add up all these complaints (the sum is 129) and divide by the number of counts that you have (24) to get the average, or $\overline{C}$, which you calculate to be 5.4. The square root of $\overline{C}$ is the standard deviation for your data. This comes out to be 2.3. Your control limits are set at 3 standard deviations ($3 \times 2.3 = 6.9$) above the average and 3 standard deviations below the average. The following are the control limit formulas with the numbers included.

$$UCL = \overline{C} + 3\sqrt{\overline{C}} = 5.4 + (3 \times 2.3) = 5.4 + 6.9 = 12.3$$
$$LCL = \overline{C} - 3\sqrt{\overline{C}} = 5.4 - 3(3 \times 2.3) = 5.4 - 6.9 = -1.5 = 0$$

(A negative control limit does not make sense, so you use ZERO as your lower control limit.)

You can now plot your monthly counts of customer complaints and add lines for the average and upper control limit. The result should look something like the chart shown in Figure 5-3.

What conclusions, do you think, can you draw from this chart? Are things getting better or are they getting worse over this time frame? What if that first data point were not present? Would you draw the same conclusion, or is that one datum skewing your perception of the performance of this work process? What about the last 7 data points on this chart? Are these points leading you to conclude anything about a trend in the data? Shortly you will find out about some tools to help you interpret what the control chart is indicating.

TABLE 5-1 Customer Complaints over a Two-Year Period

Month/Year	*Number of Defects*	*Month/Year*	*Number of Defects*
01/06	15	01/07	3
02/06	8	02/07	6
03/06	10	03/07	4
04/06	7	04/07	4
05/06	4	05/07	9
06/06	9	06/07	2
07/06	1	07/07	2
08/06	9	08/07	2
09/06	7	09/07	5
10/06	3	10/07	2
11/06	5	11/07	4
12/06	4	12/07	4

FIGURE 5-3 Control Chart of Customer Complaints

ANY Corp

Attributes Chart (C) U P NP

Business Unit: Isopropyl soapyl (Worldwide Ops) | Chart: Customer Complaints | Description: Minor & Major

Month/Yr	01-06	02-06	03-06	04-06	05-06	06-06	07-06	08-06	09-06	10-06	11-06	12-06	01-07	02-07	03-07	04-07	05-07	06-07	07-07	08-07	09-07	10-07	11-07	12-07	01-08
No. Defects	16	5	10	7	4	9	1	9	7	3	5	4	3	6	4	4	9	2	2	2	5	2	4	4	5
Sample Size																									
No./Sample																									
UCL	12.3	12.3	12.3	12.3	12.3	12.3	12.3	12.3	12.3	12.3	12.3	12.3	12.3	12.3	12.3	12.3	12.3	12.3	12.3	12.3	12.3	12.3	12.3	12.3	12.3
LCL	0	0	0	0	0	0	0	0	0	0	0	0	0	0	0	0	0	0	0	0	0	0	0	0	0
Observation	1	2	3	4	5	6	7	8	9	10	11	12	13	14	15	16	17	18	19	20	21	22	23	24	25

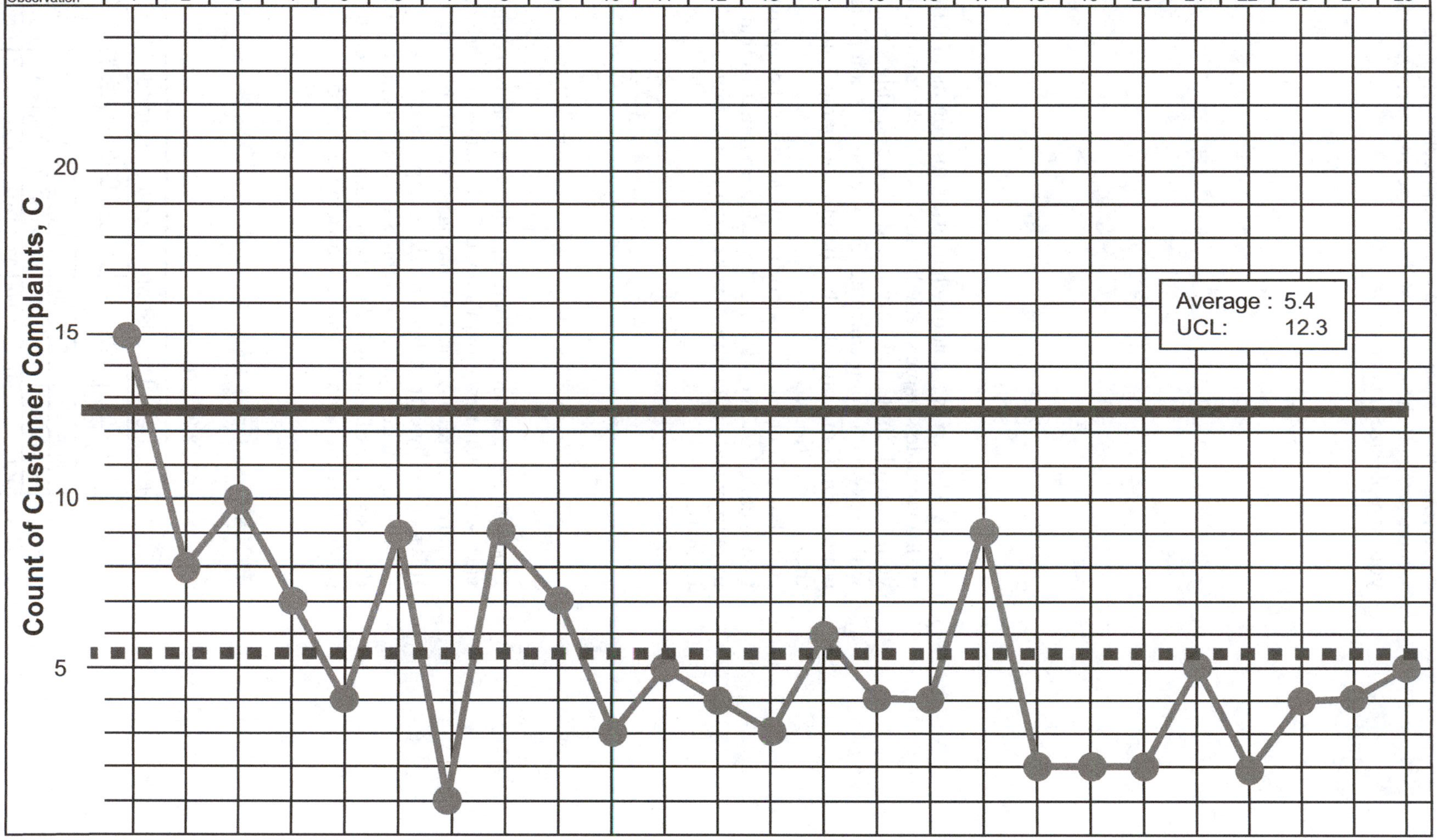

You might notice in Figure 5-3 that there is a place to collect the label for your data—in this case the month and year as well as a place to collect the count of number of defects. There are a couple of other placeholders not used. These are to collect the sample size and the number of defects per sample. These fields are needed only when your sample size is not constant.

A **U chart** is a control chart used for counts of defects (Poisson distribution) with a variable sample size. What makes a U chart more difficult to deal with is that your control limits now vary based on the sample size, and the sample size varies with every data point collected. For the sake of completeness, the formulas for the control limits are provided, but before you panic, you are not expected to perform these calculations.

For a U chart, the proportion of counts is being charted. The average proportion is called $\overline{U}$. The standard deviation is similar to that of the C chart but now you have to include a term (n) for the sample size. Figure 5-4 shows the top portion of a U chart.

$$\text{UCL} = \overline{U} + 3\frac{\sqrt{\overline{U}}}{\sqrt{n}}$$

$$\text{LCL} = \overline{U} - 3\frac{\sqrt{\overline{U}}}{\sqrt{n}}$$

Upon plugging all the numbers into the equation, you come up with control limits that vary with each new month's data. Clearly this is more complicated than any of us would care to manage on a routine basis, which is why this chart is not often employed. It is technically correct, but a hassle to implement.

Had you determined that this was the right chart to use, you would have solicited the assistance of a statistical software package to generate this chart. An example of a computer-generated U chart is shown in Figure 5-5.

FIGURE 5-4 Top Portion of a U Chart

ANY Corp

Business Unit: Isopropyl soapyl (Worldwide Ops)										
Month/Yr	01-06	02-06	03-06	04-06	05-06	06-06	07-06	08-06	09-06	10-06
No. Defects	15	8	10	7	4	9	1	9	7	3
Sample Size	600	500	550	725	350	450	550	475	625	325
No./Sample	.025	.016	.0182	.0097	.0114	.02	.0018	.0189	.0112	.0092
UCL	.0232	.0244	.0237	.022	.027	.0251	.0237	.0247	.0229	.0277
LCL	0	0	0	0	0	0	0	0	0	0
Observation	1	2	3	4	5	6	7	8	9	10

FIGURE 5-5 U Chart of Customer Complaints

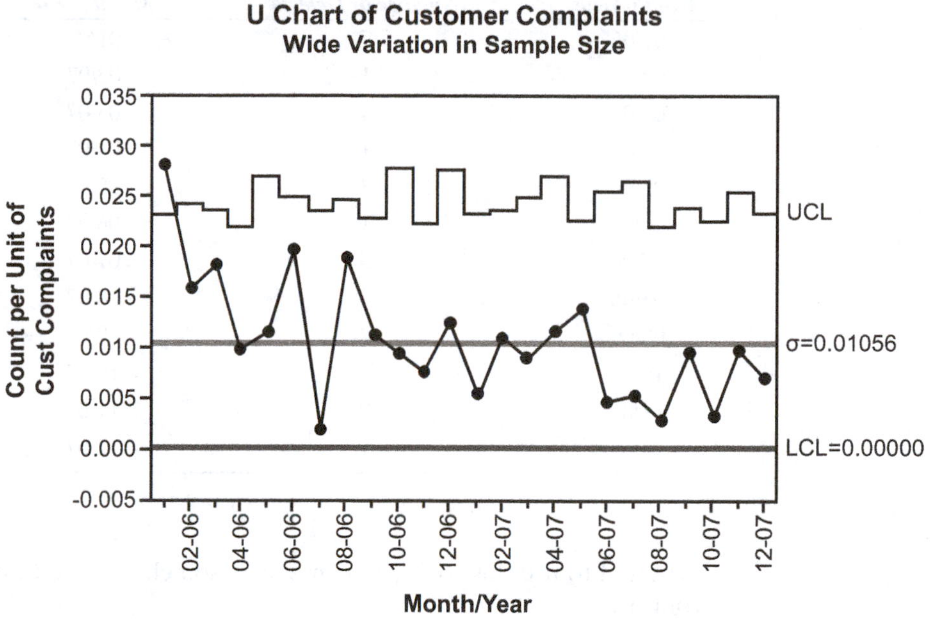

Control Charts for Binomially Distributed Data

Let us say that you are working in the supply chain portion of a chemical plant. Your job is to fill drums with a certain quantity of chemical by weight, then seal and label the drum for shipment to the warehouse. Before you let the drums go out the door, you inspect them against a checklist such as this:

- Is the net weight of the drum within 1 kilogram (kg) of the target weight?
- Are the labels applied squarely and legibly on both the top and the side of the drum?
- Is the drum in good repair? (free of scratches, dents, rust, etc.)
- Are the drum bungs secured tightly to prevent leaks?

If you note any deficiencies in the drums, then you must reject them and set them aside for rework. Shipping them to a customer would result in a customer complaint, which you are trying to avoid. In this case the data are binomially distributed. The drums are either acceptable or they are rejected. There are only two options. You are not counting how many defects exist on each drum but, rather, counting the number of defective drums. The appropriate control charts for these data would be the **P chart** if the sample size varied or the **NP chart** if the sample size was constant.

You already know that control charts for a constant sample size are easier to use, so let us start there. The data for the NP chart are shown in Table 5-2.

The centerline, or average, $(n\overline{p})$ is calculated as follows:

$$\text{Centerline} = n\overline{p} = n\frac{\text{Total Defectives}}{\text{Total Inspected}}$$

The control limits are calculated using the following formulas:

$$\text{UCL} = n\overline{p} + 3\sqrt{n\overline{p}(1 - \overline{p})}$$
$$\text{LCL} = n\overline{p} - 3\sqrt{n\overline{p}(1 - \overline{p})}$$

We are not going to go through the math for this example by hand. You need to know that this is the right tool for the job and how to use the tool; but if you are ever

TABLE 5-2 Data for NP Chart of Rejected Drums

Month/Year	*Number of Defects*	*Month/Year*	*Number of Defects*
01/06	7	01/07	5
02/06	1	02/07	2
03/06	4	03/07	3
04/06	1	04/07	5
05/06	3	05/07	1
06/06	1	06/07	0
07/06	3	07/07	1
08/06	4	08/07	1
09/06	2	09/07	1
10/06	1	10/07	1
11/06	0	11/07	3
12/06	1	12/07	0

expected to use this tool, you can expect somebody else to design it and set it up for your use.

Just so you can see the results, the NP chart is shown in Figure 5-6.

This example assumed that the sample size was constant. It did not even specify what the sample size was, but just said it was constant. If your sample size is not constant, then you must use the P chart instead. The data for the P chart are shown in Figure 5-7.

The number of drums that were filled every day varied greatly from day to day. The calculations for the P chart would be as follows:

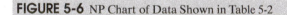

$$\text{Centerline} = \bar{p} = \frac{\text{Total Defectives}}{\text{Total Inspected}}$$

FIGURE 5-6 NP Chart of Data Shown in Table 5-2

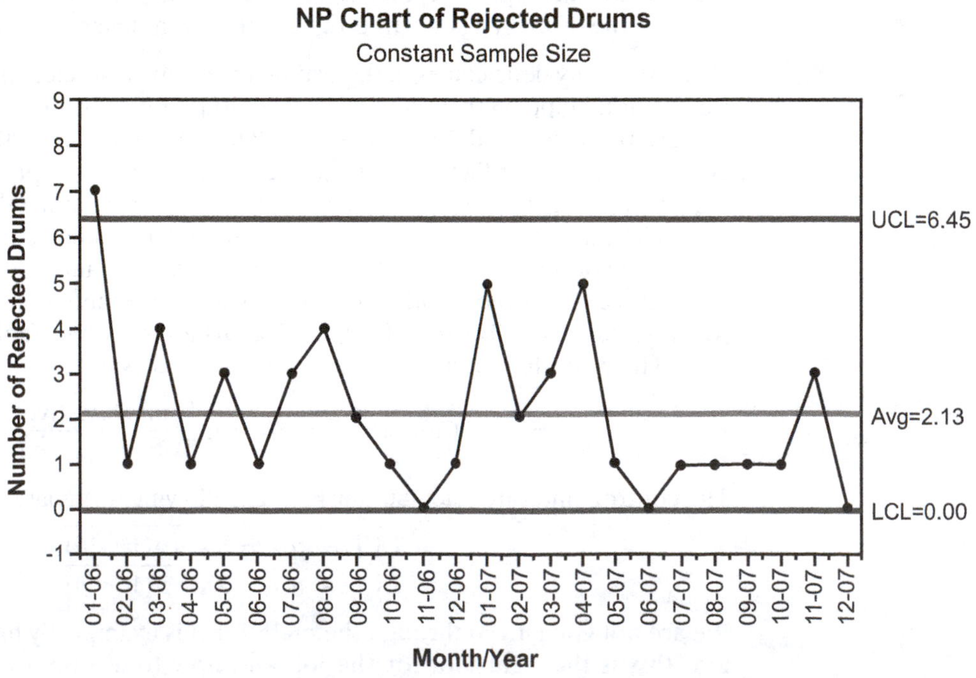

FIGURE 5-7 Data for P Chart of Rejected Drums

ANY Corp

Business Unit:										
Month/Yr	01-06	02-06	03-06	04-06	05-06	06-06	07-06	08-06	09-06	10-06
No. Defects	7	1	4	1	3	1	3	4	2	1
Sample Size	600	500	550	725	350	450	550	475	625	325
No./Sample	0.012	0.002	0.007	0.001	0.009	0.002	0.005	0.008	0.003	0.003
UCL	0.012	0.013	0.012	0.011	0.014	0.013	0.012	0.013	0.012	0.015
LCL	0	0	0	0	0	0	0	0	0	0
Observation	1	2	3	4	5	6	7	8	9	10

And the control limits would be as follows:

$$UCL = \overline{p} + 3\frac{\sqrt{\overline{p}(1-\overline{p})}}{\sqrt{n}}$$

$$LCL = \overline{p} - 3\frac{\sqrt{\overline{p}(1-\overline{p})}}{\sqrt{n}}$$

As was the case with the U chart, the P chart yields control limits that vary with each data point. The resulting control chart is shown in Figure 5-8.

Once again you find that the tool becomes difficult to employ and try to avoid this situation. Because how you define the problem determines which quality tools you will use, it is up to you to define the problem in such a way that you can effectively study the situation in as simple a fashion as possible.

FIGURE 5-8 P Chart of Data Shown in Figure 5-7

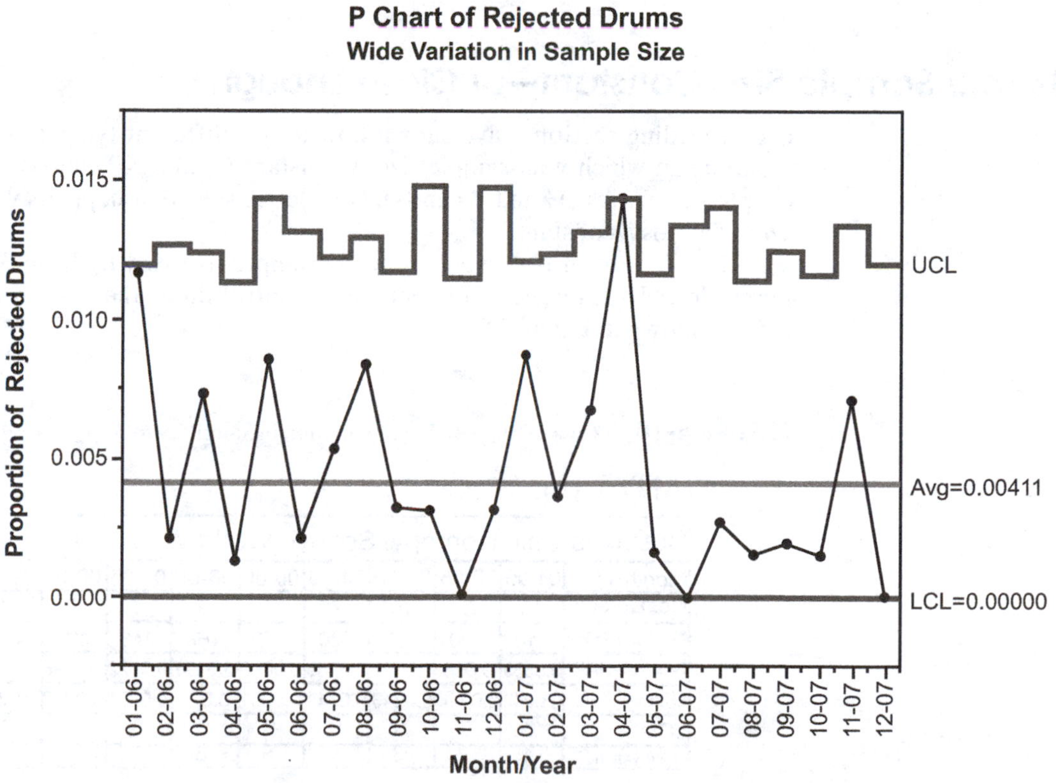

The next few pages examine the subject of sample size as well as compare these tools with each other to see what the impact on your quality might be if you used the "wrong" tool or perhaps called the sample size constant even if it really was not.

Attribute Chart Selector

Figure 5-9 summarizes which attribute chart to use in which circumstance.

FIGURE 5-9 Diagram for Selecting the Appropriate Attributes Chart

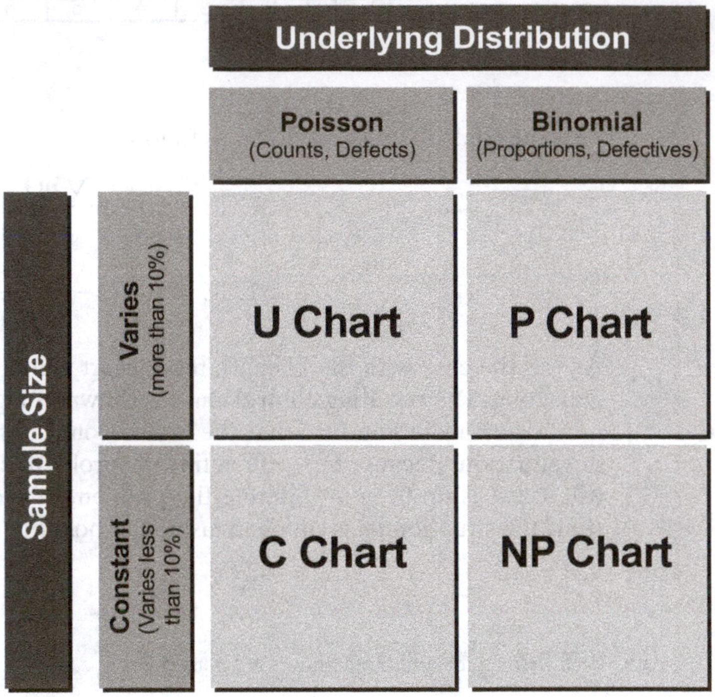

Is Your Sample Size Constant—or Close Enough?

The preceding sections have shown how to use different types of control charts for situations in which your sample size is constant (C and NP charts) versus when your sample size varies (U and P charts). Now let us see what happens when your sample size is "almost" constant.

The data from Figure 5-4 show the sample size varying from 325 to 725. This is a considerable difference and resulted in control limits that varied noticeably in the U chart shown in Figure 5-5.

FIGURE 5-10 U Chart Data with Small Variation in Sample Size

ANY Corp

Business Unit: Isopropyl Soapyl (World Wide Ops)										
Month/Yr	01-06	02-06	03-06	04-06	05-06	06-06	07-06	08-06	09-06	10-06
No. Defects	15	8	10	7	4	9	1	9	7	3
Sample Size	510	500	507	520	450	456	448	460	440	447
No./Sample	0.0333	0.016	0.0197	0.0135	0.0089	0.0197	0.0022	0.0196	0.0159	0.0067
UCL	0.0242	0.0244	0.0243	0.0241	0.0251	0.025	0.0251	0.0249	0.0253	0.0251
LCL	0	0	0	0	0	0	0	0	0	0
Observation	1	2	3	4	5	6	7	8	9	10

FIGURE 5-11 U Chart of Data Shown in Figure 5-10

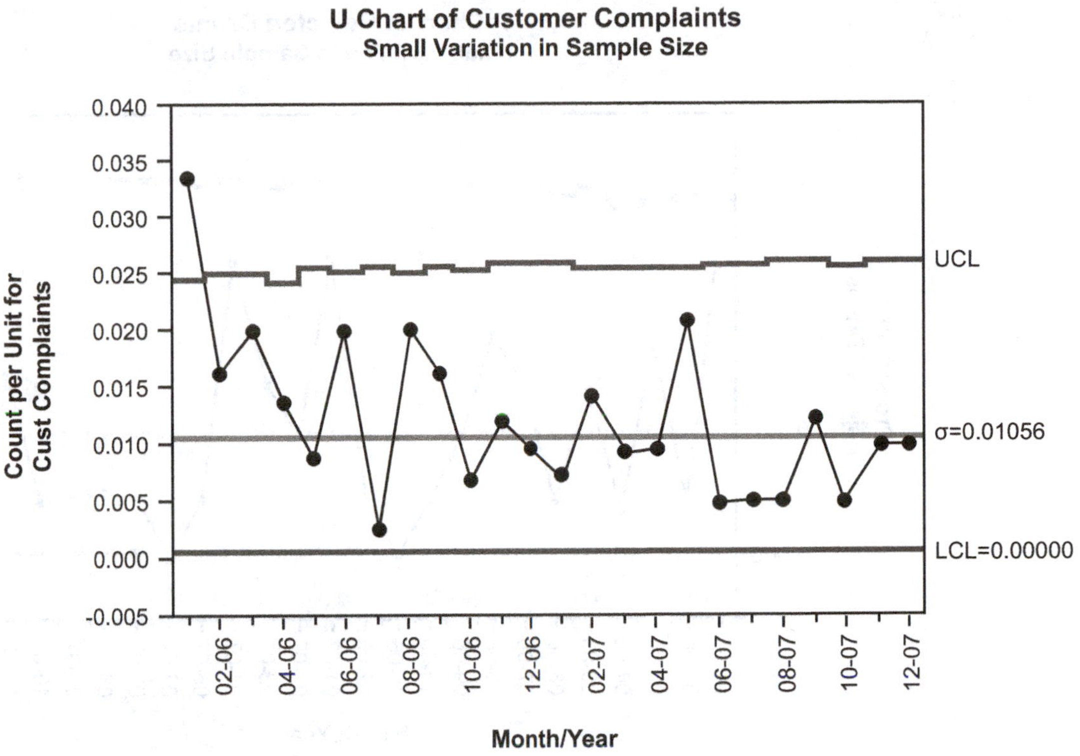

But what if your sample size data were much less variable? Would that make it easier? Consider the same counts of customer complaints but with a narrower distribution of shipments per month, which is what was used for your sample size (Figure 5-10).

Now your control limits barely vary and yield a control chart that looks a little more like what you had been using (Figure 5-11).

IN A NUTSHELL

If the sample size varies by 10% or less, you can consider the sample size to be "constant" for all practical purposes, allowing you to implement the simpler C or NP charts instead of the more complicated U or P charts.

You can see the same phenomenon when you look at control charts from the binomial distribution. The first data set had a wide variation in the sample size (Figure 5-7). This resulted in control limits that varied noticeably in the P chart shown in Figure 5-8.

Consider the same number of defects but with a less variable sample size (Figure 5-12). The resulting P chart is shown in Figure 5-13.

In both cases you see that the control limits vary little when the sample size is close to constant. For the sake of practicality, you can consider the sample size to be "constant" any time it varies by 10% or less. This allows you to use the simpler charts even when the application is not quite perfect.

FIGURE 5-12 P Chart Data with Small Variation in Sample Size

ANY Corp

Business Unit: Isopropyl Soapyl (World Wide Ops)										
Month/Yr	01-06	02-06	03-06	04-06	05-06	06-06	07-06	08-06	09-06	10-06
No. Defects	7	1	4	1	3	1	3	4	2	1
Sample Size	510	500	507	520	450	456	448	460	440	447
No./Sample	0.012	0.002	0.007	0.001	0.009	0.002	0.005	0.008	0.003	0.003
UCL	0.012	0.013	0.012	0.011	0.014	0.013	0.012	0.013	0.012	0.015
LCL	0	0	0	0	0	0	0	0	0	0
Observation	1	2	3	4	5	6	7	8	9	10

FIGURE 5-13 P Chart of Data Shown in Figure 5-12

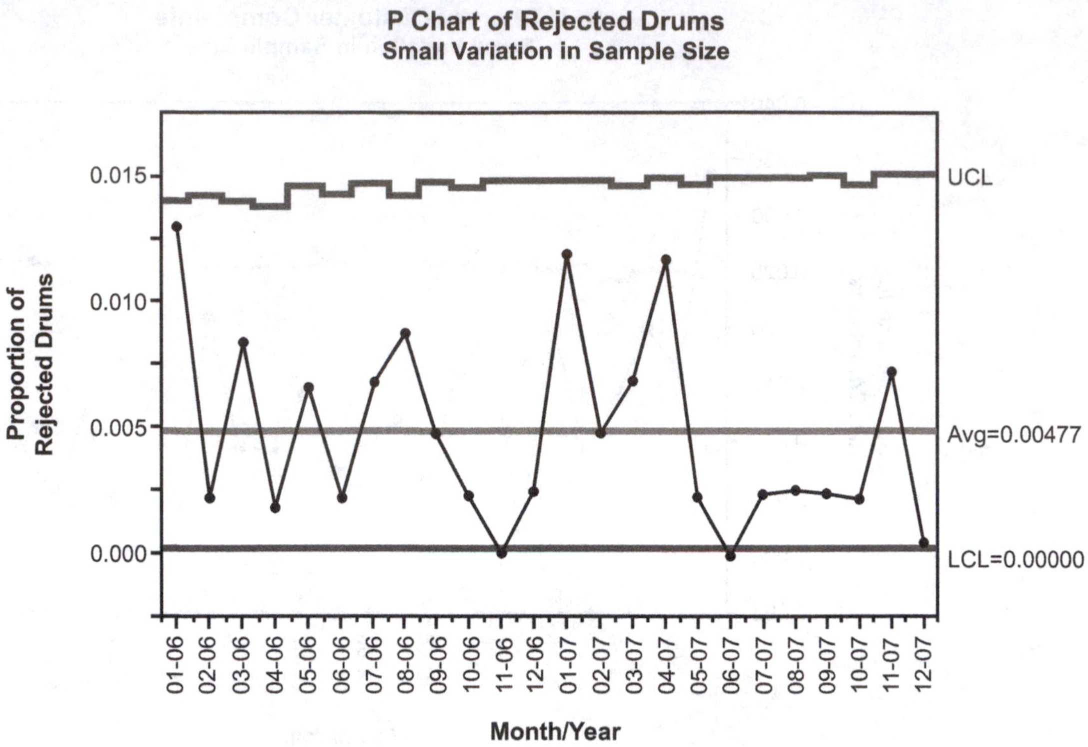

What If You Pick the Wrong Chart Type?

Here is one more topic on the subject of comparison of control charts. Even though the C (or U) chart is the technically correct chart for Poisson distributed data and the NP (or P) chart is the technically correct chart for binomially distributed data, sometimes just plotting the data and scrutinizing them using graphic tools is beneficial,

FIGURE 5-14 C Chart of Rejected Drums

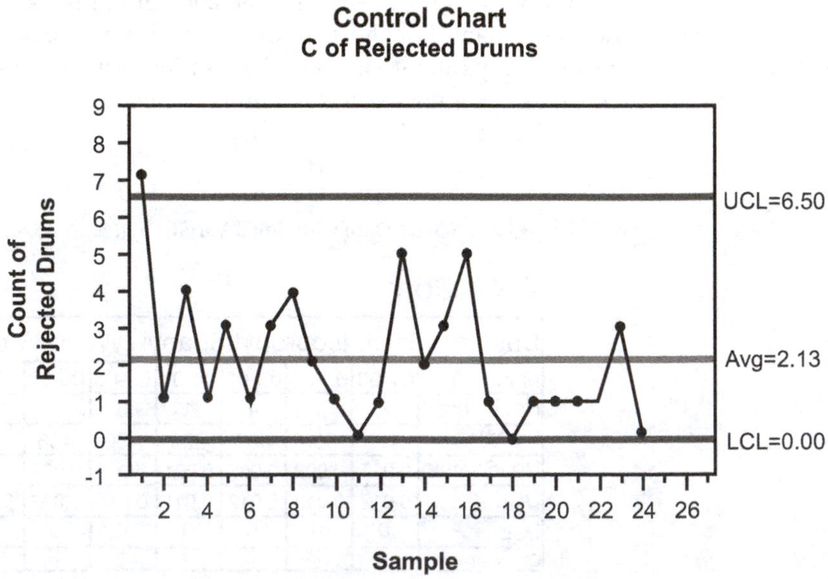

FIGURE 5-15 NP Chart of Rejected Drums

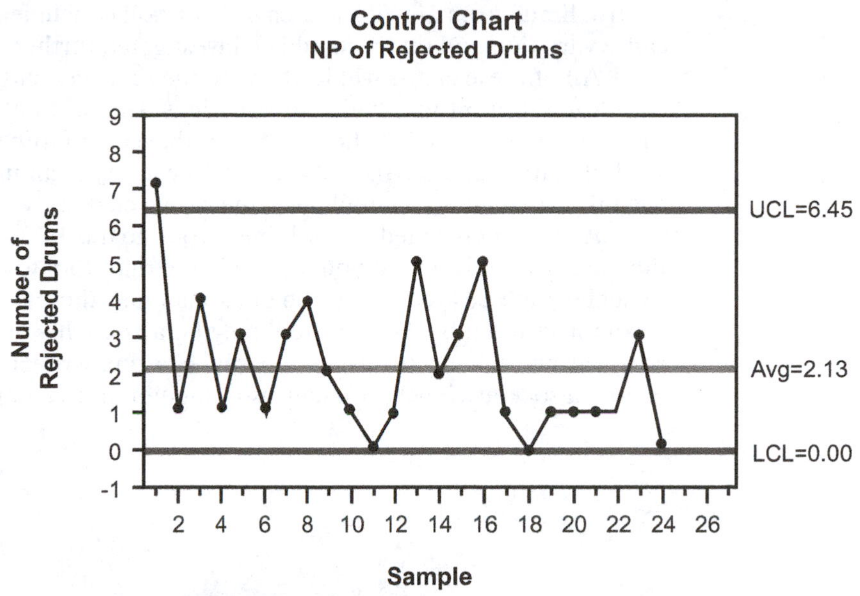

even if you do not know exactly which tool to use. You will reuse the rejected drums data set from your binomial charting example shown in Table 5-2. Plugging these data into a software package and asking for a C chart results in Figure 5-14 with an upper control limit of 6.5.

Using the exact same data but requesting an NP chart results in the control chart shown in Figure 5-15 with an upper control limit of 6.45.

Finally, putting the exact same data into an individuals control chart results in the control chart shown in Figure 5-16 with an upper control limit of 6.8.

In this particular case, the control limits calculated are all slightly different, but they are all between 6 and 7. Because you are counting the number of rejected drums, you can count a drum only as rejected or not rejected. If you count 6 rejected drums

FIGURE 5-16 Individuals Control Chart of Rejected Drums

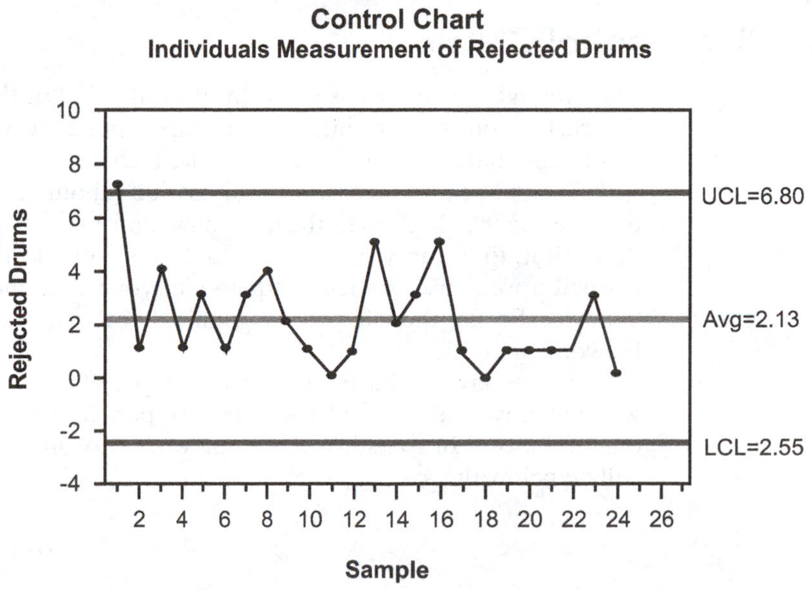

and compare with the upper control limits from any of these charts, you will conclude that this is an "okay" count. If you count 7 rejected drums and compare with the upper control limits from any of these charts, you will conclude that this is a signal of special cause variation and that it should be investigated further.

All of these charts will lead you to the same conclusion, even though they are all based on different underlying distributions. This will not always be the case, but it is surprising to some people how often it is the case. Plotting the data and applying some kind of statistical analysis will often help you make an improvement in your process, even if you might accidentally pick the wrong chart type.

As has been stated several times, your goal is to understand the various tools in the quality toolkit so that you can apply the right tool to the right job at the right time to achieve the desired improvement. Sometimes there is more than one tool that will accomplish the job. The right tool may be a wrench—but you can choose between a crescent wrench, an open-ended wrench, a box wrench, or a socket wrench. If you choose a socket wrench, you may use a 6-point or a 12-point socket.

The quality field is filled with options. Process technicians need to choose wisely from the options provided. They also need to be practical about their choice and recognize that many of the tools are really there to accomplish the same end.

Interpreting Control Charts

Even though this topic has been saved for the end of the chapter on attributes charts, this section on interpreting control charts applies to variables and attributes charts, including x-bar, individuals, C, U, NP, and P charts.

Think back to the preceding discussions about variability. Recall that in a normal distribution 99.7% of all of the data fall within 3 standard deviations of the average. By definition, then, only 0.3% of the data fall outside these control limits. You have learned how to use control charts to plot your data and detect signals of nonrandom behavior. So far, though, you have limited your signals to any points that fall outside the control limits.

Let us say you have a point on a graph. What are the odds that any given point will be below the average? There are two possibilities, and being below the average is one of the two. In statistical terms you would document the odds that any single point will be below the average as 50%.

$$\frac{1}{2} = 0.5 \qquad (50\%)$$

What are the odds that a second point will be below the average? The second point is either above or below the average. Taking the second point all by itself, the odds are still 50/50. Now, what are the odds that both the first and the second points will be below the average? In other words, what are the odds of having 2 points in a row below the average? You take the odds that the first point will be below the average and multiply it by the odds that the second point will be below the average. The odds of getting 2 points in a row below the average are thus 25%.

$$\frac{1}{2} \times \frac{1}{2} = \frac{1}{4} = 0.25 \qquad (25\%)$$

Following this same logic, the odds of getting 3 points in a row below the average are 12½%.

$$\frac{1}{2} \times \frac{1}{2} \times \frac{1}{2} = \frac{1}{8} = 0.125 \qquad (12\tfrac{1}{2}\%)$$

Extending this logic out a little further, the odds of getting 8 points in a row below the average are approximately 0.4%.

$$\frac{1}{2} \times \frac{1}{2} \times \frac{1}{2} \times \frac{1}{2} \times \frac{1}{2} \times \frac{1}{2} \times \frac{1}{2} \times \frac{1}{2} = \frac{1}{256} = 0.004 \qquad (0.4\%)$$

This means that 99.6% of the time, when you see 8 points in a row below the average, this is an indication that something is wrong. Only 0.4% of the time will 8 points in a row below the average be "normal." The reason for selecting 8 points in a row as a trigger to identify nonrandom behavior is because this closely matches the probability that a point will be outside the control limits. The reaction triggers for points **out of control** (points outside the 3 sigma control limits) have the same probability as the triggers for runs below the average.

If you were to replace the words *below the average* with *above the average* in the preceding paragraphs, you would find that the math is exactly the same. When you see 8 points in a row above the average, this is also an indication that something is wrong, 99.6% of the time. Only 0.4% of the time will 8 points in a row above the average be "normal."

Runs of 8 points above or below the average are **shifts** in the data. This is the second type of signal that something is wrong. The process has shifted from the expected average to some other value that is not expected.

A shift can be up or down from the expected average. Anytime your control chart yields a run of 8 points above or below the expected average, you need to investigate to find out what has caused your process to shift. In Figure 5-17, there is a point that almost crossed the centerline, but it is still a below-average value.

FIGURE 5-17 Example of a Shift in Data

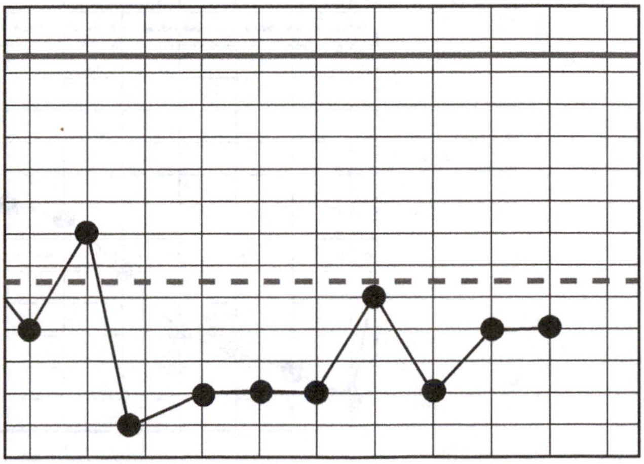

Finally, let us take a look at a similar situation. This time, instead of getting points in a row above or below the average, the process technician starts getting points that are trending up or down. Let us say you have a point on a graph. What are the odds that the next point will be above the first one? There are two possibilities, and being above the previous point is one of the two. In statistical terms you would document the odds that any single point will be above the previous point as 50%.

What are the odds that a third point will be above the second point? The third point will be either above or below the second point. Taking the third point all by itself, the odds are still 50/50. Now, what are the odds that the second point will be above the first point and the third point will be above the second point? In other words, what are the odds that 2 points in a row will be trending up? Does this logic sound familiar? The odds of getting 8 points in a row trending up are approximately 0.4%. The odds of getting 8 points in a row trending down are approximately 0.4%.

IN A NUTSHELL

Three main rules for interpreting control charts:

1. Out of control—any points that exceed the 3 sigma control limits
2. Shifts—any run of 8 points in a row below the average or above the average
3. Trends—any run of 8 points in a row that constantly increase or decrease

When you see 8 points in a row constantly increasing or constantly decreasing, this is also an indication that something is wrong, 99.6% of the time. Only 0.4% of the time will 8 points in a row moving up or moving down be "normal."

Runs of 8 points going up or going down are **trends** in the data (shown in Figure 5-18). This is the third type of signal that something is wrong. Your process is moving away from the expected average to some other value that you are not expecting. It does not matter whether your trend crosses the centerline or not. A trend this long is not normal and should be investigated.

To summarize this section, after your control chart is set up and operational, with each new point you add to the chart you should be looking for signals of special cause variation, something that indicates you should investigate nonrandom behavior in your process.

These are just a simplified set of rules for your use. There are literally dozens of control chart interpretation rules available. Some references have a lot of rules; some only a few. Some of the rules look similar but will require 6 or 7 points instead of 8 to be included in a shift or trend. Some of the rules appear to be more complicated, looking for 2 out of 3 points to be in a certain zone of the control chart. There are also rules for detecting stratification or oscillation in the data. The company you work for may have published its desired set of interpretation rules.

All of the various rules and variations of the rules exist for the same purpose—to determine when to react to the data and investigate potential problems in the process. If the cost of failure is extremely high in a certain business, then it should implement rules that have you react more quickly in order to avoid failures.

FIGURE 5-18 Example of a Trend in Data

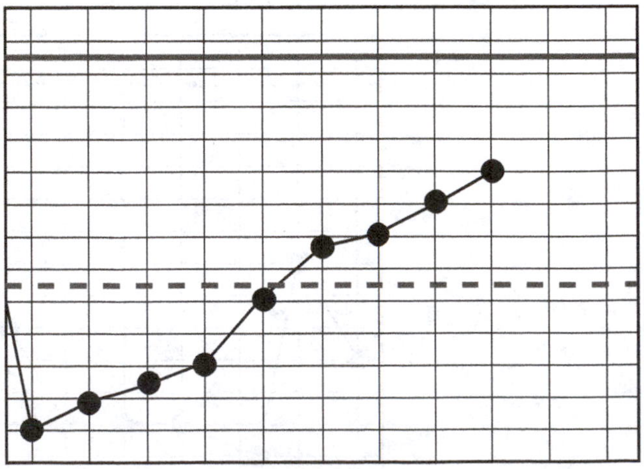

Certain businesses implement control charts with 2 sigma limits instead of 3 sigma limits. This means that they will expect 95% of their data to be inside the control limits and are willing to investigate the 5% that fall outside the limits. If the cost of failure is relatively low, then you may see rules that allow for more severe indications of nonrandom behavior before businesses expect an investigation.

Manual versus Computerized Charting

Should all control charting to be accomplished by hand, or is it okay to use computers to generate your control charts? This question has been debated in the quality field since the first personal computer hit the markets in the 1980s. Using software to perform the initial data crunching such as setting the control limits is an effective way of getting your program started quickly, and process technicians heartily endorse this methodology.

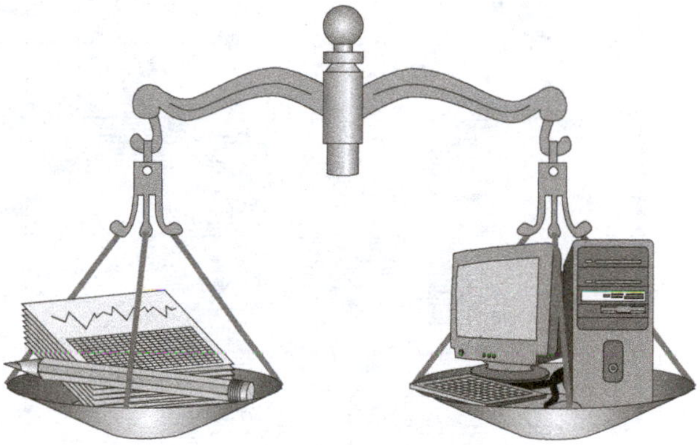

One disadvantage of using software for continuous administration of your control charting is that some people tend to dump the data into the computer, print out the charts, and then ignore the information on the charts. When you have to chart the points yourself, by hand, on a piece of paper, you are forced to spend a little time looking at the data that you are plotting.

Another disadvantage is that the true power of control charts comes from the notes captured by those using the control chart. Has the raw material changed? Make a notation on the chart so you can see the impact of this change on the process. Have you determined the root cause for an indication of nonrandom behavior? Make a note and watch the chart come back to normal. For these reasons, many organizations still require that control charts be managed manually.

Figure 5-19 is a small piece of a large control chart of a cigar lighter socket dimension from the Tokai Rika Company in Japan in the early 1980s. Notice the comments and instructions placed in the margins and below the chart. Notice how the users circled some points on the left side of the chart indicating evidence of a special cause. They also made notations below regarding what they did to fix the problem. When you see how much effort the Japanese placed on things as unimportant as cigarette lighter socket dimensions, you begin to understand how Japanese automakers took the automotive industry by storm.

There is certainly a place for computers in the world of quality; just make sure that your quality goals are being met. Sometimes, such as in the case of control charting your gas mileage, it is easier to plot a new point on a paper chart being stored in your glove box than it is to take the data back into the house, add it to the computer software, and print out an updated chart.

FIGURE 5-19 Portion of a Manually Created Control Chart.

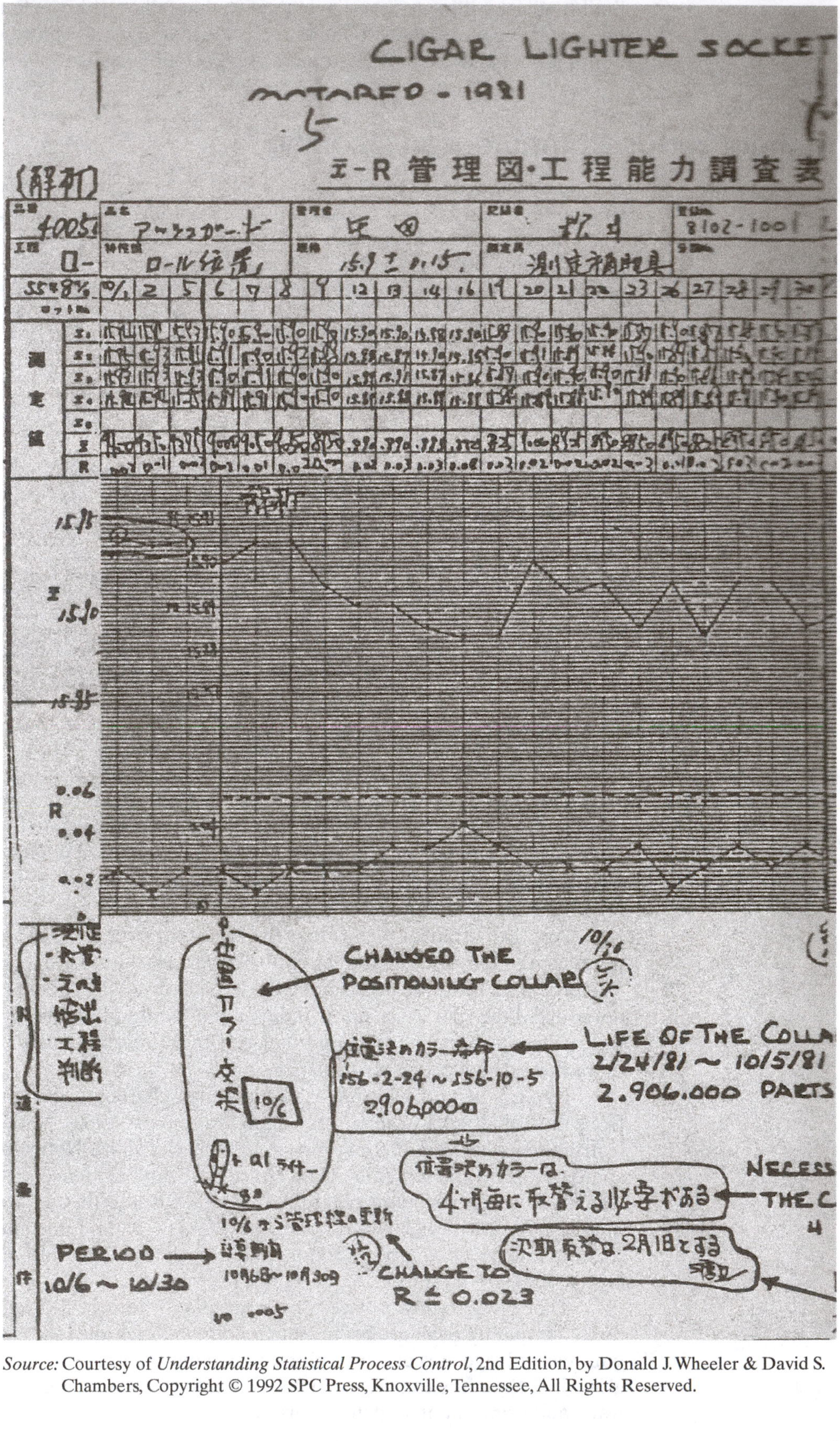

Source: Courtesy of *Understanding Statistical Process Control*, 2nd Edition, by Donald J. Wheeler & David S. Chambers, Copyright © 1992 SPC Press, Knoxville, Tennessee, All Rights Reserved.

Summary

You can construct attributes control charts so that you can apply the most effective tools to your quality improvement efforts when your data are not normally distributed.

If your data are Poisson distributed (counts of defects, safety stats, customer complaints), then use C and U charts depending on the sample size question—C charts for a constant sample size and U charts for a variable sample size.

If your data are binomially distributed, then use NP and P charts depending again on whether your sample size is constant (NP) or variable (P).

If the sample size in your data varies by less than 10%, then you can assume that your sample size is constant.

As with of the control charts examined in Chapter 4, the purpose of attribute control charts is to distinguish common cause from special cause variation. Control charts are powerful tool for accomplishing this purpose. It is often more important that you implement a control chart than it is to worry about exactly which chart you implement because they often indicate much the same thing.

When you detect special cause variation, you can investigate your process to determine the cause and remedy it. The three main rules for interpreting control charts follow:

1. Out of control—points outside the 3 sigma (σ) control limits
2. Shifts—runs of 8 points above or below the average
3. Trends—runs of 8 points going up or going down

The use of computer software to generate the initial statistics and set up your charts is absolutely supported. The decision to use computer software for ongoing plotting of control charts depends on the application of the control chart. There is a lot of value in having the charts plotted by hand, and you should not ignore this value when making the decision.

Checking Your Knowledge

1. Define the following key terms:
   a. Attribute data
   b. Binominal distribution
   c. C chart
   d. NP chart
   e. Out of control
   f. P chart
   g. Poisson distribution
   h. Shifts
   i. Trends
   j. U chart
   k. Variables data
2. C charts can be used to plot what kind of data?
   a. Measures of product quality such as purity and pH
   b. Counts of customer complaints
   c. Proportions of rejected drums
   d. Quality audit compliance statistics
3. NP charts can be used to plot what kind of data?
   a. Measures of product quality such as purity and pH
   b. Counts of customer complaints
   c. Proportions of rejected drums
   d. Quality audit compliance statistics
4. For a Poisson distribution and in a system with a constant sample size, the most correct chart to employ would be:
   a. C
   b. U
   c. NP
   d. P
5. For a Poisson distribution and in a system with a variable sample size, the most correct chart to employ would be:
   a. C
   b. U
   c. NP
   d. P
6. For a binomial distribution and in a system with a constant sample size, the most correct chart to employ would be:
   a. C
   b. U
   c. NP
   d. P

7. For a binomial distribution and in a system with a variable sample size, the most correct chart to employ would be:
   a. C
   b. U
   c. NP
   d. P
8. You assume your sample size is constant, and thus employ the simpler control charts, if your sample size varies by no more than:
   a. 5%
   b. 10%
   c. 15%
   d. 20%
9. There are three main rules for interpreting control charts. Which of these options is NOT one of those rules?
   a. Out of control
   b. Runs
   c. Shifts
   d. Trends
10. The use of a constant sample size simplifies the control charts because:
    a. Variation is bad, so you like to use only numbers that are constant.
    b. Constant sample size allows you to have only one set of control limits instead of different control limits for each point.
    c. Constant sample size does not simplify control charts.
    d. Variation in the sample size causes you to react too quickly to out-of-control data.
11. (*True or False*) The purpose of control charting is so the day staff can keep track of the shift workers' performance.
12. (*True or False*) Poisson distributions are an example of attributes data.
13. (*True or False*) Normal distributions are an example of attributes data.
14. (*True or False*) Binomial distributions are an example of attributes data.
15. (*True or False*) Control charting should always be done by hand, never by using computer software.

Activities

1. Examine the chart in Figure 5-3. Apply the three main rules of interpreting control charts. What point or points would you investigate? What do the last 7 points on this graph tell you?
2. Given the following set of customer complaint data, plot a C chart. Calculate the average ($\overline{C}$) and the control limits using the control limit formulas. Are all of these points "in control"? Do you see any evidence of special causes? A blank attributes chart template is provided at the end of this chapter.

Month/Year	Number of Defects	Month/Year	Number of Defects
01/06	3	01/07	2
02/06	0	02/07	0
03/06	3	03/07	3
04/06	3	04/07	3
05/06	2	05/07	0
06/06	1	06/07	0
07/06	2	07/07	0
08/06	3	08/07	0
09/06	1	09/07	5
10/06	0	10/07	0
11/06	2	11/07	1
12/06	2	12/07	2

Attributes Chart Template

The following page contains a blank attributes control chart template for your use in the preceding activities as well as future applications of control charts.

Attributes Chart C U P NP

Business Unit																									
Month/Yr																									
No.																									
Sample Size																									
No/Sample																									
UCL																									
LCL																									
Observation	1	2	3	4	5	6	7	8	9	10	11	12	13	14	15	16	17	18	19	20	21	22	23	24	25

CHAPTER 6

Other Basic Quality Tools

"Happy is he who can find out the causes of things."
(Felix qui potuit rerum cognoscere causas.)

~VIRGIL

Objectives

Upon completion of this chapter you will be able to:

- List and explain the steps of the plan-do-check-act (PDCA) cycle.
- Explain the purpose of brainstorming, the rules associated with it, and the circumstances in which brainstorming would be used in an industrial setting.
- Explain the concept behind the fishbone diagram, and apply that concept when capturing the results of a brainstorming session.
- Explain the purpose and benefits of work process flowcharts.
- Explain the purpose and benefits of trend charts.
- Explain the purpose of scatter plots and how they differ from trend charts.
- Explain the Pareto principle.
- Construct and interpret a Pareto chart.

Key Terms

Brainstorming—a group exercise designed to solicit a large number of ideas in a short amount of time.

Fishbone diagram—a graphic presentation device whereby ideas (or causes) are grouped into common categories such as manpower, methods, machines, and materials. Also known as an Ishikawa or cause-and-effect diagram.

Plan-do-check-act (PDCA) cycle—a cycle for continuous improvement that implements the following steps: planning, doing, checking, and acting.

Introduction

This chapter takes a break from the heavy-duty statistical stuff for a little while and discusses an overall improvement strategy (the plan-do-check-act, or PDCA, cycle) as well as some of the basic quality tools that are included in most texts on quality improvement.

Quality is not a destination—it is a journey you embark on

The PDCA cycle is, as the name implies, a continuous improvement cycle. It goes on forever. Quality is not a destination—it is a journey you embark upon. Your competitors are constantly improving their processes, so the day you stop improving yours is the day you start falling behind. As Jim Lamar put it in *Quality Management in a Wellness Program*, "Quality is not a thing we do. It is the way we do things."

The basic quality tools outlined in this chapter include brainstorming, fishbone (Ishikawa) diagrams, work process flowcharts, trend charts, scatter plots, and Pareto charts. If to this list you add histograms, which were covered in Chapter 2, and designed experiments, which will be covered in Chapter 7, you have a collection of tools sometimes referred to as the "elegant eight."

Each of these tools plays a part in achieving your overall goal of improving your processes so that you improve your products and services to the customer. If you satisfy your customers, they will continue to buy from your company preferentially, so that your company will make money and stay in business.

Brainstorming is a fairly straightforward technique for capturing the ideas and thoughts of a group of people on a given subject. This chapter applies the concept to the world of quality improvement. Ishikawa diagrams, more commonly referred to as fishbone diagrams, provide a mechanism to capture the results of the brainstorming exercise and even categorize the results.

Work process flowcharts are exactly what they sound like—a flowchart of the steps in your work process. Although this may seem trivial as first glance, it never ceases to amaze us how many improvements are identified just by assembling all of the people who perform steps in the work process and asking them to map it out for us. And after having spent a couple of chapters learning all about control charts, the trend chart section should be a breeze for you. Think of a trend chart as a control chart without the control limits. Scatter plots are graphs of two variables or parameters against each other.

The final tool to discuss in this chapter is somewhat statistical in nature—the Pareto chart, which is a type of histogram. With these tools in your toolbox, you will be well prepared to support your company's quality improvement efforts in your role as process technician.

Plan-Do-Check-Act (PDCA) Cycle

Walter Shewhart proposed that quality improvement is accomplished through a plan-do-check-act cycle. A **plan-do-check-act (PDCA) cycle** is a cycle for continuous improvement that implements the following steps: planning, doing, checking, and acting. Some companies may refer to it as the plan-do-study-act cycle. The concept is straightforward.

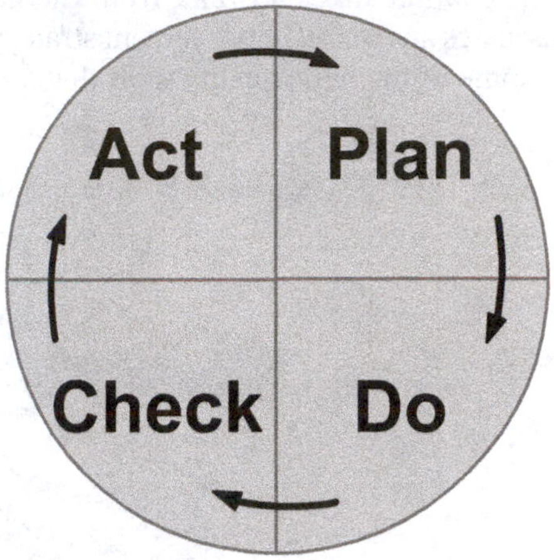

First you have to plan your attack. What is the current situation? What data exist to support the issue? What does the process look like?

Once you have studied the situation and analyzed the data, you can do something about it. In the do phase you implement solutions to accomplish improvements to your system. This is the action phase of the cycle.

In the check phase you reanalyze your data to validate that the improvement worked. Did the improvement eliminate or only reduce the problem? How much of an improvement did it make? Is further improvement needed or have you done enough?

From these questions, you decide how to act. If you have completely eliminated the problem or reduced it to an acceptable level, then you can start again in the plan phase to work on the next issue.

If your first pass through the cycle failed to address the issue or made a reduction but did not reduce the problem to an acceptable level, then you will need to revisit the plan phase to decide what your next approach will be. Either way, your review of the results of your initial improvements should always culminate in your starting over again.

Continuous improvement means that you are always looking for ways to make your processes and products better. You can never stop moving forward or you will most certainly be left behind. We have yet to come across a company or an operating unit that did not have a single area of opportunity for improvement.

The PDCA cycle is not the only strategy for accomplishing quality improvement. When you get to Chapter 10 and discuss Six Sigma, you will notice a lot of similarities between the two. Think of the PDCA cycle not as a tool to employ but as a guideline for the overall steps to follow while on your continuous improvement journey.

In your role as a process technician, you may be asked to participate on an improvement project for which the issue has already been identified. For example, you may serve on a team to reduce or eliminate unplanned shutdowns of the process. In that case, the job of the improvement team is to do the following:

Plan—collect the data around the unplanned shutdowns, analyze the data to see what the most common cause of shutdowns might be, brainstorm possible solutions, and select the most effective one.

Do—implement the selected solution, perhaps all at once—perhaps beginning with a pilot or test case to make sure the solution really works. Implementation should include updating of the documented procedures and drawings to ensure the problem stays fixed.

Check—once the solution has been implemented, you have to validate the results. Did you get all or most of the expected improvement? Were there other results obtained as a result of your implementation that were not expected? Sometimes, okay often, your processes are made up of many interconnected parts, and changing something on one end will cause a change that you were not expecting somewhere else.

This is an especially critical role for process technicians. As the ones out in the field day and night, you are the most likely to notice a small change in the process that may not be readily apparent through the normal control systems readouts. In the process industries, we have seen many, many, many instances in which a process technician that really knew the equipment could provide an early warning of pending problems because of his or her awareness of how the equipment sounded and felt. Obviously you have to trust the instrumentation for your process. You should also learn to "feel" the process because it can sometimes tell you when things are happening that your instrumentation does not measure or even tell you that the instrumentation needs attention.

Act—how you react to the results obviously depends on the results. If the problem is fixed, then you can move on to the next problem. If the problem is not fixed to the satisfaction of the business, then you need to go back to the drawing board and take another crack at it. Do not be discouraged if the first pass at any given problem does not completely eliminate it. Often there are several contributing factors to an issue, and your initial solution may have addressed only one of them. In that case, your action is to revisit the plan phase and analyze the data again. This time, because you have already made some improvement, you have more and better data to analyze.

Sometimes you learn that what you thought would address the problem did not. Even this scenario can be classified as a success if you take the time to incorporate the learnings into your policies, procedures, and drawings.

Thomas Edison was fond of saying that there were no failed experiments; for with each experiment that did not achieve the desired result, he had eliminated one of the possible options and was thus narrowing down the field. The importance of keeping your policies and procedures up-to-date will be discussed more fully in Chapter 12.

In each phase of your journey, you use the appropriate tools to help you along the way. In the plan phase, your control charts may provide excellent data to show how your process was behaving just prior to the unplanned shutdown.

A work process flowchart may be an excellent tool to ensure that everybody on the improvement team has the same level of understanding of what your process is doing. Brainstorming potential root causes and potential solutions is a great way of getting ideas on the table. Fishbone diagrams can be used to organize potential root causes. Pareto charts are designed to identify the vital few causes versus the trivial many.

The do phase is the action phase—just implement the selected solution. The check phase can employ many of the same tools. Your control charts should measure the improvements. A new Pareto chart will let you know whether the same causes are still the vital few, or if the particular causes you addressed are now part of the trivial many. The act phase is a decision point—either you move on to the plan phase for the next opportunity, or you recycle to the plan phase to take another run at the current opportunity.

Brainstorming

Whether you are trying to generate a list of quality opportunities to solve, a list of potential root causes to a certain opportunity, or a list of possible solutions to a validated root cause, brainstorming is a tool to help you get ideas on the table. **Brainstorming** is a group exercise designed to solicit a large number of ideas in a short amount of time. Keep in mind that these options are just the opinions of the participants, and data will be needed to determine which option(s) to pursue further. Most quality texts cover some variant of brainstorming.

Brainstorming can be conducted in a structured or an unstructured manner. It can be accomplished in small groups or in larger groups. In some teams, the exercise can be accomplished as a group exercise, whereas other teams will benefit greatly from having a dedicated facilitator lead the exercise. It is most likely that when you are called upon to participate in a brainstorming exercise as part of an improvement team or a root cause analysis team, specific instructions and guidelines will be set in place for the exercise. What is included here is one example of how the activity might take place.

How to brainstorm:

1. Assemble a group of 3 to 9 participants.
2. Include some subject matter experts as well as some participants who are not experts in the subject at hand.
3. Have the facilitator reviews the process to be followed, the rules for brainstorming, and the problem statement.
4. Clear off a section of the conference room or control room wall.

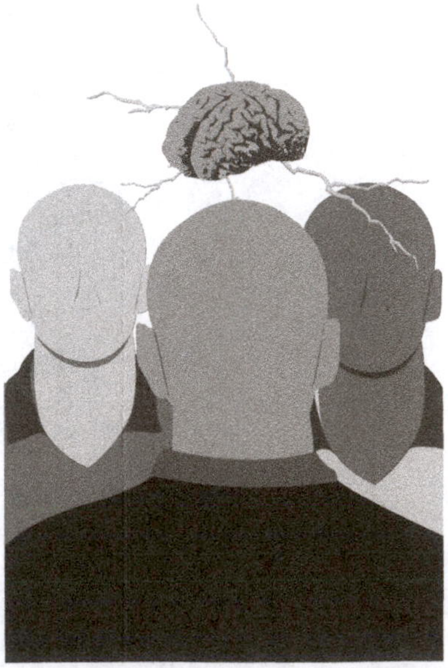

5. Pass out Post-it notes and pens to each participant.
6. Set a 5-minute time limit, and have each participant write down as many ideas as possible. Perhaps set a minimum of 10 ideas per person—one idea to a Post-it note; no talking.
7. Have each participant read his or her ideas and post them on the wall at the end of the time limit for everyone to see. Do not allow any discussion or criticism of the ideas; just voice them and post them.
8. Group together any ideas that are duplicates of ones already posted as each participant reads his or her ideas out loud.
9. Give everyone a fresh set of Post-it notes once all the ideas from round one have been posted, and instruct the group to generate 10 more ideas in the next 5 minutes. It is perfectly okay to build off of ideas already posted.
10. Have the participants read their ideas out loud one at a time and post them on the wall.
11. Run through the process again if everyone was able to quickly come up with 10 ideas. Keep pushing the process until the participants are drained of any ability to contribute further. (One way we have heard this phrased is "brainstorm to rebellion.")

It is not uncommon to see a group of 5 people generate a list of 100 ideas in less than an hour using this approach.

Some rules for brainstorming:

1. Strive for quantity.
2. No idea is a bad idea.
3. Encourage "out of the box" thinking.
4. No discussion is allowed—just post the answers as given.
5. No judging or criticizing of others' ideas is allowed.
6. Building upon an idea is encouraged.
7. Keep the process moving quickly.

Fishbone Diagrams (Ishikawa Diagrams)

One way of capturing the results of your brainstorming activities is to organize the ideas on a fishbone diagram (shown in Figure 6-1), sometimes referred to as an Ishikawa diagram after Dr. Kaoru Ishikawa, who first proposed its utility. A **fishbone diagram** is a graphic presentation device wherein ideas (or causes) are grouped into common categories. It is easy to see where the nickname "fishbone" diagram came from.

Imagine that you were looking for a small set of categories that would capture every idea ever generated. Dr. Ishikawa suggested that causes of a problem were related to either manpower, methods, machines, or materials. Somewhere down the road somebody added the category of environment as a way of capturing causes outside of humans' control, such as hurricanes and earthquakes. Others might argue that your machines and materials should be designed to handle expected natural causes, but leave that argument to them for the time being.

Let us fill in an example. A business unit held a brainstorming session during which dozens of ideas were proposed regarding what causes unplanned shutdowns in the unit.

FIGURE 6-1 Example of a Fishbone Diagram

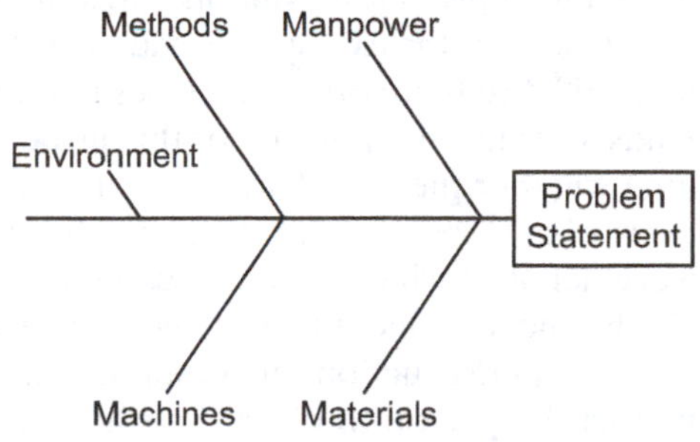

FIGURE 6-2 Example of a Completed Fishbone Diagram

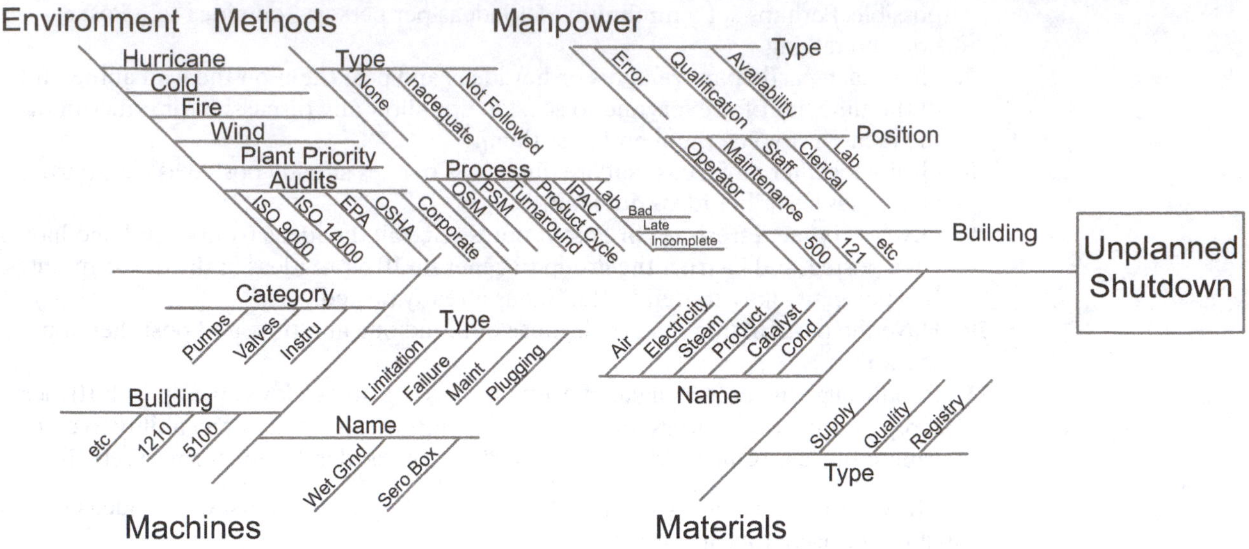

Some of the ideas were the catalyst or product type being produced. Other ideas were related to the utilities used in the manufacture of the product. The people brainstorming grouped all of these ideas under the Materials section. They could classify some shutdowns as due to quality problems, regulatory problems, or supply problems—all also materials issues. Some shutdowns were due to the specific machinery being run. Others were related to human error and further subcategorized by type, position, and building where the people worked. The end result of this classification exercise is shown in Figure 6-2.

Flowcharts

Flowcharts are an excellent method for ensuring that everybody possesses the same understanding of any given issue. The purpose of a flowchart is to provide a graphic representation of a process or a portion of a process.

Take four shift workers, put them in a conference room with the staff person over the operation, and ask them to draw a flowchart of any given work process. Each person in the room "knows" how to do the job, yet each one may approach the job from a slightly different perspective. If everyone in the room is allowed to contribute his or her perspective, then not only will the resulting flowchart of the work process be the most accurate and most complete flowchart possible, but also it may shed some valuable light on where improvement opportunities exist.

Just to illustrate the point, Figure 6-3 shows an example of a high-level flowchart for changing a flat tire. Diamonds are used to represent questions or decision blocks. Rectangles are used for process steps. Circles are often used as pointers to connect portions of flowcharts that cannot fit on a single page. If you were to consult enough different sources, you would find there are dozens of symbols that you can use to make flowcharts specific. Most people find that the majority of flowcharts can be accomplished with just a few symbols and still meet the basic needs of the users.

Consider the example provided. Is this detailed enough for your purposes? Is it too detailed? Are there other processes that tie into this one? The answer is, of course, it depends. Because the purpose of this flowchart is not specified, it is not possible to determine whether the flowchart meets the purpose. If the purpose was to describe in broad terms what steps to go through, then this is probably fine. If the purpose was to describe every detail of what tasks to accomplish in order to get the tire changed by somebody who has never done it before, then this example is probably lacking in necessary detail.

As an instruction for somebody that has never done it before, we would want to instruct the person to block the wheels and where to place the jack before raising the

FIGURE 6-3 Example of Flow Chart

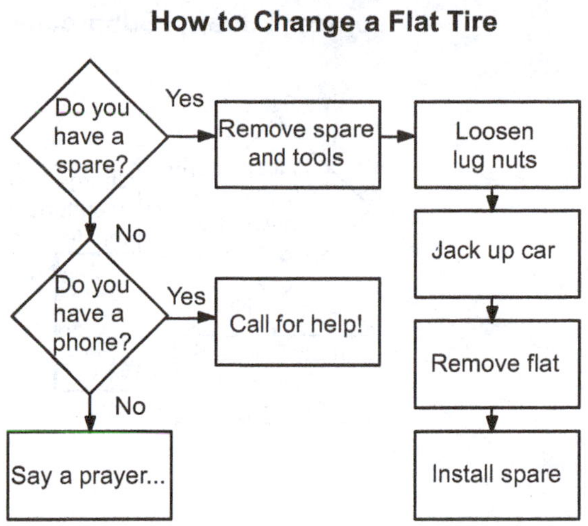
How to Change a Flat Tire
Do you have a spare?
Yes
Remove spare and tools
Loosen lug nuts
No
Jack up car
Do you have a phone?
Yes
Call for help!
Remove flat
No
Say a prayer...
Install spare

car. We would want to tell him or her to hand tighten the lug nuts when installing the spare and then to tighten them evenly to ensure proper seating of the rim on the hub. Perhaps we would want to start with a high-level flowchart such as this one and then provide a detailed flowchart for each of the major steps. This is how we typically see flowcharts implemented in the processing industries.

Here is another example. First, start with the overall process for how to make isopropyl soapyl (shown in Figure 6-4). For each of these large process steps, there are detailed work process steps.

You may be in a position to benefit from seeing the entire process in broad terms. You may be given the job of charging the raw materials, in which case you would benefit more from the detailed work process steps (shown in Figure 6-5).

Another benefit of flowcharts is that they provide a framework for analyzing the process. In this example, which is, of course, completely fictitious, the technical staff, maintenance staff, and operations staff were assembled to discuss unplanned events. The goal was to discover the reason for the recent downtime and to reduce these events to manageable levels.

With this flowchart on the wall in front of the team, each of the unplanned events of the last 12 months was discussed. Nearly every person in the room had experienced one unplanned event, but nobody had been there for all of them. By marking which process steps were involved in the historical unplanned events, they were able to determine that over 80% of their problems were related to the "heat to 150 degrees C" step. The flowchart

FIGURE 6-4 Process Flow for Isopropyl Soapyl

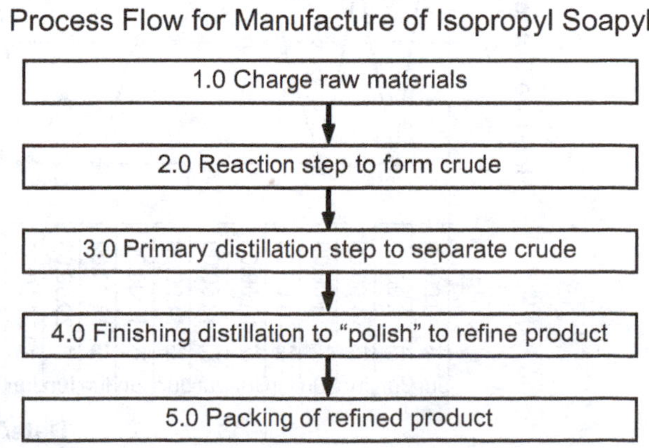
Process Flow for Manufacture of Isopropyl Soapyl
1.0 Charge raw materials
2.0 Reaction step to form crude
3.0 Primary distillation step to separate crude
4.0 Finishing distillation to "polish" to refine product
5.0 Packing of refined product

FIGURE 6-5 Example of Detailed Work Process Steps

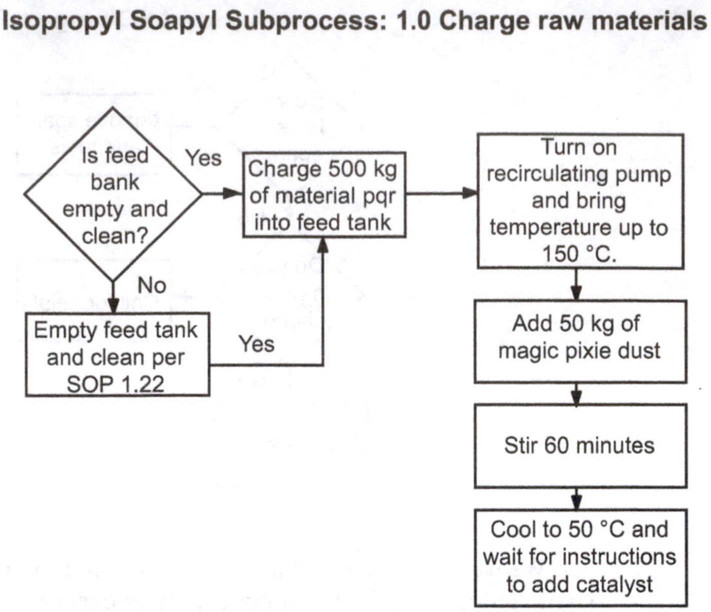

becomes a simple data collection and data analysis tool, allowing process technicians to see where the problems are occurring so they know where to direct their improvement efforts.

Trend Charts

Because preceding chapters have already covered control charts, trend charts will be a brief topic. Trend charts are simply plots of data against time to check for trends.

Consider the case of the isopropyl soapyl plant. The operators there check the moisture content of the final product stream every 2 hours. What does a quick glance at this trend chart in Figure 6-6 tell you? Are there any obvious signs that the process is trending up or down? None are visible without further analysis.

FIGURE 6-6 Example of a Trend Chart

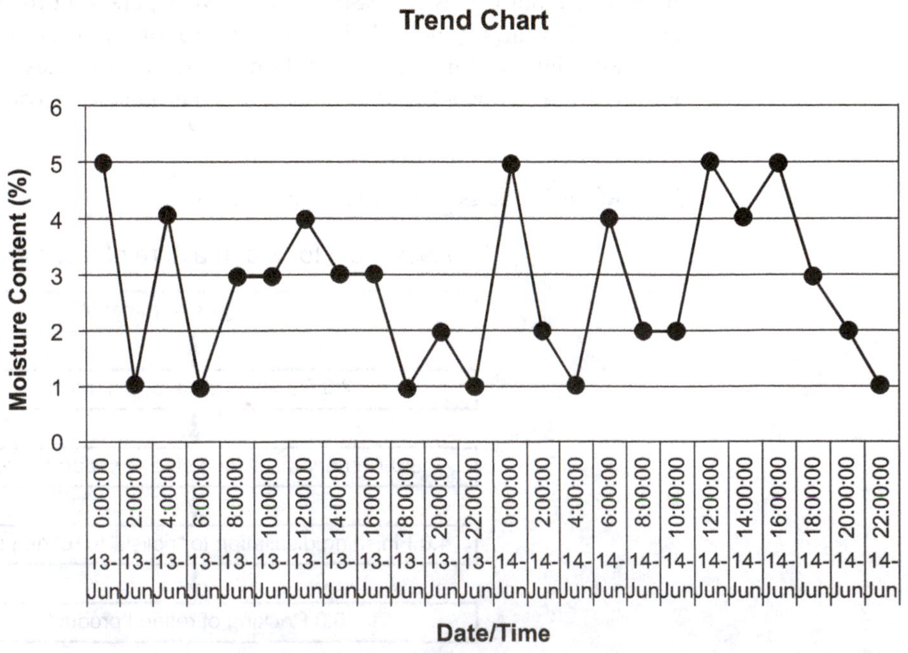

If this were an issue of ongoing concern, you would use these data to establish control limits and then monitor this parameter in an ongoing manner so that the control chart would alert you to out-of-ordinary performance. The benefit of trend charts is that they are quick and easy to plot—no statistics. If you do not know what you are looking for or how important the various parameters are, you can quickly plot a lot of variables, then visually check to see which ones seem to deserve more exploration.

Trend charts are one form of a time series chart, meaning the data are plotted against time. They show how the process changes over time. They are different from histograms, which are snapshots of how a process behaved during a specified time frame. The data on the histogram can tell you something about the distribution of your process.

The data on a trend chart can tell you how your process behaved from point to point to point. Each one has value. Each one tells you something different. As one old statistics professor used to comment, "Never underestimate the value of throwing the data up on the wall in graphical format and just staring at it for a while. You never know what you might see when you take the time to look."

Scatter Plots

Scatter plots plot two parameters against each other. For example, you could plot the same quality parameter that interested you before, like moisture content, against the temperature readout in the distillation column. Instead of telling how the moisture changed with time, the plot would tell how the moisture changed with column temperature.

In the example of a scatter plot shown in Figure 6-7, can you predict at what temperature you can achieve the lowest moisture content? Hopefully you said no. In this example, you can see that moisture contents of 1% are achieved at 50 degrees and 60 degrees and at many temperatures in between. In Chapter 7 correlations will be discussed in more detail. For now it is enough to know that scatter plots are graphs of two variables or parameters against each other.

FIGURE 6-7 Example of a Scatter Plot

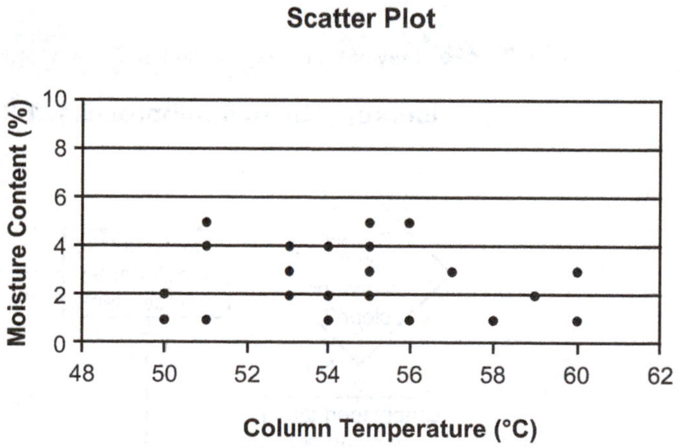

Pareto Charts

Pareto charts are named after a prominent Italian economist named Vilfredo Pareto. He was the first to note that 80% of the land was owned by only 20% of the people. These were the vital few people who controlled 80% of the wealth. The other 80% of the people were the trivial many. Combined they owned only 20% of the country's wealth.

This principle is used extensively in the quality world to describe the relationship between problems and causes. You can use this concept to prioritize your improvement efforts.

Your job is to discover which of the root causes are causing 80% of your problems. You can then work to eliminate those causes first, then reevaluate the data to see where your next priority should lie.

The improvement team went through all of the unplanned events and just marked which step of the process was active when the event occurred (Figure 6-8). As you have already learned, the "heat to 150 degrees C" step is where most of the problems were occurring.

Had you just taken raw data and charted it instead of making marks on the flowchart, you would have obtained a Pareto chart (shown in Figure 6-9). As you can see, a Pareto chart is a bar chart showing the frequency of occurrence for each of the causes or categories. The categories are sorted by count (frequency), so that the biggest numbers always show up on the left side of the graph. An added feature to most Pareto charts is the cumulative percentage curve, which has a scale on the right-hand side. This shows how many and which categories to group together to reach the 80% mark.

Present these data in this format to your management, and you will have no trouble at all convincing management that you need to improve the heat-up phase of the raw material charging operation. Put in the right format, the conclusions are obvious.

FIGURE 6-8 Tally of When the Unplanned Event Occurred

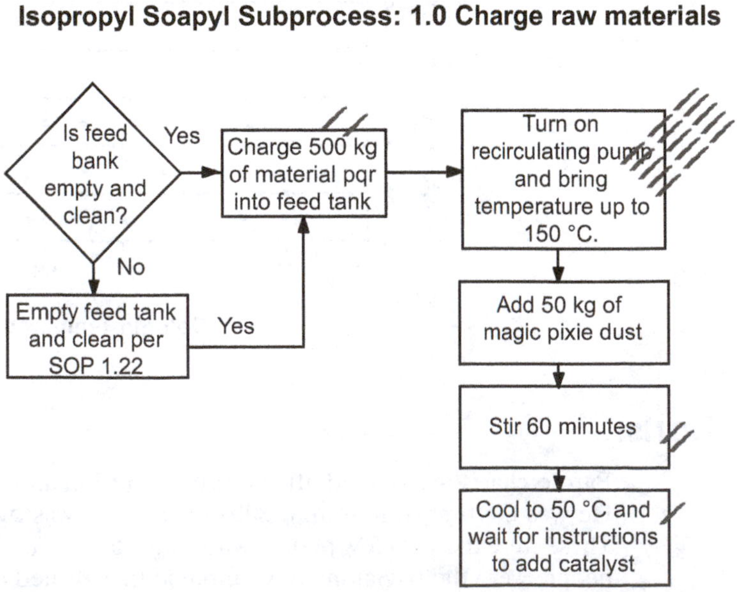

FIGURE 6-9 Pareto Chart of Unplanned Events from Figure 6-8

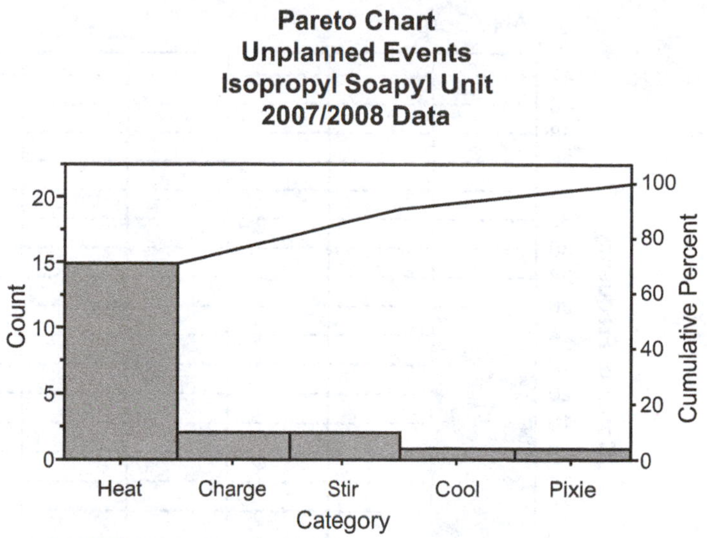

A Pareto chart is generated as follows:

1. Collect your data.
2. Sum up the number of incidents per cause.
3. Sum up the total number of incidents.
4. Divide the number of incidents for each cause by the total number of incidents, and multiply by 100 to get the percentage for each cause.
5. Sort the results so you always plot the highest numbers to the left.
6. Plot your results.

Figure 6-10 shows an example worksheet for steps 1–4, and Figure 6-11 shows the final Pareto chart.

Note that we chose to put the "unknown" category at the right-hand side of the listing of categories, even though it occurred more frequently than the bullets and arrows categories. Because we cannot fix what we do not know, there is no reason to include the unknown in the priority list of things to fix. In this example, how many of the categories would fall in the general range of the 80% rule? Hopefully you concluded only the first two. If you were to address the thorns issues and the normal wear and tear issues, you could eliminate approximately 80% of the flat tire problems.

FIGURE 6-10 Example of a Pareto Chart Worksheet

Pareto Chart Worksheet

Flat Tire Causes

Category	Tally	Subtotal	Percent
Nails	II	2	6%
Normal Wear/Tear	IIIIIIII	8	23%
Cold Weather	II	2	6%
Bullets	I	1	3%
Arrows	I	1	3%
Thorns	IIIIIIIIIIIIIIIIIII	19	54%
Cause Unknown	II	2	6%

GrandTotal= 35

FIGURE 6-11 Example Pareto Chart of Data from Figure 6-10

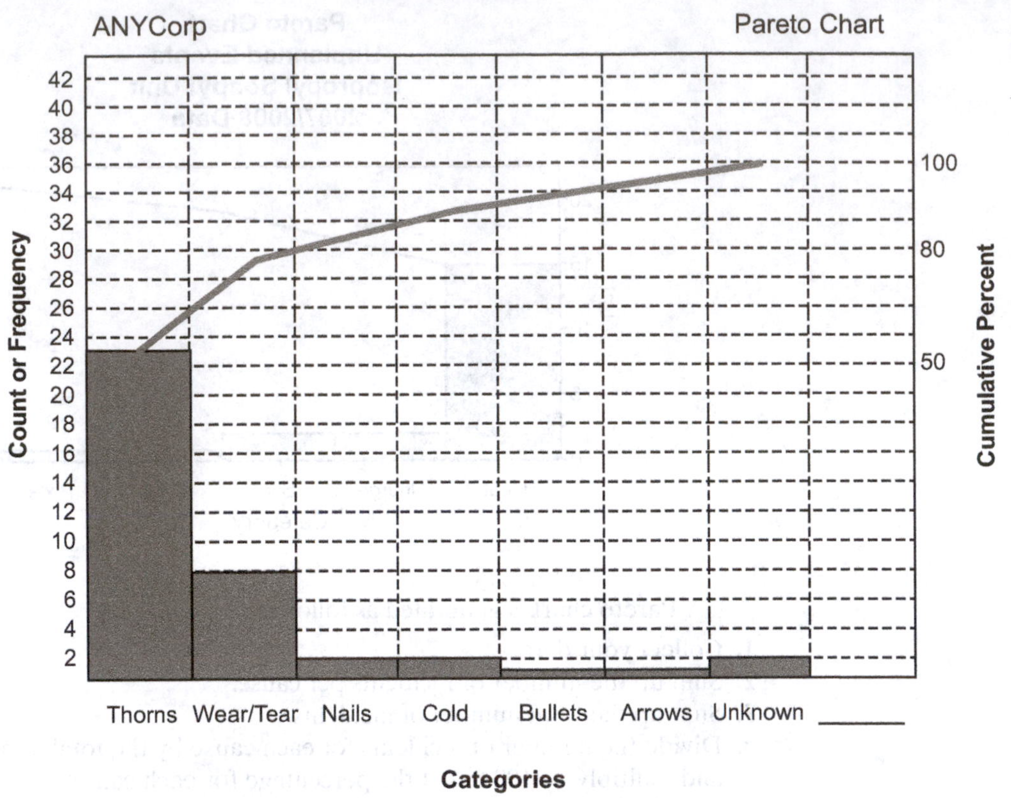

Although it is true that we occasionally experience flat tires due to nails, these are not the priority.

Summary

This chapter has provided a glimpse into the use of a wide variety of quality improvement tools. The overall strategy for quality improvement is one of continuous improvement. You plan out your improvement effort; you do that work using the tools you have learned; you check the results of your work to see whether you accomplished what you set out to accomplish; and finally you act upon the results of your process check and start the planning phase all over again. The plan-do-check-act (PDCA) cycle is at the heart of all good quality-improvement efforts. During the PDCA cycle, you employ one of the many tools at your disposal to accomplish the specific goals of that part of your quality improvement effort.

To gather ideas regarding what needs to be worked on or to generate a list of potential root causes to explore further, you might use the brainstorming technique. This is extremely common in the process industries, and process technicians provide a valuable contribution to the effort. A fishbone diagram can be used either to guide the brainstorming effort or as a mechanism to organize the results into categories for further exploration.

Once you have a list of possible root causes, you will be expected to collect data to either validate or invalidate that this truly is a contributing cause. This is accomplished through trend charts, scatter plots, control charts, histograms, or whatever tool is the most appropriate for this specific question. If you were looking to see whether a parameter was deteriorating with time, you would use a trend chart. If you were looking to see whether a parameter was statistically stable, you would use a control chart. As you learned in Chapters 4 and 5, the exact control chart that best fits the need will depend upon the type of distribution underlying the data. If you needed to know how capable your process was of meeting specifications, then you would use a capability graph.

A Pareto chart is the best tool to use when you are determining the relative contribution to an overall problem of the various causes being studied. The Pareto principle states that 80% of the problems are due to only 20% of the causes. To get the most from your quality improvement, you need to fix the vital few causes that are responsible for the majority of your problems. By knowing and applying the right tools, you help ensure the success of your quality improvement efforts.

Checking Your Knowledge

1. Define the following key terms:
   a. Brainstorming
   b. Fishbone diagram
   c. Plan-do-check-act (PDCA) cycle
2. Name three (3) basic graphic shapes used in flowcharting, and define what each shaped is used for.
3. The Pareto principle is also known as:
   a. The 80/20 rule
   b. The 30/70 rule
   c. The golden rule
   d. The 60/40 rule
4. A group technique for gathering large quantities of ideas in a short time span is:
   a. Pareto charts
   b. Brainstorming
   c. Designed experiments
   d. Group dynamics
5. Which of these options is a graphical device for grouping causes together?
   a. Pareto chart
   b. Scatter plot
   c. Fishbone diagram
   d. Group dynamics
6. Which of these options is NOT typically included on a fishbone diagram?
   a. Man
   b. Methods
   c. Materials
   d. Motors
7. According to the Pareto principle, the 20% of the causes responsible for 80% of your problems are called:
   a. The trivial many
   b. The vital few
   c. The misunderstood
   d. The Pareto pack
8. According to the Pareto principle, the 80% of the causes responsible for 20% of your problems are called:
   a. The trivial many
   b. The vital few
   c. The misunderstood
   d. The Pareto pack
9. Which tool is used to show changes over time?
   a. Histograms
   b. Scatter plots
   c. Trend charts
   d. Fishbone diagrams
10. Pareto charts have scales on both the left and the right side. What is the scale on the left axis used for?
    a. Frequency
    b. Cumulative percentage
    c. Time
    d. Weight
11. Pareto charts have scales on both the left and the right side. What is the scale on the right axis used for?
    a. Frequency
    b. Cumulative percentage
    c. Time
    d. Weight
12. (*True or False*) Trend charts show the relationship of two variables to each other.
13. (*True or False*) Scatter plots are used to show the effect of time on a parameter.
14. (*True or False*) Fishbone diagrams can be used for guiding brainstorming activities as well as categorizing the results of brainstorming activities.
15. (*True or False*) Brainstorming activities are best carried out as a group exercise.

Activity

1. The college you are attending has experienced a reduction in enrollment over the last few years and has asked the Process Quality class for assistance.
   a. Conduct a brainstorming session to generate a list of potential reasons for this decline.
   b. Organize the results from your brainstorming exercise into a fishbone diagram.
   c. Review the fishbone diagram as a group. Have each member of your workgroup "vote" for the three potential causes on the fishbone diagram that he or she believes to be the most likely culprits. Place tally marks on the diagram as you go to show which causes are garnering the most votes.
   d. Construct a Pareto chart of your results.

Pareto Chart Template

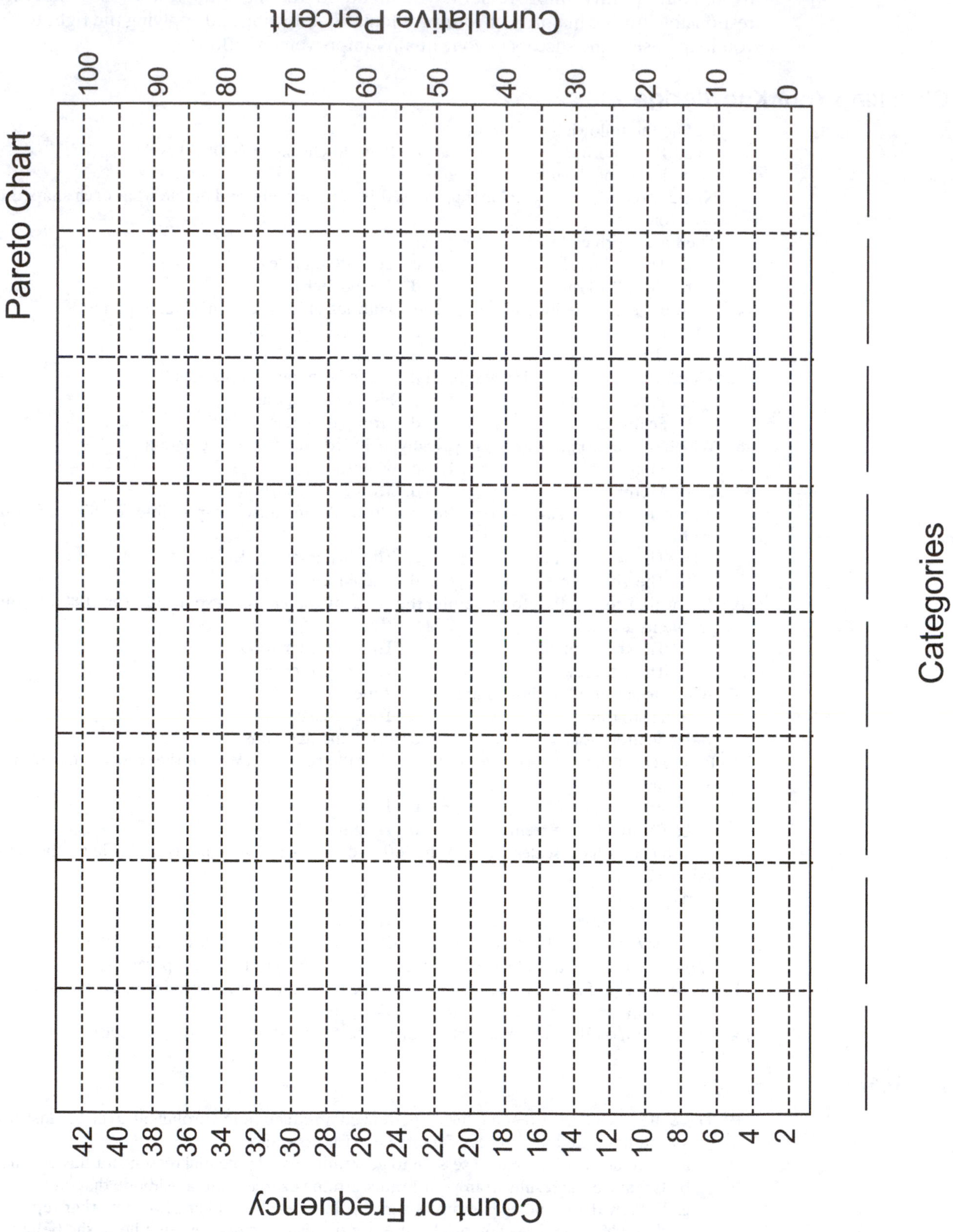

CHAPTER 7

Designed Experiments

"Life is the art of drawing sufficient conclusions from insufficient premises."

~SAMUEL BUTLER

Objectives

Upon completion of this chapter you will be able to:

- Explain the importance of modeling a process.
- Determine the direction of a correlation (positive or negative).
- Determine the strength of a correlation using the *R*-squared value.
- Explain regression and the types of regression.
- Explain how designed experiments generate data that allow you to model a process.
- Explain the process technician's role in designed experiments.

Key Terms

Correlation—the mutual relationship of two or more things; the degree to which two or more attributes tend to vary together.

Performance model—the mathematical model that predicts how your process will perform based on the input variables.

Regression—the statistical analysis that generates a mathematical representation of the correlation between two or more variables.

Introduction

Welcome to the last chapter on statistics in this textbook. As mentioned in Chapter 6, designed experiments are the eighth tool in the collection referred to as the elegant eight. Are you ready for more good news? Even though designed experiments are definitely a statistical topic, discussion of it in this textbook is strictly in conceptual terms. You will not need a calculator, and any formulas you see are just for understanding.

The statistical quality tools covered in the first six chapters are mainly used for looking back into the history of the process to see how you have done. You used historical data to perform the capability studies that define "normal" for your processes. Based on what your process has done in the past, you learned how to use control charts to monitor your process's performance in the present. The next step in your statistical quality pilgrimage is to look ahead into the future.

In order to predict the future performance of your process, you need to have a model of your process. This does not mean a plastic model that shows what your process looks like, although that type of model is also useful; it means a performance model. This type of **performance model** typically looks like a mathematical formula that predicts how your process will perform based on the input variables. With a process model in your hands, you can predict the outcome of changes you make to the process control scheme. Without a model, you are basically flying blind and hoping everything will turn out okay.

The question then becomes, How do we generate a model? The answer is, by using data from the process. Here is one of the cardinal rules of modeling: your model is good for predicting process performance only for the range of data that you used to calculate the model. In other words, if you ran your process between 100 psi and 110 psi and calculated a model for how much product your process generated, you cannot trust that this model is useful at 150 psi. The model has been validated only between 100 and 110.

It is quite common for your processes to be held in tight control; consequently, you may not have any data at 150 psi. You might extrapolate what should happen as you go outside the range of 100 to 110, but you do not know what will happen. This poses a bit of a problem.

In order for your model to be useful, you need to collect data over as broad of a range as is safely possible, but running your processes at widely varying conditions is completely the opposite of the consistent operation you try to achieve in order to produce a quality product. To address this issue, use designed experiments, which are short, logically designed experimental runs whose purpose is to collect the greatest amount of useful data over the broadest range possible with the least amount of waste.

This chapter introduces you to the topic of correlation and shows how to determine the direction and strength of correlations. It briefly discusses the use of regression analysis, which is the mechanism for turning data into models, and then covers designed experiments and the role of the process technician in making them work. A small-scale Roman catapult will be used to demonstrate the concepts of designed experiments in a manner that can be applied in the classroom.

Correlation

Correlation is derived from "co" and "relate," meaning to relate together. A **correlation** is the mutual relationship of two or more things, or the degree to which two or more attributes tend to vary together. Let us introduce you to a mathematical formula.

$$Y = f(x)$$

FIGURE 7-1 Graphs of the U.S. Population and the Brown Pelican Population

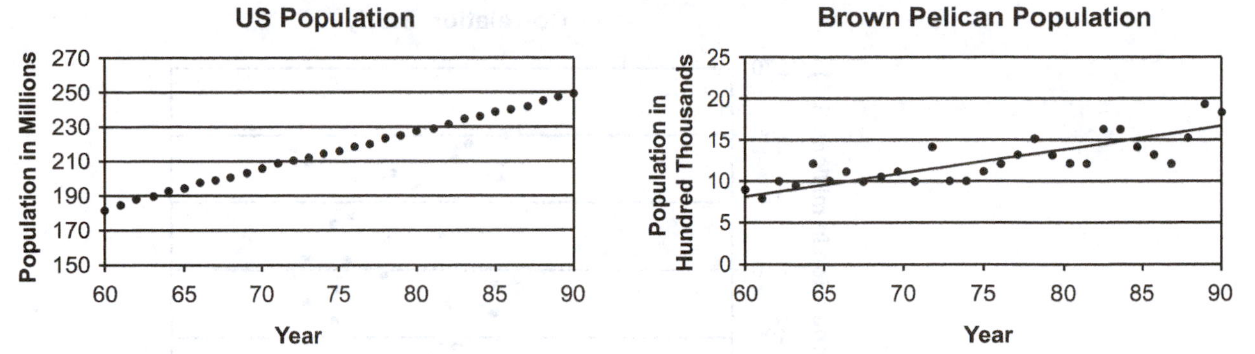

This is read as "*Y* is a function of *x*." *Y* is your process performance, the output of your process. *X* is one of the inputs into the process. Let us say that your process is boiling water on the stove. *Y* is the temperature of the water in the pan. *X* is the heat being applied to the pan. As you turn up the heat applied to your pan, you expect to get hot water. The temperature of your water is a function of the heat applied to the pan. Is the heat applied to the pan the only variable that affects the heat of the water? No. How much water is in the pan? How cold was the water when you put it in the pan? How long have you been applying heat to the pan? Each of these variables could be measured and your formula updated to be more precise. The new formula might take this form:

$$Y = f(v, w, x, z)$$

which you would read as "*Y* is a function of several variables." In this example, the heat of the water is a function of how much water you have, how cold the water was when you started, how much heat is applied, and how long you have been applying the heat. Have all of the possible inputs been named? Assume you do name another input, as no model is completely perfect. There is almost always another variable that might help make the model better. One of the most respected industrial statisticians of our era, George E. P. Box, is quoted as saying, "All models are wrong; some models are useful."

Consider the graphs of the human population and the population of the common brown pelican in the United States from 1960 to 1990 shown in Figure 7-1.

Each one appears to be increasing, so they must be correlated. In order to check for a correlation, the graph shown in Figure 7-2 was created.

FIGURE 7-2 Correlation of the Brown Pelican Population and the U.S. Population

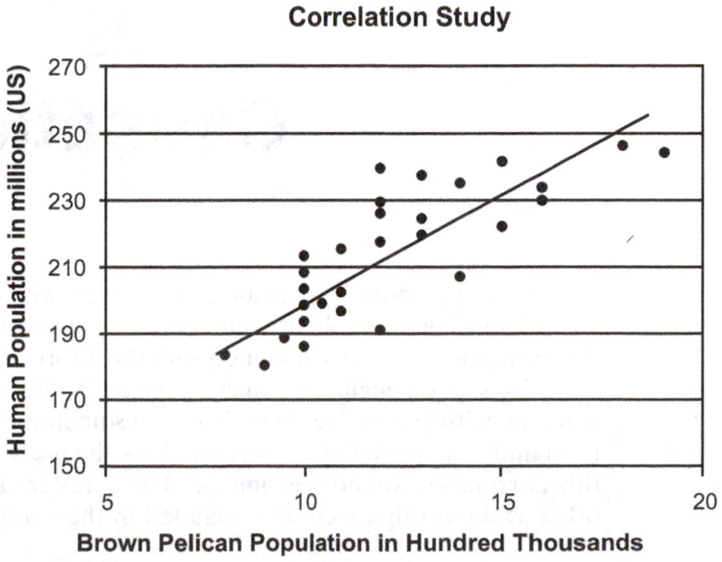

FIGURE 7-3 Correlation of the Wood Stork Population and the U.S. Population

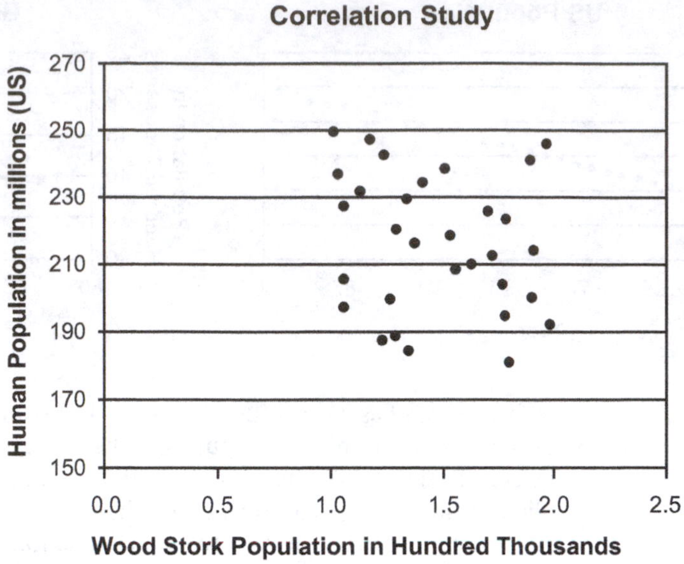

It looks like there is a relationship. You can draw a line through the data that show the population of humans increasing during this 30-year time span just as the population of pelicans was increasing.

Let us now check the population of humans versus the wood stork, shown in Figure 7-3.

There is not much of a pattern in this graph. In fact you would have a tough time drawing any conclusions at all from this. There does not seem to be any correlation between the human population and the stork population. What do you conclude? Humans are not delivered by storks but by pelicans.

Correlation does not equal causation. Drawing a model of human population shows a correlation to pelican population. That does not mean that pelicans deliver babies. It does not mean that they do not, but it does not mean they do. In this case, both populations were on the rise during this time frame, so the data correlate.

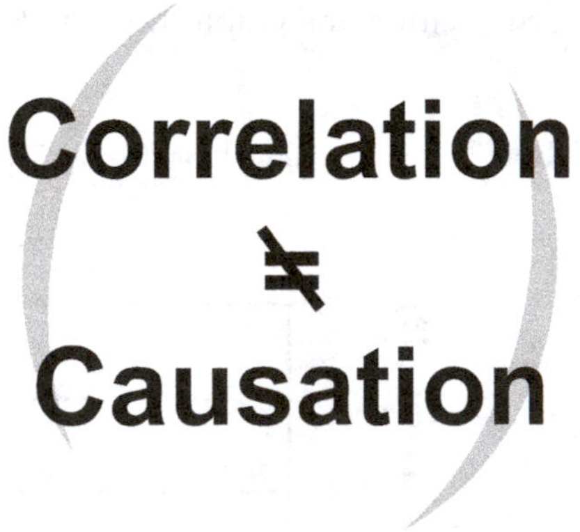

Blindly performing statistical calculations with a computer does not make you any more knowledgeable about your process. It is not the performing of the calculations that is important; it is what you do with the information you gain from the calculations.

Sometimes engineers run experiments, gather data, and run statistical analyses to come up with impressive charts. To learn something from these analyses, everyone has to sit and stare at them for a while and ask themselves how meaningful they look. Does this correlation sound reasonable? Have any laws of physics been violated? Were other variables that were not included in the study changing at the same time? The process technicians play a key role in understanding what the results from these types of experiments mean. They are there all night long. They see things that the day staff does not see.

DIRECTION OF CORRELATION

Now that you know what a correlation is, and what it is not, let us take a look at a couple of characteristics of correlations. Look at the direction of the correlation. If both properties go up together, a line starts at the lower left and goes up and to the right. These variables are said to be positively correlated.

Direction of Correlation

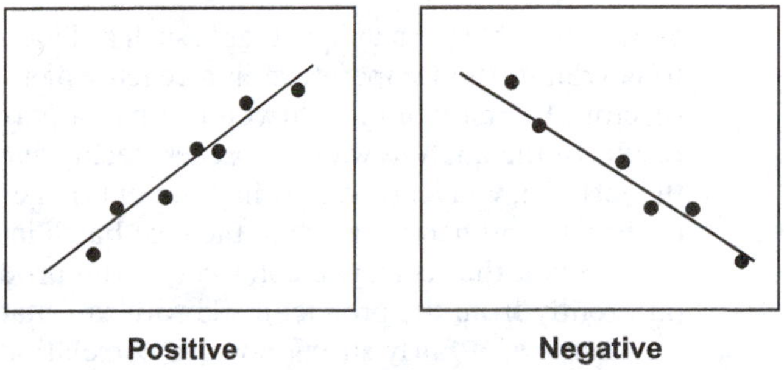

Using the same example, if you want to boil water, you have to increase its temperature. Therefore, as the temperature of the water increases, the amount of time required to boil will be reduced. These variables are negatively correlated. Negatively correlated variables will result in a chart that goes down and to the right.

STRENGTH OF CORRELATION

You can visually tell how strong a correlation is by looking at the graphs. Let us look at the correlations between human population versus the pelican and stork again.

Strength of Correlation

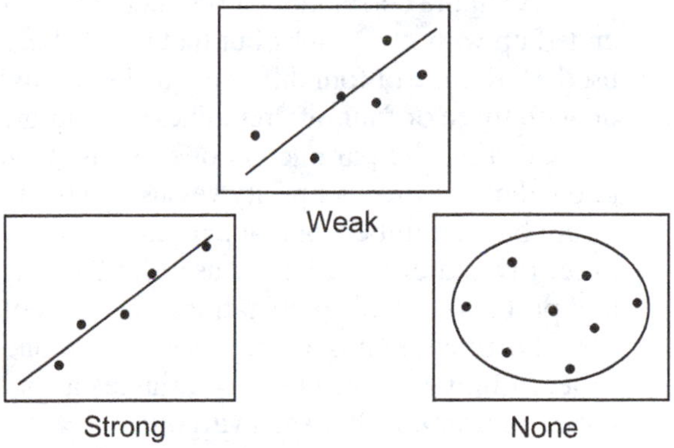

The data from the pelican correlation make a more or less straight line (Figure 7-2). The data from the stork correlation make a scatter plot (Figure 7-3). When you run the formal statistical analysis, you obtain a correlation coefficient that is a number between -1 and $+1$. A value of $+1$ means a perfect correlation in the positive direction. A value of -1 means a perfect correlation in the negative direction.

The closer to 1 the correlation coefficient is, the stronger the correlation. The closer to zero the correlation coefficient is, the weaker the correlation. As a general rule of thumb, what you hope to find is a correlation coefficient that securely establishes a link between the variables, perhaps something in the 0.75 or higher range. A correlation coefficient of anything less than 0.50 is not considered useful.

FIGURE 7-4 Graph of Temperature and Purity

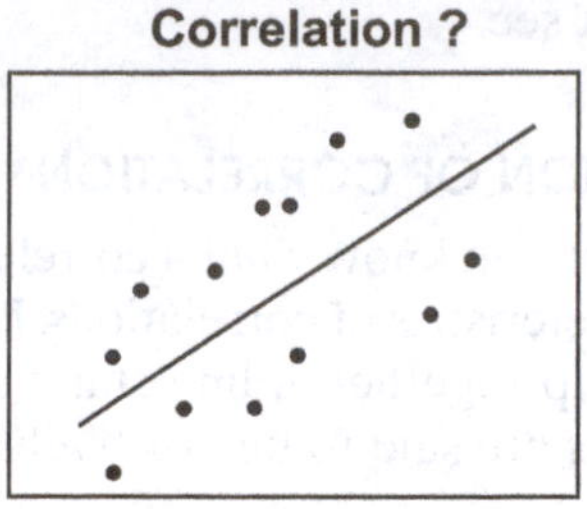

Dual Correlation

Sometimes there is a unique graph, such as Figure 7-4. In this case you were attempting to correlate the temperature in a reactor against product purity coming out of the reactor. At first you thought you just had a poor correlation; but while reviewing the results of the analysis with the process technicians who were working the board during the test run, you learned they had run out of a certain raw material right in the middle of the run and had to switch to the next batch in the warehouse.

Given that clue, you color coded the product made with raw material batch A differently from the product made with raw material batch B. What you could clearly see then was a fairly strong positive correlation between the product purity and the reactor temperature (Figure 7-5), but the correlation was slightly different when using different batches of the same raw material. Had you not reviewed the test results with the hourly staff, you might not have discovered this phenomenon.

At the least it would have a taken a lot more work to determine what had happened. In effect, you learned during this experiment that the product purity was a function of both raw material purity and reactor temperature. The next step was to examine the quality information on this raw material to see what was different between the two batches.

As it turned out, you went on to prove that one certain characteristic that was significantly different was the variable causing the difference in your process performance. In this case, you were trying to develop a model for $Y = f(x)$, where x was the reactor temperature, and instead found that you needed to be developing a model for $Y = f(x, z)$, where x was the reactor temperature and z was the raw material purity.

A simple correlation of product purity to reactor temperature was not possible. You ended up with two graphs, but that is only because only two batches of raw material were used. Had three or four different batches of raw material been used, you would have ended up with three or four different lines on your graph. Clearly there must be a better way.

Testing the parameters one by one would take forever. Even worse, you already knew that the product purity versus reactor temperature line was not the same with different batch numbers of that particular raw material. What if the lines were different with other raw material variations as well? There could be an infinite number of raw material and process control combinations that you could spend your entire career testing.

Fortunately, it is not necessary to test one variable at a time. In fact, it is much better to learn how to test multiple variables at one time for the specific purpose of finding these interactions between variables.

FIGURE 7-5 Correlation Determined from Figure 7-4

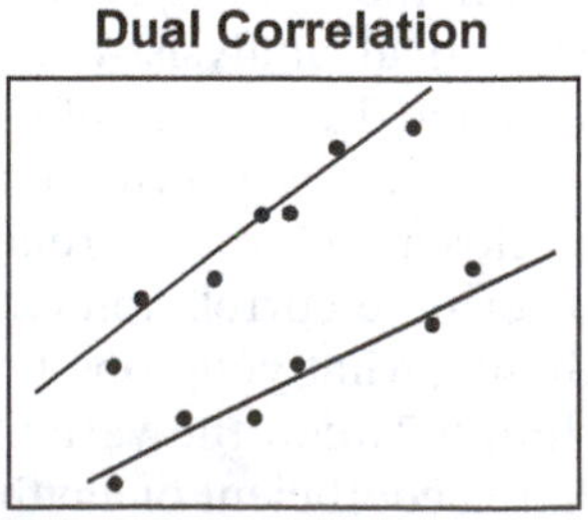

Regression

The fitting of a line to the experimental data that have been collected is called regression. **Regression** is the statistical analysis that generates a mathematical representation of the correlation between two or more variables. Many types of regression analyses are used in different situations. The end result is that somebody plugs the data into a computer, and the computer cranks out an analysis that has to be interpreted. As has been discussed already, without the human interpretation of what the data mean and what the analysis appears to be indicating, you might draw all kinds of wrong conclusions, such as determining that babies are delivered by pelicans instead of storks.

Simple regression is the correlation of a single x (input) to a single Y (output). It is the determination of the formula $Y = f(x)$.

Multiple regression is the correlation of many xs (inputs) to a single Y (output). It is the determination of the formula $Y = f(w, x, z \ldots)$.

There are times when a simple regression is all that is needed. In the process industries, it is more common to have complex operations with dozens of variables to study, so multiple regression analyses are most often employed.

As mentioned earlier in this chapter, the problem with using historical data to predict future performance is that your historical data may exist in too narrow of a range. In order to overcome this issue, you can design experiments to collect a small amount of data over a wide range for each of the inputs and analyze those data to determine where to operate.

Designed Experiments

Let us take a look at some data from a chemical process. Your process technician filled in a log sheet every 2 hours for the last couple of days (Table 7-1). Available data include the reactor temperature, the reactor pressure, and the yield, which is the measure of your process output.

TABLE 7-1 Log Sheet

Rx_Temp	*Rx_Press*	*Yield*
50.4	6.8	87.5
52.6	9.3	93.2
51.3	8.9	90.9
53.0	7.3	91.7
52.7	6.3	90.4
52.9	5.2	89.5
52.8	6.5	90.7
52.8	7.4	91.5
51.1	6.7	88.4
50.8	8.6	89.8
51.5	9.3	91.6
51.2	5.7	87.6
50.3	8.2	88.6
51.4	7.3	89.4
50.1	9.8	89.9
51.3	8.7	90.6
52.6	6.6	90.5
50.2	9.5	89.8
52.0	5.5	88.5
50.3	9.6	90.1
50.8	6.4	87.7

FIGURE 7-6 Temperature and Yield Correlation

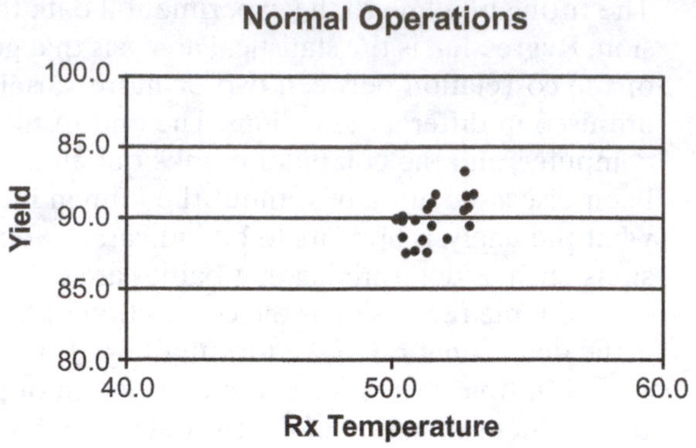

The reactor temperature varied from around 50 to 53 degrees C during this time frame. The reactor pressure varied from around 5 to nearly 10 psi. Under these conditions your process produced between 87 and 92 pounds of product per hour.

Plotting the yield versus the temperature to see whether there was a correlation (Figure 7-6) did not result in much evidence of correlation. You tried plotting the yield versus the pressure as well (Figure 7-7). It looks like the same yield results no matter what pressure is run.

Everybody knows that the temperature and pressure in the reactor are the control handles for the process, so where is the correlation? First, it is not evident because you are looking at one variable at a time and these variables are related. Second, as noted in the preceding section, these input data are collected over a small range, the normal operating range.

The conclusions drawn from analysis of the data are only as good as the data being analyzed. To address the first reason, the data were plotted on a three-dimensional graph (Figure 7-8) to see what the impact of yield was from both temperature and pressure.

Unfortunately, these kinds of graphs are hard to read. What is needed is a way to take the individual points and plot them in a plane so you can see what the overall system looks like. Fortunately, this can easily be done with statistical software. The yield versus pressure and temperature graph looks like that shown in Figure 7-9.

What you see is a solid plane that represents your yield as a function of both pressure and temperature, called a surface plot. Plotting either of these variables against yield did not show how the process operates. Plotting both of them against yield gives you a clear picture of the process.

In this simplistic case, a flat plane results because the variables both influence yield, but they are not actually correlated with each other. In real life, these variables

FIGURE 7-7 Pressure and Yield Correlation

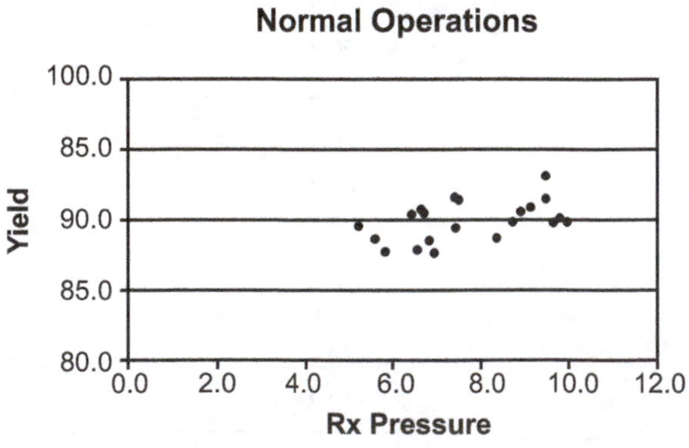

FIGURE 7-8 Three-Dimensional Graph of Pressure, Temperature, and Yield

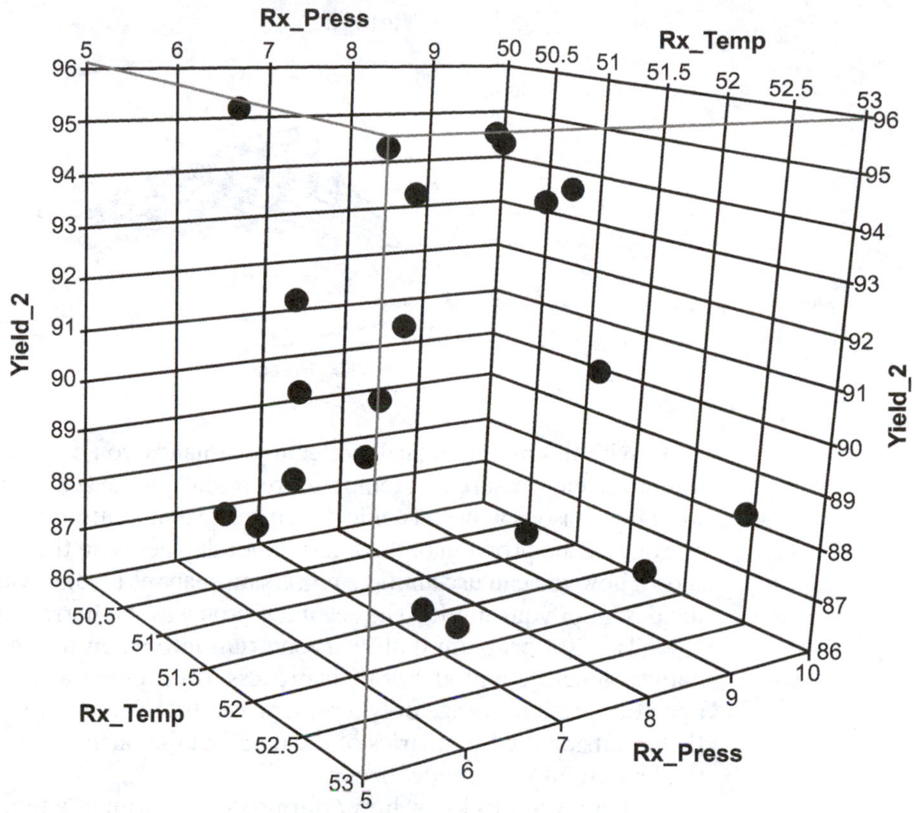

are very much related to each other as well as to final yield. When input variables are related to each other, the result is a surface plot that is not flat, such as that shown in Figure 7-10.

What this means is that not only is yield a function of pressure and temperature, but the effect of pressure varies with the temperature, and the effect of temperature varies with pressure. Using simple correlations will not give an accurate model of your process. Multiple regression analyses will provide a mathematical model of your process performance. As stated earlier, for a multiple regression you need a model that shows Y as a function of many xs, or $Y = f(w, x, z)$. The exact model for the curved response surface in Figure 7-10 follows:

$$\text{Yield} = (\text{Rx\_Temp}^2 + 2(\text{Rx\_Temp})(\text{Rx\_Press}) + \text{Rx\_Press}^2)/35$$

FIGURE 7-9 Plane Representation of the Three-Dimensional Graph Shown in Figure 7-8

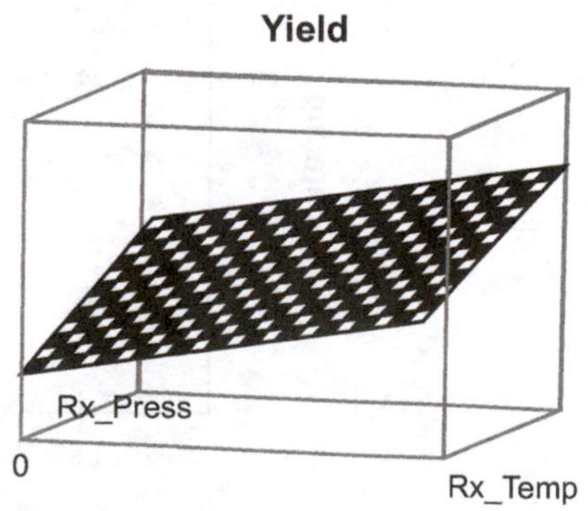

FIGURE 7-10 Realistic Plane Representation of the Three-Dimensional Graph Shown in Figure 7-8

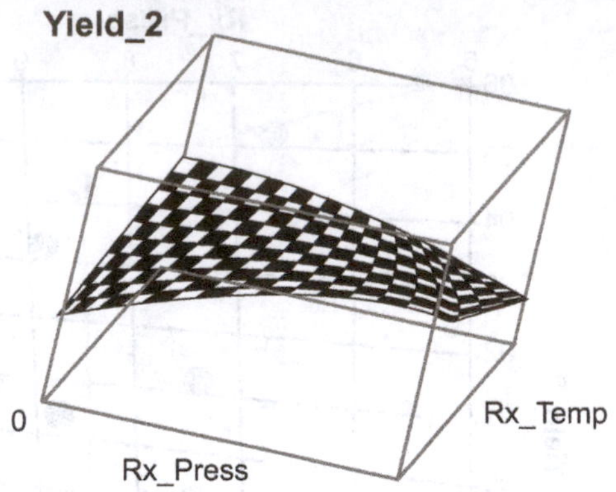

With this mathematical model in your hands, you can predict what the yield will be for any given pressure and temperature. Right? Not exactly. As already mentioned, there were two reasons why you could not correlate temperature and pressure to yield. The first reason was the problem of the interrelationship between the input variables. We have just shown how you can use multiple regression analyses to take both variables into account in the design of your model. The second reason was the narrow range of *x*s in your model.

This is a problem that everyone runs into when trying to use existing operating data to generate a model of their process. Your temperature range was only 3 degrees and your pressure range only 5 psi. You can try to extend the value of your model a little bit outside the boundaries of the data, but the farther you get from the input values, the less useful your model becomes.

If you want to know how your process will run at a temperature of 30 degrees or 60 degrees, you need data in that range. If you want to know how your process will run at 2 psi or 20 psi, you need data in that range. If you want a good model of your process that predicts how it will perform at a wide range of temperatures and pressures, you need data across the entire range.

Basically you need to build a box that encompasses the data range that you are interested in looking at further. This is a 2^2 factorial design (shown in Figure 7-11); it consists of two variables at two levels. This model will require four experimental runs to develop. You will want to set your low and high values for the experiments at the extreme edges of where you believe you can run safely. If the reactor vessel is rated for only 15 psi, it would not be in your best interests to run the experiments at 20 psi. This is about the simplest designed experiment that can be shown.

FIGURE 7-11 Example of a 2^2 Factorial Design

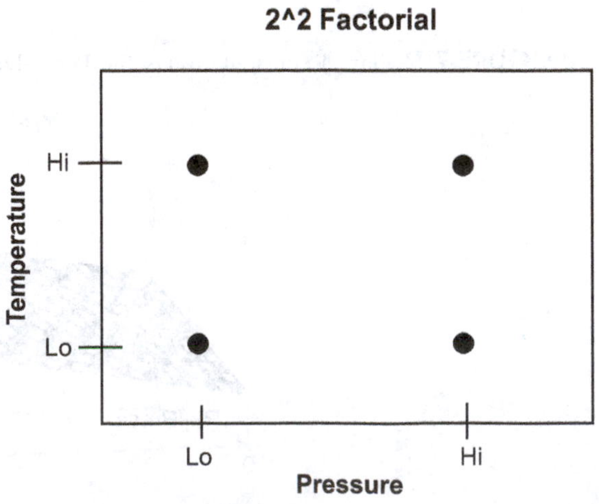

FIGURE 7-12 Example of a 2^2 Factorial Design with Centering

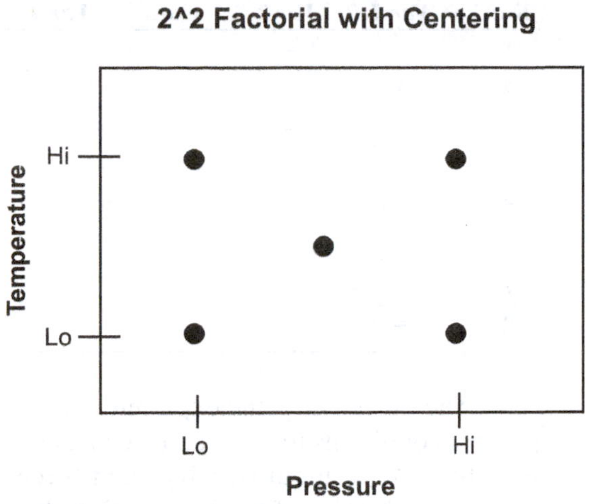

The only problem with this particular design is that all of the data are at right angles with each other. This may mask any curvature in the model due to interactions between the variables. To make sure you see this curvature, add a center point to your design (Figure 7-12).

If you tried to run your process at every possible combination of temperature and pressure, it would be necessary to collect literally hundreds of data points to build your model. By structuring your experiments like the ones shown here, you can develop a mathematical model with just four or five experimental runs. If you are concerned about the normal variation in the process, you might repeat the experiments in blocks to get two data points at each set of conditions. Designed experiments are a powerful tool for understanding and predicting your process' performance.

When the engineering staff set up a designed experiment, they will hopefully include process technicians in the discussion about how far the unit can be pushed in all directions. During those discussions, you may see a form such as the one in Figure 7-13, which is used to both describe the experiment as well as provide a place to collect the results.

In Figure 7-13 there are five experimental runs. A "+" sign has been used to indicate the high settings and a "−" sign to indicate the low settings. Run number 5 is the zero value or center point of the design. You can see that each combination of high and low values for both pressure and temperature is accounted for.

Processes are often complex with many input variables. There may be readings and controllers for multiple temperatures at multiple locations in the process, plus the various flow rates, pressures, and other readings. To make matters even more interesting, there is often more than one output (response) variable that you are interested in. You might desire the creation of a model that predicts yield, purity, and some major impurity.

FIGURE 7-13 Example of Form Used in a Designed Experiment

+ = 15 psi	+ = 65 C
0 = 10 psi	0 = 50 C
- = 5 psi	- = 35 C

Run	Rx_Press	Rx_Temp	Yield
1	+	-	
2	+	+	
3	-	-	
4	-	+	
5	0	0	

TABLE 7-2 Sample Input Sheet for a Designed Experiment

Run	*Var-1*	*Var-2*	*Var-3*	*Var-4*	*Response-1*	*Response-2*	*Response-3*
1	−	−	−	−			
2	+	−	−	+			
3	−	+	−	+			
4	+	+	−	−			
5	−	−	+	+			
6	+	−	+	−			
7	−	+	+	−			
8	+	+	+	+			

The objective of the experiment may be to define a model that provides the best process conditions to meet all three desired outputs. An input sheet for an experiment with four major input variables and three outputs might look like the one shown in Table 7-2. The rest of the information that you would need is what the variables are, where they are measured and controlled, and what constitutes a high value and low value for each variable. It would be wonderful to draw this design out for you in a graphic form, but drawing in four or five dimensions has not yet been mastered.

In the discussion of simple regressions, it was noted how to use the correlation coefficient to gauge how good the correlation was. In that case it was easy to draw a line that represented the relationship between the single input variable and the output variable. When you are running experiments that have four, five, six, or even more input variables, it is impossible to draw the relationship on a graph. Instead, the result of the analysis is a two-dimensional graph that plots the actual values obtained versus the predicted values obtained. If this graph makes a straight line, there is a good correlation. If this graph makes a shotgun pattern, there is not a good correlation.

Figure 7-14 is an example of the output graph for Yield-2. Notice a few things on this graph. In addition to the single solid line that represents your model, there are a couple of dashed lines on each side of the model. These are the prediction limits for the model. When you see these prediction limit lines curve sharply away from the center, as they do in this example, it means that you cannot be confident in your correlation except right in the center.

This model is pretty good at a yield of about 100, but it is poor at 95 or 105 and even worse at 90 or 110. This is a result of having all of your data clustered in a tight range. If you had used a designed experiment to generate this model, your prediction limits would more closely "hug" the model line. The other thing you might notice is the RSq value, which in this example is equal to 0.41. The *R*-squared (R^2) value is always a

FIGURE 7-14 Sample Graph Generated from the Yield-2 Example

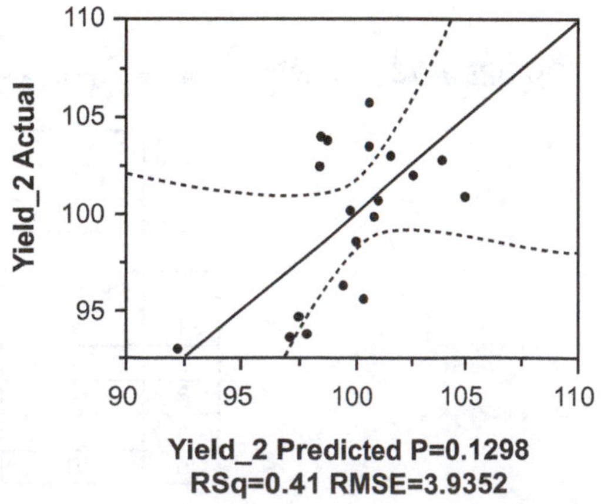

number between zero and 1. Just like with the correlation coefficient, the closer to 1 the R^2 is, the better the model. The closer to zero the R^2 is, the less useful the model.

Again you desire models that have an R^2 value of 0.75 or greater. This model gave you an R^2 of 0.41, which is not good enough. The R^2 value tells you the percentage of the variability in the data that the model explains. This model explains only 41% of the variability in the data. You need either more data, data spread across a larger range, or more variables in the model in order to make this model useful.

Earlier 2^2 factorial experiments were mentioned as being 2 variables at 2 levels. A 3^2 factorial is 3 variables at 2 levels each. Sometimes you have reason to design experiments that are conducted at more than 2 levels, so you may have 2^3 or 3^3 experiments. If you ask about the type of experimental design being conducted, you may hear about these plans plus several others, such as the following:

- Full factorial designs
- Fractional factorial designs
- Central composite designs
- Equiradial designs
- Box-Behnken designs
- Mixture designs

The only point in mentioning these is to give you an appreciation for the complexity of the topic and the diversity of what you may be exposed to later. There is no way you can become an expert in the field of experimental design in one short chapter, but it is important to appreciate your role in the success of the experimental runs.

Role of the Process Technician

What is your role? What is expected of the process technician during the execution of experimental designs? As you might expect, the answer is almost too obvious to state, but it is still offered here. You will be expected to do the following:

- Monitor process conditions during the run.
- Take copious notes about what you observe.
- Gather the requested data.
- Report all of the preceding information accurately and precisely.

In our experience, process technicians have done a great job of monitoring process conditions and gathering the requested data. However, because they did not understand the nature and value of designed experiments, technicians often did not capture notes about observations during experiments and pass them on to the day staff.

Think back a few pages to the discussion of the dual correlation. A wrong conclusion would have been drawn had a process technician not made mention that there was a raw material change in the middle of the experimentation. Sometimes your observations will not be related to the experiment. Sometimes they will. The only way to know is to capture everything and pass that information on.

Another failure that has been seen occasionally is that sometimes information gets reported, but not as accurately and precisely as desired. In many organizations the salaried and hourly staffs work together well, but unfortunately in a few these staffs have yet to establish a high level of trust and understanding. In those cases, a scenario might arise in which the shift process technician was instructed to run the unit at 50 degrees and 10 psi, but due to circumstances outside his or her control, the process technician was able to achieve only 48 degrees and 9.5 psi. These things happen.

It is not a major failure that the exact experimental conditions were not met. The problem comes up when, rather than risk a confrontation with the day staff, the process technicians report that they ran 50 degrees and 10 psi as requested and these conditions gave a yield of x. When the results of the experiments are put into the statistics software to generate the response surface plots and other analyses, the analyses do not work out as well as expected because the model was assuming the results were due to 50 degrees but the process really operated at 48 degrees. The model was assuming the results were due to 10 psi but in reality it operated at 9.5 psi.

The output from your multiple regression analysis of data collected from designed experiments is only as good as the data you put into the system. The conditions at which you are instructed to run are desired conditions. If you cannot reach them for any reason, document what you really achieved so that the model will still be useful. Communicate any issues with the day staff so they can have the option to make modifications as needed to maintain symmetry in the design.

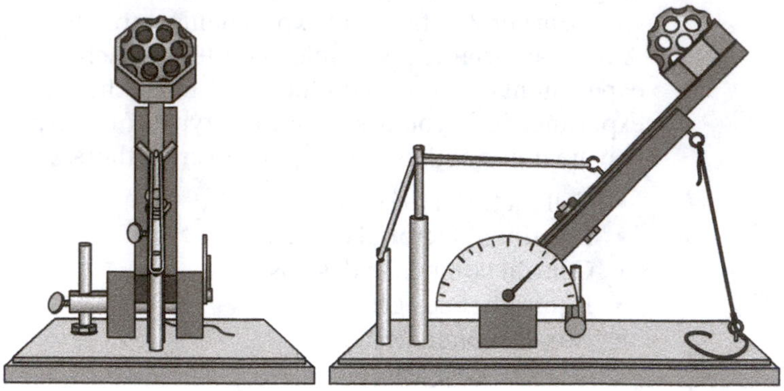

Roman Catapult Example

Instead of building a large process to simulate a designed experiment, you are going to use a catapult. The output of your process is the launching of your ball across the room.

There are at least two measurable outputs from your process: how far the ball travels and how high the ball travels. You can easily measure the travel distance with a tape measure and the height by launching the ball over a barrier. Can you think of another possible process output?

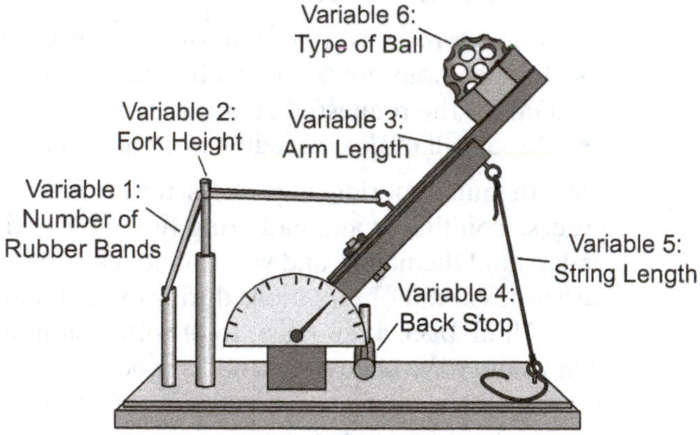

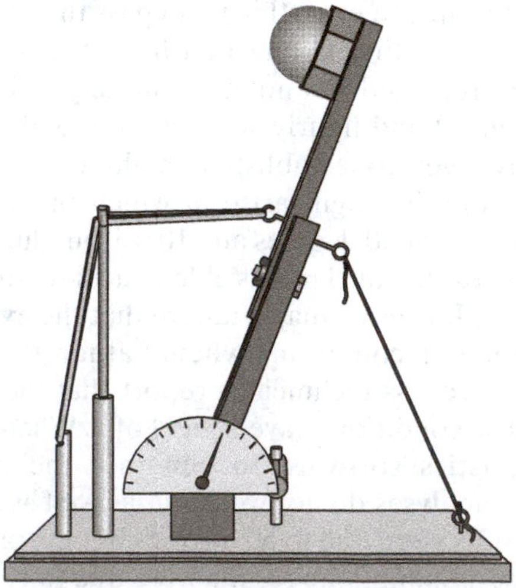

In medieval times the catapulted objects were required to travel the requisite distance to the target and clear the top of the castle wall. The process had to be modeled with multiple output variables as well.

There are several input variables for your process as well. You can put a varying number of rubber bands on your catapult. Different brands and different sizes of rubber bands can be used. Just as in your processes, raw materials are not always constant and consistent.

Your second input variable is the height of the fulcrum. You have interchangeable fulcrum posts and each one is adjustable. Next you have the length of the catapult arm. The arm may come with a measurement system included; otherwise, you will have to measure the distance from the pivot point to the ball holder.

Your next two variables are the start angle and stop angle of your catapult. Your process comes equipped with a backstop that allows you to start your catapult at an incline. It also comes with a string that can adjust how far your catapult will move forward. With a protractor you can measure the angles at which to start and stop your process.

Finally you have the ball itself. You can use Ping-Pong balls, plastic golf balls, waffle balls, or just about anything that will fit in the ball holder. Each different ball will have a different mathematical model for how to make it fly a certain distance and obtain a certain height.

During one of the first exercises with this example apparatus, one team of students discovered that when they randomly placed their ball in the holder and launched it, they received the same basic result ± 3 inches. When they took care to place the ball in the holder so that the seam was oriented exactly the same every time, they received the same accuracy but with ± 1.5 inches of precision. This was a classic case of the process technician (the student) discovering a source of variability that was previously unknown or at least undocumented by the engineering staff (the instructor).

At this point, you might be able to make an attempt at setting up a set of designed experiments to learn how each of these variables affects the process output and how each of these variables interrelates with the others.

Upon completion of the technical training in designed experiments, students are limited to eight launches of their catapult. They are then challenged to conduct the experiments of their choosing to hit a certain distance target. After a maximum of eight shots, the teams are judged to see whose model is the most effective in hitting the target. With success in a single output behind them, the next challenge is to hit the distance target while also clearing a minimum height.

Summary

By now you are convinced that designed experiments are an extremely powerful statistical tool for gathering minimum amounts of data that provide maximum amounts of useful information. Designing your experiments with purpose instead of collecting random data ensures that you cover as large a range as possible for your input variables in order to make your model as useful as possible.

The data from designed experiments are used in a multiple regression analysis to generate mathematical models of your process's performance. A process performance model such as this enables you to customize your product offerings to your customers when needed as well as make quick adjustments to your process to keep it running on spec and consistent.

Even without the technical knowledge regarding how to perform the regression analyses, you can determine the relative strength of the model simply by looking at the R^2 values for the model. For simple regressions, you can tell the strength of the model and the direction of the correlation by examining the correlation coefficient.

The role of the process technician is to monitor the process during the experimentation, take notes regarding any changes in the process during the experimentation, gather the requested data, and report all information accurately and precisely. Any fudging or rounding off of the input or output readings may make the data sheet look good but will defeat the purpose of generating a useful process model.

Checking Your Knowledge

1. Define the following key terms:
   a. Correlation
   b. Performance model
   c. Regression
2. The model for a simple correlation is often written as:
   a. $Y = x(f)$
   b. $Y = f(v, w, x, z)$
   c. $Y = f(x)$
   d. $Y = mx + b$
3.

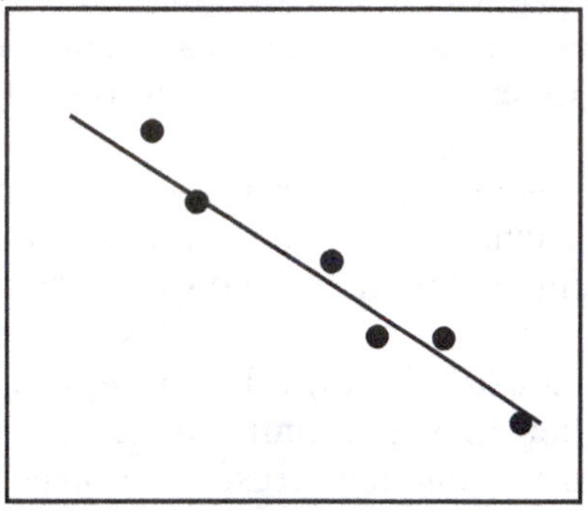

   a. This is a positive correlation.
   b. This is a negative correlation.
   c. This is a zero correlation.
   d. This is a hyperbolic correlation.
4.

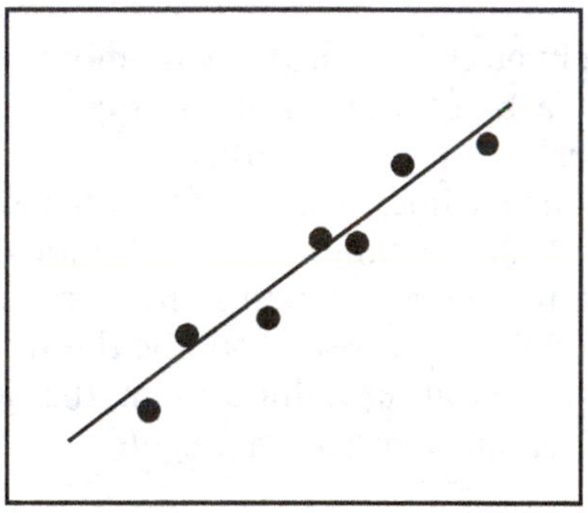

   a. This is a positive correlation.
   b. This is a negative correlation.
   c. This is a zero correlation.
   d. This is a hyperbolic correlation.
5.

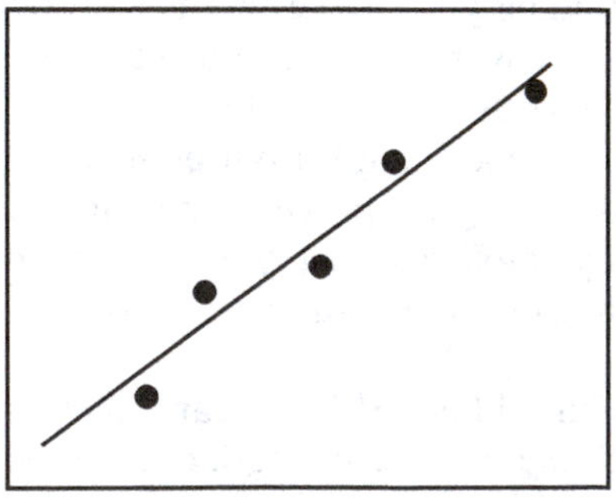

   a. This is a moderate correlation.
   b. This is a strong correlation.
   c. This is a zero correlation.
   d. This is a correlation coefficient.

6.

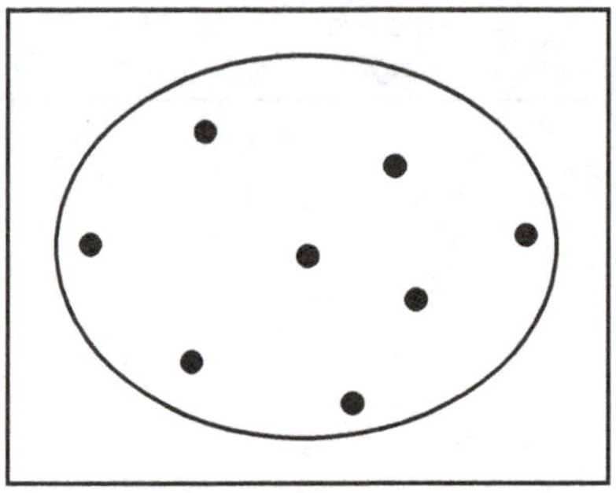

a. This is a moderate correlation.
b. This is a strong correlation.
c. This is a zero correlation.
d. This is a correlation coefficient.

7.

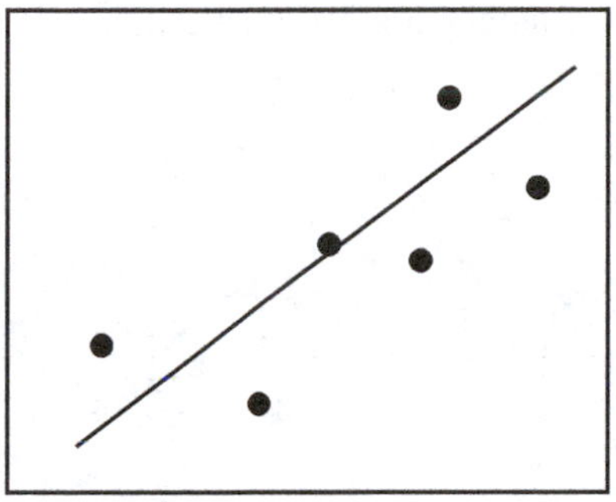

a. This is a moderate correlation.
b. This is a strong correlation.
c. This is a zero correlation.
d. This is a correlation coefficient.

8. A good rule of thumb is to desire a correlation coefficient of at least:
   a. .10 to .25
   b. .30 to .45
   c. .50 to .65
   d. .75 to .90
9. The R^2 value is best defined as:
   a. The square of the correlation coefficient
   b. The percentage of the variability in the data explained by the model
   c. The percentage of the variability in the model that is explained by the data
   d. The desired target value for all designed experiments
10. (*True or False*) Correlation does not equal causation.
11. (*True or False*) If you cannot achieve the targeted conditions, do the best you can and write down that you hit them exactly.
12. (*True or False*) Process technicians play a key role in the success of designed experiments.
13. (*True or False*) Only one output variable (response) can be studied at a time in designed experiments.
14. (*True or False*) A perfect correlation has a correlation coefficient of ZERO, which tells you that there were ZERO deficiencies in the model.
15. (*True or False*) Taking notes about what you observe in the process during the experimental runs is critical to the success of designed experiments.

Activities

1. The instructor will divide the class into teams to demonstrate the use of the catapult as a designed experiment exercise.
2. Here is a potential design for your catapult. What values would you use for the high and low values for each variable?

Experiment Number	Arm Length	Fulcrum Height	Start Angle	Stop Angle	Distance
1	–	–	–	–	
2	+	–	–	+	
3	–	+	–	+	
4	+	+	–	–	
5	0	0	0	0	
6	–	–	+	+	
7	+	–	+	–	
8	–	+	+	–	
9	+	+	+	+	

3. Imagine that you have a barbeque pit. Identify the key input and output characteristics that you might consider including in a designed experiment to model the performance of your pit.

CHAPTER 8

Root Cause Analysis

"Don't cross the bridge until you get to it."

~ENGLISH PROVERB

Objectives

Upon completion of this chapter you will be able to:

- Explain the purpose of root cause analysis.
- Explain the processes technician's role in root cause analysis.
- Compare and contrast the most commonly applied root cause analysis methodologies (i.e., Kepner-Tregoe, Five Whys, and Apollo).
- Recognize and identify cause-and-effect diagrams, interrelationship digraphs, and current reality trees.

Key Term

Root cause analysis (RCA)—the analysis of data to determine the true root cause of a problem. Also called root cause investigation.

Introduction

This chapter takes a break from the statistical subjects and focuses your attention on quality topics that are more procedural in nature. This chapter deals with root cause analysis, which, as the name implies, means analyzing the information to determine the true root cause of a problem or incident. This means digger deeper than just pointing to the first contributing factor.

It is almost shocking how often when a problem occurs, people jump to conclusions about what caused the problem, completely ignoring any real analysis of the data or events surrounding the situation. Yet, in order to improve your processes, you must make sure to eliminate the root cause, not just contributing causes or factors to the problem. There will be further discussion of the importance of eliminating root causes in Chapter 10 on Six Sigma.

This chapter presents several different styles of root cause analysis such as Kepner-Tregoe, Five Whys, and Apollo. It also outlines the role of the process technician in supporting these efforts. There are a few tools commonly used in root cause analyses, or RCA, that you should be able to recognize by sight, such as cause-and-effect diagrams, which were already presented in Chapter 6. In addition, you are introduced to a couple of new tools such as interrelationship digraphs and current reality trees.

As has been the case in several other topics that have been covered and will be covered in future chapters, what is important is that you get the job done, not that you apply a certain tool. As long as the methodology gets you to the root cause, we support it. There are lots of tools in your toolbox; you just need to use an effective tool to get the job done.

Purpose of Root Cause Analysis

When something unplanned, unexpected, and undesired happens in your process, you need to identify the true root cause so that you can eliminate the root cause and thus prevent future recurrence. Which methodology you employ will depend on the company that you work for and what is in vogue at the time. If you fail to conduct a thorough **root cause analysis** (analysis of data to determine the true root cause of a problem), you may end up spending a lot of time, money, and resources to fix the wrong thing.

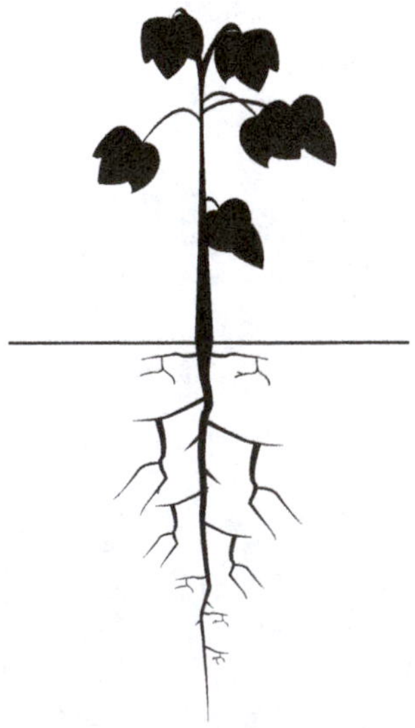

Think about your car or truck. It is making a rattling noise that is driving you crazy. You spend half of your day in the garage working on the problem and discover the area where the rattle is coming from, so you tie things off and stuff towels around the area to stop the rattling noise.

Unfortunately, you addressed the symptoms but not the root cause. Perhaps a bolt or bracket has vibrated loose and is allowing a hose to swing back and forth and make noise. You find the loose bracket and replace it. Is this good enough? What caused the bracket to come loose in the first place? Maybe it was a poor design. Maybe your car has over 150,000 miles on it and the bracket came loose from normal wear and tear. But perhaps there is a deeper cause. Maybe the engine is vibrating more than it used to because of a shaft imbalance. Perhaps the flywheel has lost a tooth and is no longer balanced.

Sometimes a problem is obvious, but oftentimes a more thorough investigation into the problem is needed to determine the true root cause. How deep do you have to dig? It is hard to complete a root cause investigation with a couple of people in an hour or two.

The purpose of root cause investigations or root cause analyses is to identify the underlying reason(s) behind a problem. There may be more than one root cause. There may be a combination of root causes with mitigating circumstances. Things that do not cause a problem in the winter may cause a problem in the heat of the summer. You must evaluate all of the relevant data, by sorting through all of the data, in order to find the combination of causes and circumstances that allowed your problem to manifest itself. Then and only then can you apply corrective action to ensure that this problem will not bother you again.

Role of the Process Technician in Root Cause Analysis

As a process technician, you are on the front lines of the process industries. You are there where the action is around the clock. Process technicians will be involved in any root cause analyses that involve the operation. Modern plants have distributed control systems that record thousands of pieces of information every hour. These computerized systems hold a wealth of raw data but are oftentimes devoid of information. They will never replace a human being's ability to see what is happening around the plant and establish relationships between those events and the readings on the computer.

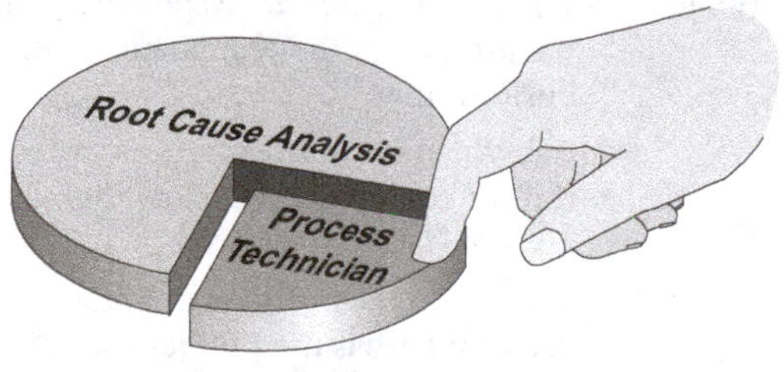

The discussion has already mentioned that the scope of a root cause analysis may be broad or narrow, may consume many weeks or just a few hours, and may involve many people or just a few. An informal analysis is often just one person talking to a few others to document the issue.

A formal root cause analysis typically involves a team of people assigned the role of understanding an issue in depth. This team should include the individuals involved in the issue as well as individuals that were not involved. It should have a mix of technical experts as well as hands-on experts. The makeup of the team will depend on the nature and complexity of the issue.

As a process technician, you can expect to participate on many of these teams during your career, sometimes as an ad hoc member called in to answer just a few questions, sometimes as a full-fledged team member. However you are asked to participate, it is in your best interests to help the team put all the pieces together so the right issue gets fixed the right way to prevent recurrence.

Kepner-Tregoe

(For more information, visit **www.kepner-tregoe.com.**)

Dr. Charles Kepner and Dr. Benjamin Tregoe were researchers employed by the Rand Corporation in the 1950s conducting research in decision-making breakdowns at the Strategic Air Command. Finding that rank was less critical to successful decision making than were rational thinking and the use of logical processes, they developed a methodology to guide individuals through a logical thought process for gathering, organizing, and analyzing information before making decisions.

The Kepner-Tregoe company that they founded is still going strong and still supporting industry, having celebrated 50 years of success in 2008. Obviously, this chapter cannot cover in depth in only a few pages what Kepner and Tregoe cover in an entire book, but here is a brief overview of how they approach root cause analyses.

The methodology they employ is unique in its approach to determining the root cause of a problem or, as they call it, a deviation. Their process is, as are most processes, broken down into several steps. The first step is to outline the details of the deviation. To define the problem, the root cause analysis team answers the typical questions such as who, what, when, and where.

The unique part of the Kepner-Tregoe (K-T) approach is that it also asks the negative of each of these questions. Who was not involved? Where did the problem not occur? What did not happen? The terminology that approach employs is to document the "is" versus the "is not." Table 8-1 outlines the types of questions that may be asked in the problem definition step.

The next step in the K-T process is to develop hypotheses as to the possible cause(s) of the deviation. For each hypothesis, the K-T process looks for distinctions and changes that might give clues about the root cause. A possible root cause that you identify might be based on the differences between the "is" and "is not" documentation

TABLE 8-1 Example of Questions That May Be Asked in the Problem Definition Step

	Is	*Is Not*
What	What is the defect? Be specific. Name the product or the equipment or the business unit.	What is not defective? Again, be specific. What products or equipment or businesses were not affected?
Who	Can you name a specific person or shift or contractor company that was involved?	Who (person, shift, company) was NOT involved?
Where	Where did the problem manifest itself? Where is the problem seen? Is the defect on a certain location on the product or in a certain geographic location?	Where is the problem not occurring? Where is it not seen? What other places might you have expected to see the problem but did not?
When	When did the problem start? When did it stop? Did the timing coincide with another event such as a shift change or a change in the raw materials?	When was the problem not a problem? When else might the problem have been seen but was not?
Severity	How bad was the problem? How many units of product were affected? How many dollars did it cost?	How many other units might have been affected but were not? How severe might the problem be if left unattended?

TABLE 8-2 Example of a Completed Kepner-Tregoe (K-T) Process

Deviation Statement: Three bad batches of product were produced in the plant in September.

	Is	*Is Not*
What	Three batches of dibutyl futile failed the water specification.	All other specification parameters were normal.
Who	The IBF Line II production team.	A specific shift, as these batches were made over the course of several days with several shifts involved.
Where	Problem was with dibutyl futile manufactured in the IBF Line II facility.	Affecting IBF Line I.
When	The batches were the last three in the campaign. Starting September 12 and no end date. The plant shut down to prevent further off-spec.	A problem in August or at the beginning of the campaign, which started on September 2.
Severity	Three batches failed and the plant shut down. In addition to discarding these batches, we are losing money by not running.	All products produced in the plant.

generated in the first step. What you are doing here is using the data from the first step to generate a small list of potential causes that you have reason to suspect.

The third step in the K-T process is to test these hypotheses to see whether they logically explain every "is" and "is not" that is listed. This narrows down the amount of work that has to be accomplished in the final step of the process by eliminating potential root causes logically without the need for data collection.

The fourth and final step in the K-T process is to validate the potential causes as root causes by collecting data. The team will have to decide what the easiest, cheapest, and quickest way to collect these data might be, but without validation, you never really know whether you are fixing the right item. Table 8-2 shows a completed K-T in order to clarify how this methodology is used to solve problems.

With this information in hand, the root cause analysis team noticed that the company had switched to a new lot of a certain raw material on the same day that the problem began. Team members then developed the hypothesis that the problem was caused by a bad batch (lot number X15) of a raw material called gotchagoin. They noticed that lot X15 of gotchagoin was used in the manufacture of all three of these batches and was not used on Line I at all. This lot of raw material was not staged for use until September 12, when the previous lot ran out on Line II.

It was common to all of the shifts. It appeared to meet all of the criteria of the "is" and "is not" definitions, passing the company's logical review. Every other theory that the team came up with failed to satisfy the "is"/"is not" criteria. The last thing for the team to do is to validate or invalidate this potential cause with data. How would you suggest the team proceed?

The obvious way to proceed might just be to make another batch of dibutyl futile with gotchagoin X15 and a batch with lot number X14, which was discovered to have made good product prior to the problem.

Management might have a problem with this validation method because if the team is right, the company is going to make yet another bad batch of product that will cut even further into its profits. Another way might be to send samples of lots X14 and X15 to the laboratory for testing to see whether something is different about them. A few lab tests may be quicker and cheaper than making commercial-sized batches of product. The team might even contact its supplier of gotchagoin and ask whether it saw any differences in the batches. Enlisting the support of your suppliers is an excellent way of understanding your own process and helping them to understand yours. It may be that

FIGURE 8-1 Five Whys

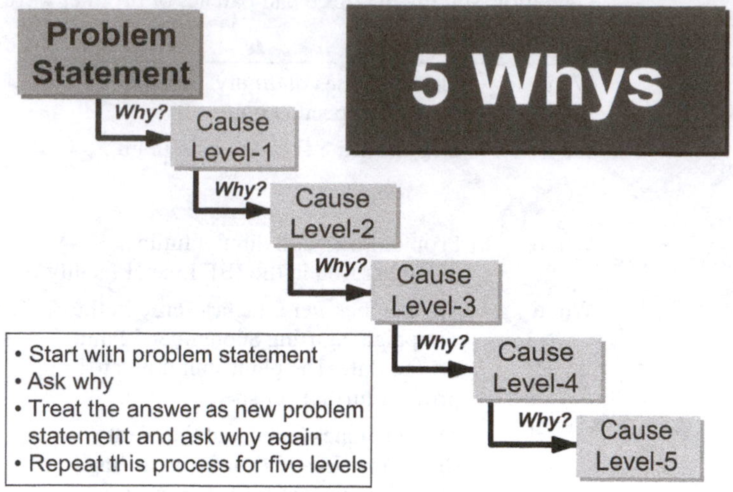

the supplier noticed a difference but did not think it would be important to you. If in doubt, ask. It is easy to undercommunicate but almost impossible to overcommunicate.

Five Whys

Let us switch gears now and look at one of the most commonly employed root cause analysis techniques applied in the process industries today. It is a simple application of root cause analysis called the Five Whys (shown in Figure 8-1). As always, start with a problem statement. The root cause analysis team asks the question, Why did this occur? and documents the response. It may be that data are needed to answer the question. Remember: "In God we trust—all others bring data." The first why question is now treated as the new problem statement, and the team asks the question, Why did this occur? This process is repeated five times to determine the root cause of the original problem statement.

Let us take a look at an example of how this might work itself out in a plant (Figure 8-2). Your department tracks product quality on every batch it makes. During the month of September it had a bad month and made three batches of "bad" product that failed the specifications and could not be sold to customers.

FIGURE 8-2 Practical Example of Five Whys

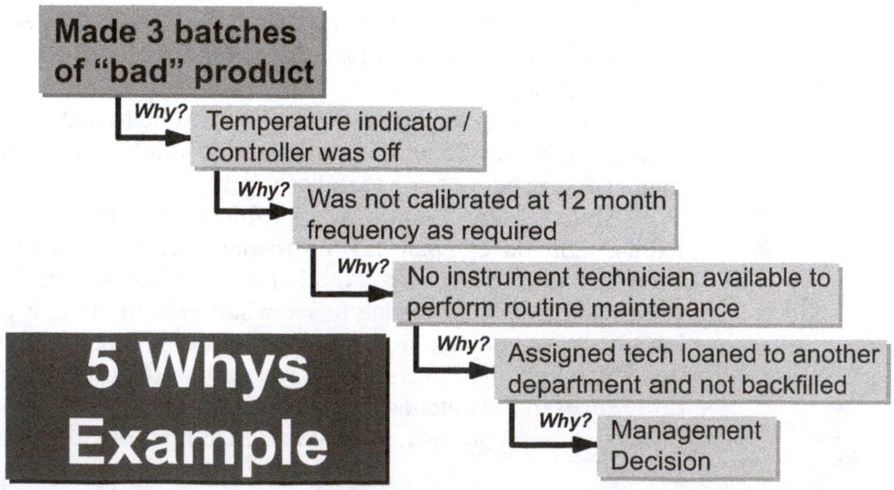

The cost of making this bad product must now be absorbed out of your profits. The boss is not happy and charters a root cause analysis team. The team documents the problem statement and conducts a brainstorming exercise. Somebody mentions that the particular failure your department encountered is indicative of a temperature problem in the plant. You check it out and, sure enough, your temperature indicator is not reading correctly.

Your department thought it was running at 100 degrees C, but it was really running at 110 degrees C. This difference is enough to explain why you made bad product. In some companies the team might stop here. It found the root cause and fixed it. If you do that, you might miss an opportunity to prevent recurrence. Instead, the team members treat the temperature controller issue as the problem statement and ask themselves, Why was the temperature controller off? With a little investigation, you find out that the device is scheduled for routine calibration on a 12-month frequency but has not been calibrated for over 18 months.

Again you dig deeper, asking, Why was the device not calibrated on schedule? You find out that you no longer have an assigned instrument technician. Why not? Your technician was loaned out to another department and no backup was identified. Notice that this statement is really two answers. Reassigning the instrument technician to another department for a special project or shutdown is not an unnatural occurrence. The part that hurts is that no provisions were made to backfill that position in the meantime. The team asks, Why not? one more time and management embarrassingly admits that it just made the decision and accepted the risk.

A couple of things to take note of:

1. You may have to ask why four, five, or six times, but eventually you will get to a real root cause. The number of levels is not fixed at five; that is just a rule of thumb to get you going and make you think deeper than you may have otherwise.

2. You need to have data to make your decisions. Intuition is okay to guide you, but gather the data to validate your gut. Sometimes this means the RCA team has to get together and generate a list of desired information, then dismiss for a day or two while the data are gathered, then come back together to review the information and plan the next steps. Sometimes there is more than one answer to the question why. In this case the diagram will get more complicated as each one of the new branches is further explored. This is normal. Do not think that every problem you encounter can be solved with a few quick questions.

Apollo

(For more information, visit **www.apollorca.com.**)

Another widely implemented approach to root cause analysis is the Apollo root cause analysis methodology. The Apollo root cause methodology is well ingrained in process industries with clients such as Hercules, Valero, Huntsman, Rohm and Haas, and Dow Chemical.

As was the case with the Kepner-Tregoe methodology, it is difficult to cover anywhere near the details that the methodology deserves in just a few pages of this text. This section is an attempt to familiarize you with the basic approach so you will be ready to take your place in industry and participate in root cause analyses as needed.

In at least one way, the Apollo root cause analysis methodology (shown in Figure 8-3) is similar to the Five Whys approach just covered in that you treat the answer to each why question as the new problem statement and repeat the process to keep digging deeper and deeper into the issue. In the Apollo methodology, you do not ask why. Instead you ask, Caused by? The answer to each Caused by? question is not a single response but a consideration of the action(s) and condition(s) that existed to cause the problem to occur.

There may be more than one action and more than one condition that merged to allow any given problem to occur. If the beauty of the Five Whys methodology is its

FIGURE 8-3 Apollo Cause-and-Effect Diagram

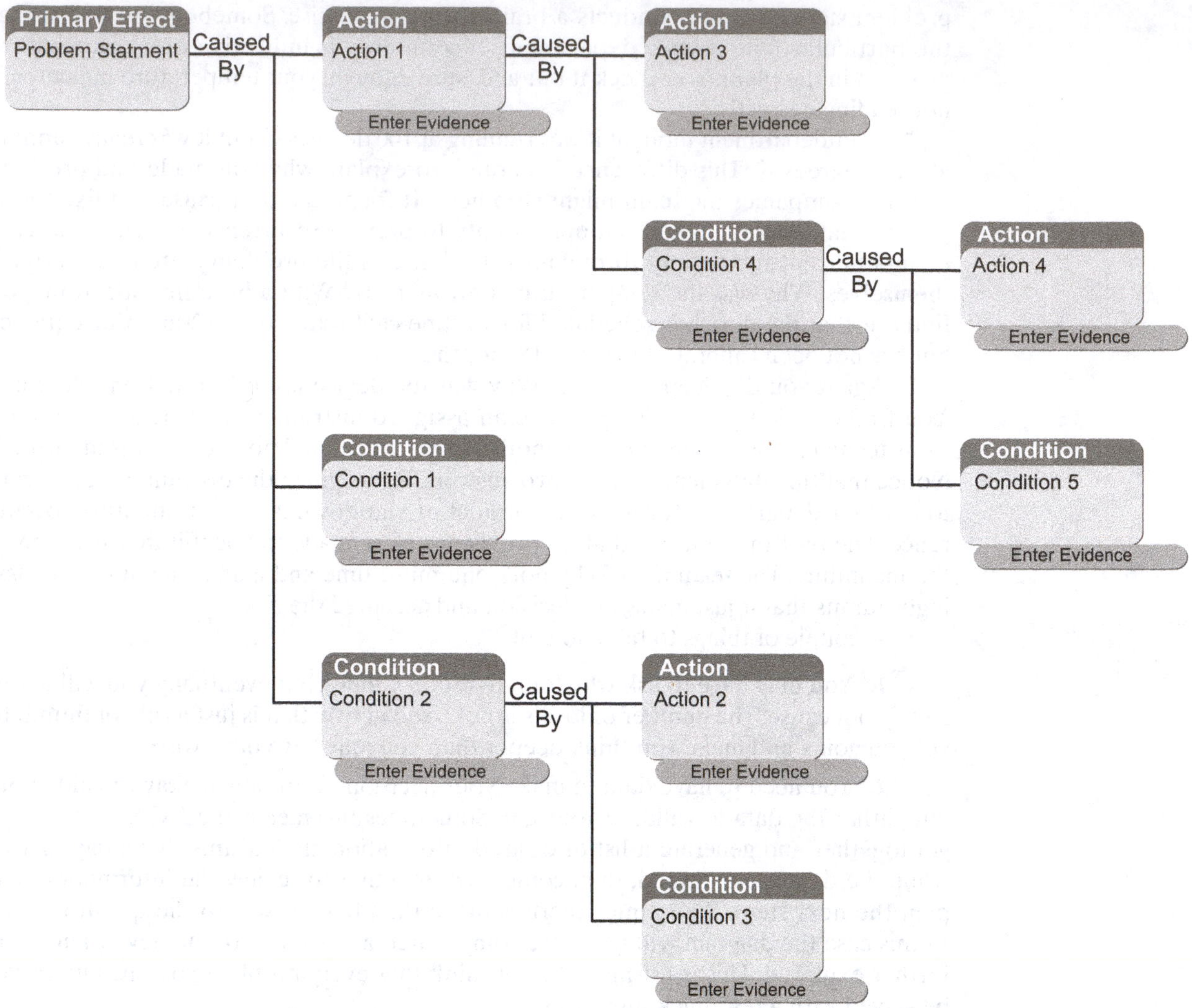

Source: Courtesy of Apollo Associated Services, LLC.

simplicity, then the beauty of the Apollo approach is the recognition that most problems are not that simple; many interconnected actions and conditions worked together to cause the problem to exist.

Consider the example of the employee who broke his leg, shown in Figure 8-4. The action that the employee was taking was walking across the supply room floor. The conditions of the floor were found to be wet and dark. The employee was also carrying a box of supplies from the supply room. You have at least two actions (walking and carrying) and two conditions (wet floor and dark room) to consider.

Each one of these four causes must now be examined as if it were the problem statement. What caused the employee to be walking through the supply room? What caused the floor to be wet? What caused the room to be dark? What caused the employee to be carrying a box of supplies?

For each of these problems, seek to identify the actions taken and the conditions that existed. The room was dark because two of the four lights had burned out and had not been replaced yet. What caused the bulbs to not be replaced? Lack of manpower? Unreported problem? The floor was wet because of the action of the person who had mopped the floor. The condition: No "Wet Floor" sign had been posted and no communication had been issued. What caused that failure?

FIGURE 8-4 Example of a Completed Apollo Cause-and-Effect Diagram

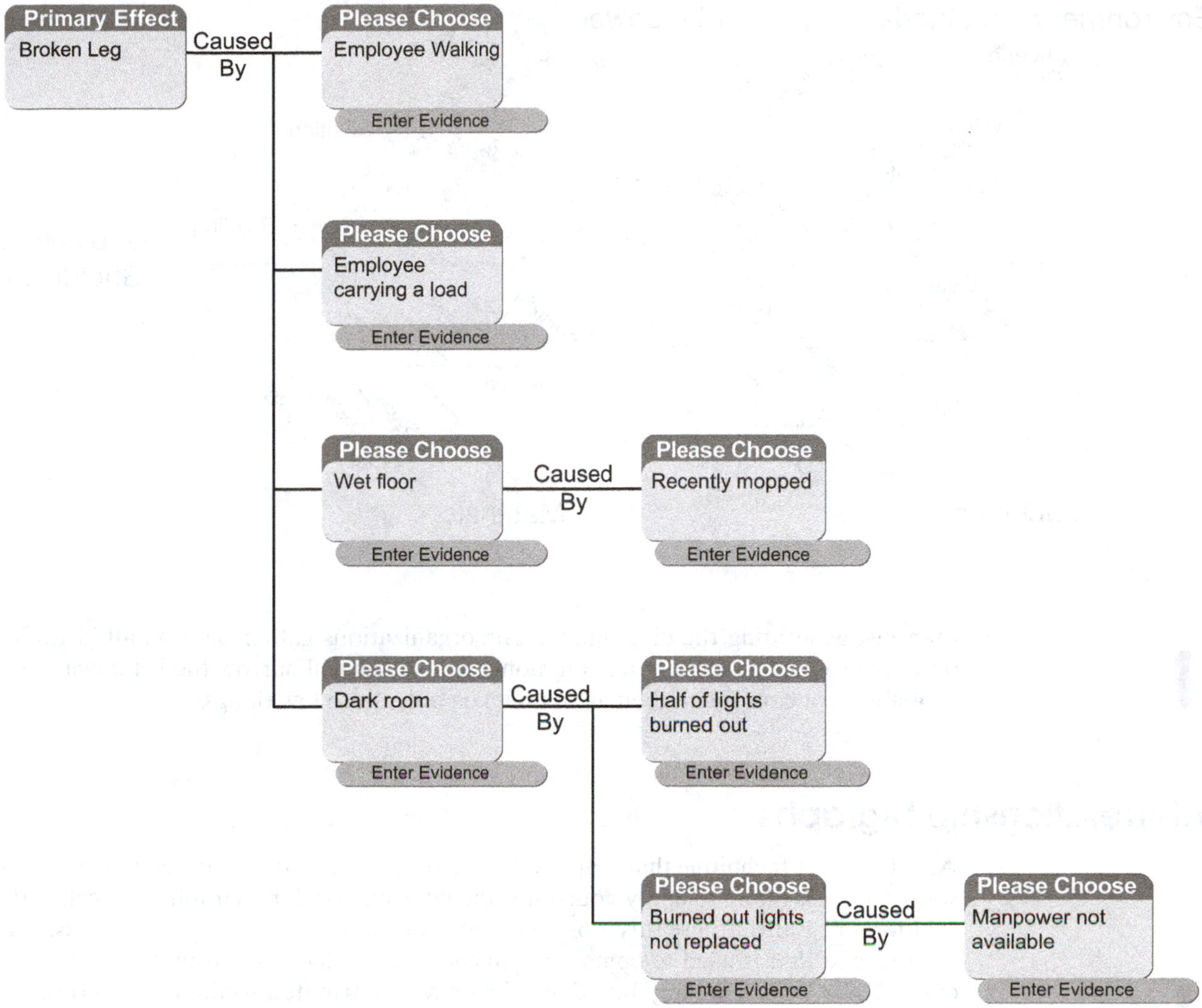

Source: Courtesy of Apollo Associated Services, LLC.

As you can see, the cause-and-effect diagram for something as simple as an employee's slipping on a wet floor can be fairly large. In the end, you do not identify a single root cause, not even a single action with a single condition. Note that there were several actions taken and several conditions that existed that all contributed to the problem. Mitigating one or more of the conditions and/or actions might have been enough to prevent the employee's broken leg.

Cause-and-Effect Diagrams

Cause-and-effect diagrams were discussed in Chapter 6 and are also called fishbone diagrams or Ishikawa diagrams (shown in Figure 8-5). With this technique, you start with a problem statement and then brainstorm potential causes. Each cause is then placed on the diagram in the most appropriate category. Commonly used categories are manpower, methods, machines, and materials.

At this point you have a list of potential root causes, not a root cause. If all you do is generate diagrams of all possible causes, you end up with diagrams on the wall and no idea how to fix the problem. From this list of potential root causes, you must gather data to either validate or eliminate the potential cause. This is obviously more tedious

FIGURE 8-5 Example of a Fishbone Diagram

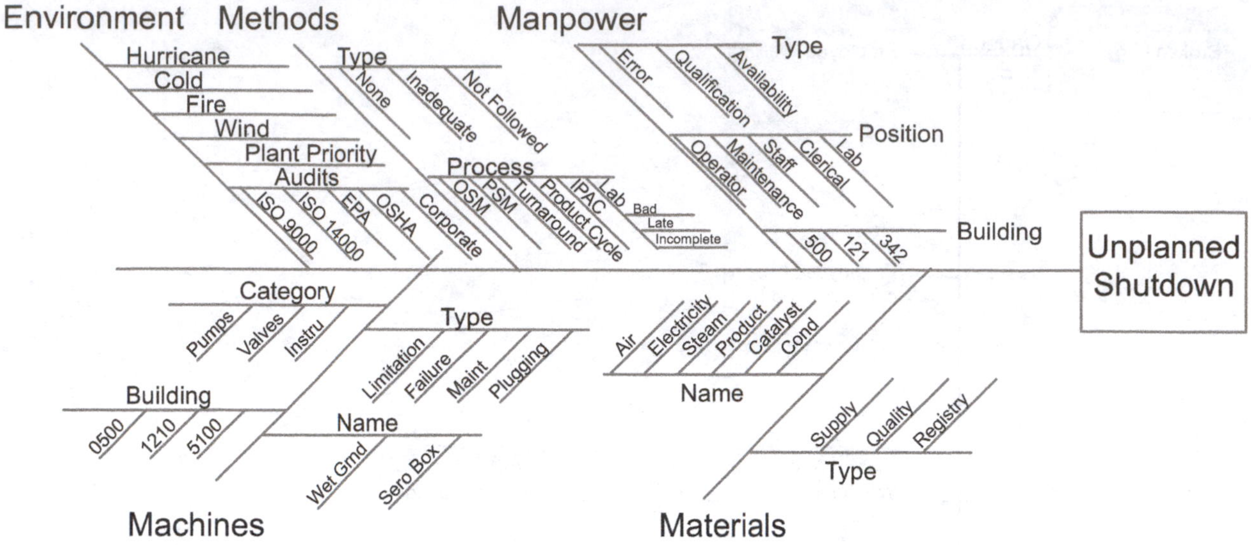

than just generating the diagram. Certain organizations gather data on all identified potential root causes, but more commonly individuals will narrow the list down to the probable root causes and then gather data on only those few things.

Interrelationship Digraphs

Another RCA technique that is applied in the process industries but that you would be somewhat less likely to apply yourself is the interrelationship digraph (also called the relations diagram). To use this tool, you basically conduct the same type of brainstorming exercise that is used to come up with contributing factors and plot them all on a chart. Draw lines from any box that might have contributed to the other boxes, and even sum up the number of inputs and outputs seen.

The result is a diagram that shows how each of the pieces of this puzzle might be related to the other pieces. The "root cause" that needs to be addressed is the box that has the most influence on the other boxes. Figure 8-6 shows an interrelationship digraph for the employee who slipped in the supply room and broke his leg.

FIGURE 8-6 Example of an Interrelationship Digraph

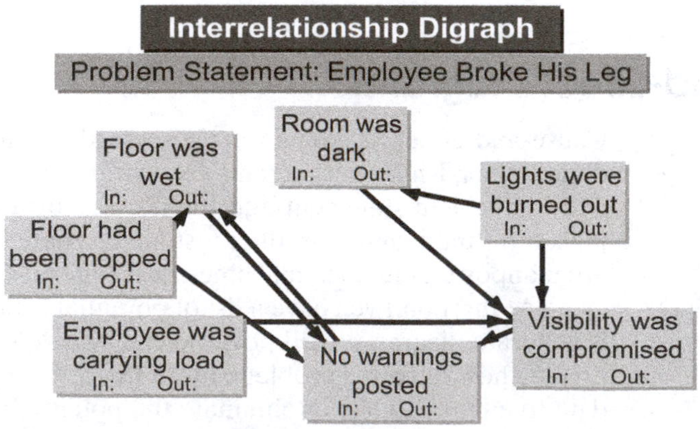

FIGURE 8-7 Example of a Current Reality Tree

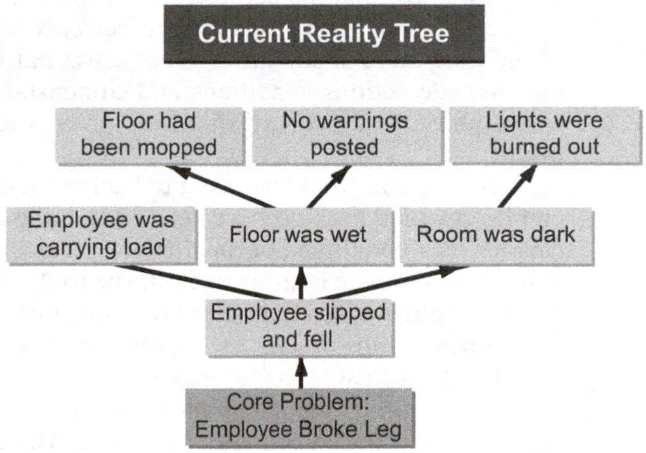

This tool is less often employed in the manufacturing departments, seeing more use in groups of people that are more visual and perhaps less technically oriented, such as human resources, but you should recognize an interrelationship digraph when you see one.

Current Reality Trees

The last example of a root cause analysis tool is the current reality tree. As was the case with the interrelationship digraph, this tool is covered briefly. This methodology is similar to that of the some of the other RCA tools examined thus far.

The most obvious difference is that it is oriented vertically, so that the root problem starts at the bottom and the root causes show up in the top of the chart. Figure 8-7 shows a current reality tree for the employee who slipped in the supply room and broke his leg.

Timelines

No matter which root cause analysis methodology you use, start your investigation by putting together a timeline of events. It can be a simple text list of dates, times, and events, or a graphical depiction of events, but get the information together as quickly as possible when gathering the baseline information for your investigation.

It is common to have people make comments such as, "This happened about the same time as this problem, so they are probably related," only to find out that the problem occurred before the event, so they could not possibly be related as cause and effect. In the previous example, you might find out that the floors had indeed been mopped causing them to be slippery, but that this occurred 3 hours before the slip, in which case the floors would have had plenty of time to dry. The timeline in Figure 8-8 shows the events related to the employee who slipped in the supply room and broke his leg. There is no substitute for good data, and a timeline makes for good data.

FIGURE 8-8 Example of a Timeline

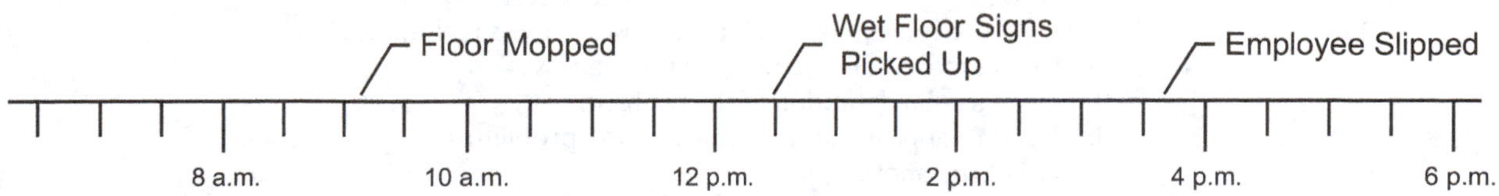

Summary

You now know that the purpose of root cause analysis is to determine the true underlying cause of a problem so that its recurrence can be prevented. Recognize that most of the time there is not a single root cause but a list of contributing root causes that may include actions, conditions, and circumstances. Many root cause analysis teams will require the direct participation of process technicians in order to understand the problem.

Before you would be asked to lead or facilitate a root cause analysis team, your employer would provide more detailed training on its particular methodology, but this chapter may be all the information you receive in root cause analysis before you are asked to participate in such an effort. The root cause analysis methodologies most commonly employed in the process industries include the Kepner-Tregoe approach, the Five Whys approach, and the Apollo RCA approach. Each methodology brings a slightly different twist to the analysis.

The Kepner-Tregoe approach uses the "what it is" versus the "what it is NOT" approach to understanding the problem. The analysis of these contrasting bits of information helps to narrow down the list of potential root causes to only those that can satisfy all of the documented criteria.

The Five Whys approach suggests that for each cause that can be identified, you treat that cause as a problem and ask yourself why this new problem occurred. You must repeat this approach at least five times to drive down to the true root cause.

The Apollo root cause analysis approach builds on the Five Whys methodology by evaluating each potential cause for both the actions and the conditions that resulted in the cause. This recognition that there are usually combinations of multiple actions and multiple conditions that contribute to a problem is realistic and explains to some extent the popularity of this approach in the process industries today.

Each of these approaches requires a solid, detailed definition of the problem and the validation of the suspected causes before you can declare the true root cause. A root cause that has not been validated is not a root cause at all but, rather, a potential cause.

Finally, there was a brief review of a few other root cause analysis tools that are applied, although not as commonly, in the process industries to ensure you can recognize these for what they are. This list of additional techniques included the cause-and-effect diagram, the interrelationship digraph, and the current reality tree.

Checking Your Knowledge

1. Define the following key term:
   a. Root cause analysis (RCA)
2. Which of the following root cause analysis techniques is best described as using the contrast between what is versus what is not in its methodology?
   a. Five Whys
   b. Apollo
   c. Kepner-Tregoe
   d. Current reality tree
3. Which of the following root cause analysis techniques is best described as a chain of multiple questions digging deeper and deeper into the cause of the problem?
   a. Five Whys
   b. Apollo
   c. Kepner-Tregoe
   d. Current reality tree
4. Which of the following root cause analysis techniques is best described as seeking out the combination of actions and conditions that converge to cause a problem?
   a. Five Whys
   b. Apollo
   c. Kepner-Tregoe
   d. Current reality tree
5. Which of these is common to all root cause analysis techniques?
   a. Need data to validate the true root causes
   b. Need a solid definition of the problem
   c. Need to apply a team approach to the problem
   d. All of the above

6. The following is an example of which root cause analysis technique?
   a. Kepner-Tregoe
   b. Cause-and-effect diagram
   c. Current reality tree
   d. Interrelationship digraph

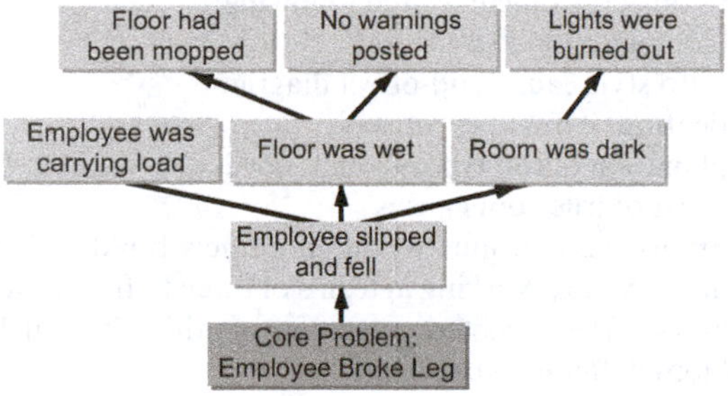

7. The following is an example of which root cause analysis technique?
   a. Kepner-Tregoe
   b. Cause-and-effect diagram
   c. Current reality tree
   d. Interrelationship digraph

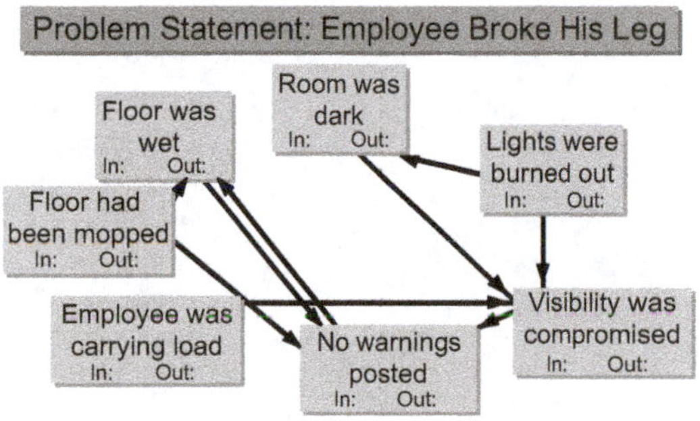

8. The following is an example of which root cause analysis technique?
   a. Kepner-Tregoe
   b. Cause-and-effect diagram
   c. Current reality tree
   d. Interrelationship digraph

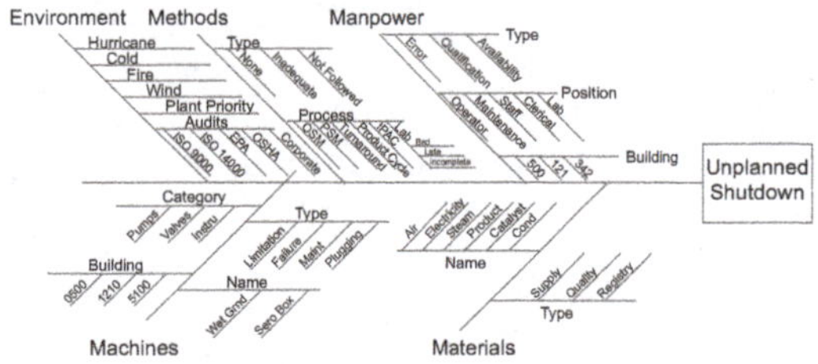

9. (*True or False*) Process technicians will rarely be called upon to participate in root cause analyses.
10. (*True or False*) Determining the true root cause is essential if you are to prevent the recurrence of the problem.
11. (*True or False*) Root cause analysis tools are relatively new and as yet untested in the process industries.
12. (*True or False*) A timeline of events is often useful in reconstructing a problem, so that you know when a problem occurred and when it did not.
13. (*True or False*) Validation of the suspected cause(s) with data is essential to determining the true root cause.
14. (*True or False*) There may be more than one root cause to any given problem.
15. (*True or False*) Graphical tools such as the cause-and-effect diagram can be used to guide a root cause analysis team.

Activities

1. Several weeks ago, a mandatory evacuation was required for your area. Unfortunately, traffic conditions were horrible and it took you almost 8 hours to evacuate. Working in teams of three to five students, conduct an Apollo root cause analysis for this problem. The expected outcome from this effort includes the following:
   a. A problem definition statement
   b. An Apollo style cause-and-effect diagram
   c. Identification of possible cause(s)
   d. An explanation of the types of data needed to validate these possible causes
   e. A short list of true root causes
2. The process industries continue to see more new builds taking place in foreign countries than in the United States. Working in teams of three to five students, conduct a Kepner-Tregoe root cause analysis. The expected outcome from this effort includes the following:
   a. A problem definition statement
   b. An analysis of "is" versus "is not"
   c. Hypotheses of possible causes
   d. An evaluation of your hypotheses against the "is" and "is not" criteria
   e. A listing of probable root causes (those that meet the "is"/"is not" criteria)
   f. An explanation of the types of data needed to validate these probable causes
   g. A short list of true root causes

CHAPTER 9

Customer Quality

"Don't find fault, find a remedy."

~HENRY FORD

Objectives

Upon completion of this chapter you will be able to:

- Explain the importance of customer service.
- Describe the supplier → process → customer relationship in terms of a flowchart.
- Explain the differences between internal and external customers.
- Explain the differences among specifications, requirements, and expectations.
- Describe an effective customer service interaction.

Key Terms

Expectations—desires of the customer that might not prevent the product from working as intended but would cause dissatisfaction if not met.

Requirements—the criteria that describe the traits that a product or service must possess in order to function as intended. Performance characteristics of your product.

Specifications—the subset of the requirements that the customer puts into the transaction contract. May also be the translation of the requirement into a measurable output for the purpose of tracking compliance.

Introduction

This piece of the quality puzzle is all about customers. Customers are the reason businesses exist. Without customers you have no business. For most readers of this textbook, this chapter should be an easy one. After all, every day you all play the role of a customer, so you all should be customer experts.

Think about the last time you went to the grocery store or the hardware store. For just a few minutes, set product quality aside and look at the service you experienced. Was the parking lot well marked and spacious? Was the store comfortable? How well marked were the aisles and shelves? How clean were the restrooms? Were the employees helpful? Did each and every person that you talked to form a part of your impression of that establishment? Some played a greater role than others, but all of them were part of your shopping experience.

Sometimes customers come to visit their suppliers. They tour the control rooms and talk to process technicians. Along the way they are taking mental notes about cleanliness, knowledge, friendliness, and so on. Your behaviors will form a part of their impression of your company.

This chapter discusses the importance of customer service and describes your process in terms of a flowchart whereby you receive inputs from your supplier, add value by virtue of your process, and then sell your outputs to your customer. This discussion will cover the differences among internal customers and external customers and outline the differences between specifications, requirements, and expectations. Finally, what makes an effective customer service interaction is discussed. Naturally, all of this is brought close to home by showing you how a process technician will play the role of customer service representative.

The Importance of Customer Service

The age-old question is, Which came first, the chicken or the egg? In the world of quality the question has been reframed as, Which comes first, quality or profits? Best in class companies believe that if you take care of the customer first, then profitability will naturally follow. The importance of customer service cannot be overstated. If profits are your primary focus, then you will make short-term decisions that increase profits this quarter.

In the quality world companies are taught that profits are the outcome of doing other things right: making the right product, shipping that product in the right container to the right place and with the right paperwork attached. Companies constantly work to improve their processes to make them the best. If they do these things well, the companies should make money. The primary focus has to be the customer because the customer has the money. The company has to provide a product that convinces the customer to give it that money.

Every contact between you and your customers, every pound of product that you make, every transaction that you conduct should be focused on customer satisfaction. Too often customers are ignored, treated as a nuisance, or treated with such disdain that they choose to look elsewhere. You should never give a customer an excuse to shop elsewhere.

In the quality election, the customer has all the votes. This is not to say that the customer is always right, but the customer is always the customer and has earned the right to be wrong. In all honesty, it is hard to be nice to some customers. Some are just plain difficult. We have all seen examples of customers who complained about a product even though nothing was wrong. Sometimes they complain because they think something might be wrong. This may even be a ploy to help negotiate a better price. After all, they cannot praise your company and product as the best in the world and then ask for a low price.

If a customer does file a complaint, the response needs to be a rigorous root cause investigation (RCI), as described in Chapter 8. If your company is at fault, then you obviously need to make it right. If you are not at fault, then this is your opportunity to share the RCI results with the customer and impress the customer with the lengths to which your company will go to improve its processes. Either way, keeping the customer happy must be your goal if you want to stay in business.

The response to the customer complaint needs to be thoughtful, be respectful, and represent the entire company. For example, a customer complaint comes in regarding drum labels being messed up on the last shipment. You respond that this is not a manufacturing issue, so you do not have any way of dealing with the problem, and suggest the customer talk to the shipping department instead. Do you think this will satisfy the customer? After all, you have helped the customer understand where the problem lies. The customer does not care who messed up but just wants it fixed and some assurance that it will not happen again. Your job is to convince the customer that you care, assure the customer the problem will be dealt with swiftly, and improve your process to prevent future occurrences. The appropriate attitude to adopt when customers voice dissatisfaction is, "Thank you for bringing this improvement opportunity to our attention."

Another response that does nothing to make a customer feel good is the response, "That's not my job." The attitude is, I am just a process technician; customer relations is not my job. However, everybody that comes in contact with a customer, directly or indirectly, is a customer service representative.

When it comes to customer satisfaction, how good is good enough? Do you want a 90% supplier rating? How about a 95% rating? Perhaps a 99% rating? Is there such as thing as putting too much effort into this process? Let us do the math. If the electricity were on at your house 99% of the time, it would be turned off for approximately 1 day every quarter. Would that work for you?

Let us use a doctor dropping a baby analogy (calculations shown below). If 99% of the babies born are not dropped, then 1% of them are dropped. This means that, in 2005, 42,000 babies would have been dropped. This is a failure rate of one baby every 12 minutes all year long. Good enough? A 99.99% success rate still equates to a baby's being dropped every single day and twice on Sunday. This is still not good enough. Nothing short of perfection is good enough. You do not want any babies dropped. You do not want any unhappy customers. You must adopt the zero defects attitude. As one executive put it, "If you do not shoot for 100% you are tolerating mistakes and you will get what you ask for."

Do the Math

Birthrate = 14 births per 1,000 population

Population = 300,000,000

Births = 4,200,000/year

1% of 4,200,000 = 42,000 (99% success)

(42,000 per year ÷ 365 days per year ÷ 24 hours per day = approximately 5 per hour or 1 every 12 minutes)

0.1% of 4,200,000 = 4,200 (99.9% success)

(one every 2 hours)

0.01% of 4,200,000 = 420 (99.99% success)

(8 per week = one every day and twice on Sunday)

Nothing short of perfection, that is your goal when it comes to customer satisfaction. It is built on more than just your product quality. Customer satisfaction is built upon the overall experience of doing business with you. Have you ever been to a department store or grocery store and noticed several employees standing around talking to each other but not to customers? Ever wished they would take the time to find out what you needed and help you find it? When you are the customer, what is important to you?

- Price?
- Product quality?
- Service during and after the sale?
- Ease of transaction?
- Selection?
- Availability of technical information?
- Attitude and personality of the sales force?

If it is important to you as a consumer, it must also be important to your industrial customers.

SPC = Supplier → Process → Customer

In earlier chapters the concept of statistical process control, or SPC, was covered in some depth, as these concepts form the foundation for most quality improvement processes. Here is another suggestion of a meaning for this acronym: supplier → process → customer. Whatever job you perform, this is your process. Your process produces an output that goes to a customer. Your process requires inputs that come from a supplier. From your supplier's perspective, you are the customer. From your customer's perspective, you are the supplier. The point is that the role you play in this chain of events depends upon your perspective.

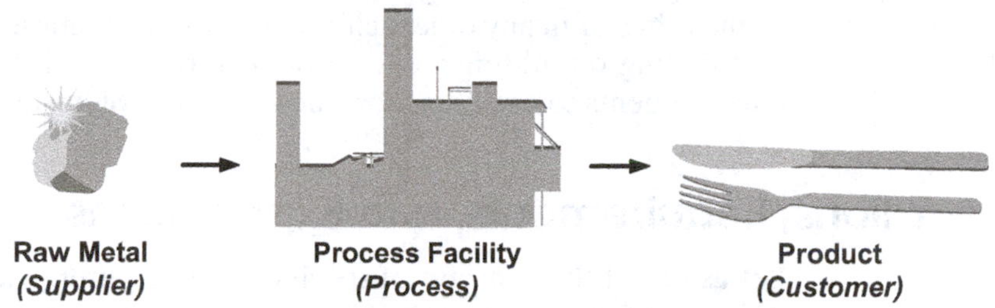

Each of these parties needs something in terms of their relationship. Customers have many needs, some of which are communicated and some of which are not. They may have needs that they do not even know about. They also have expectations that may exceed what they really need. At the least, all customers need the product that they asked for, delivered on time, to the right location, with all of the proper paperwork. In order to fulfill those needs, the process that supplies the customer must understand what product is being asked for, when the product is needed, where it is needed, and what kind of paperwork is required. Without knowledge of what the customer needs, you have little chance of meeting those needs. Later in this chapter there will be more about the differences between needs and expectations.

Internal Customers and External Customers

If the person or organization that receives the output from your process is not part of your company and is paying for the product or service, then this is an external customer. Many times your customer is part of the same organization and is an internal customer. Should you be concerned with pleasing an internal customer just as you would an external customer? Is doing so of the same importance as pleasing the external customer? No, but you must remember that the person or organization is still your customer and that this is an important relationship to manage. You, your internal suppliers, and your internal customers are all in existence for the same reason and that is to serve your external customers.

Beware of Silo Mentality

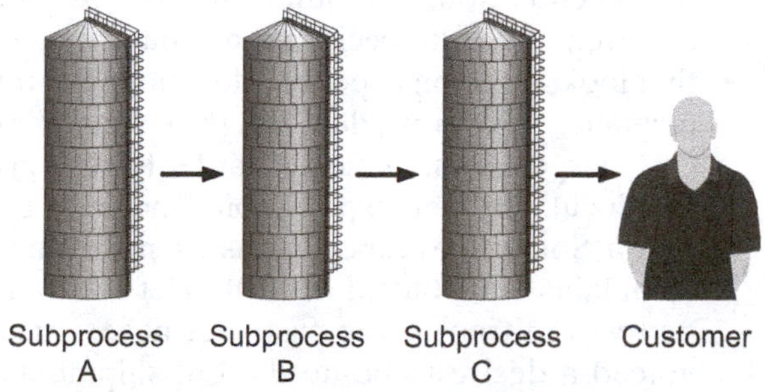

If you keep your internal customers happy but collectively fail to satisfy the external customer, you all lose. Sometimes organizations overemphasize the importance of internal customers and in doing so create wonderfully efficient processes for the various subprocesses within their companies. They optimize each phase of their operation internally, which is a good concept unless taken too far. If your internal processes focus solely on the internal customers and you lose sight of the external customer, you may suboptimize the overall process to the external customer. We call that phenomenon a silo mentality, which is thinking about only one small portion of the overall operation.

An excellent book on this subject is called *Theory Why*, by John Guaspari. The boss in this book solves the "riddle" of quality by getting employees to remember why they do what they do. You do not improve quality by improving productivity in manufacturing or by improving morale in human resources or by improving technology in engineering, or in any other achievement of functional excellence. You improve quality by focusing on adding value to the external customer. Otherwise you make great improvements to each function and still go out of business.

Specifications, Requirements, and Expectations

Let us revisit the concepts of specifications, requirements, and expectations, which are defined as follows:

1. **Requirements**—the criteria that describe the traits that a product or service must possess in order to function as intended. Performance characteristics of your product.

2. **Specifications**—the subset of the requirements that the customer puts into the transaction contract. May also be the translation of the requirement into a measurable output for the purpose of tracking compliance.

3. **Expectations**—desires of the customer that might not prevent the product from working as intended but would cause dissatisfaction if not met.

Your customer buys dibutyl futile from you. This product is a "filler" in the customer's application and takes up space in its final product mix. The customer has a requirement that this product flow through a 2-inch unloading hose at a certain rate under ambient conditions in order to unload the truck in a reasonable time frame. The customer set specifications around a trait of your product that can be used to measure your ability to meet its requirements. In this case, the customer set a viscosity specification, which is a way of measuring how easily a liquid flows. It may be that the customer came to you and asked about unloading times, and you helped with the setting of the specification. It is often true that customers know what they think they want but are not knowledgeable enough in the technology to understand how to translate their requirements into specifications.

Another chemical, isopropyl soapyl, is used as a cleaning agent. The customer uses this particular chemical to wash certain organic polymers away in its process. The requirement is that the isopropyl soapyl be an effective cleaning agent, but this is a difficult thing to measure in a lab. You set a purity specification based on testing that shows your isopropyl soapyl does an excellent job of dissolving organic polymers as long as the product is at least 95% pure.

Understanding the customers' requirements and translating them into meaningful specifications are important. If your company does a lousy job of translating requirements into specifications, you end up with customers that are receiving product that looks good on paper but does not perform as they expected. Any gap between performance of your product and the customer's expectations is a quality problem.

Expectations are even harder to get a grip on than requirements. Customers may not document their expectations. They may not even be able to articulate every expectation. Sometimes expectations are not even known until one is not met. The customer complains that you did not call to let the company know that the product had shipped and was in transit. The customer never made this contact a specification and never voiced a desire to be notified of shipments made, but because its other suppliers routinely made this phone call, the customer naturally expected that you would, too. Once the expectation has surfaced, work with the customer to define requirements and specifications to keep it happy.

There is a large home improvement store down the road. It carries everything under the sun. Because this nationwide chain buys supplies in railcar quantities, it gets a great volume discount, some of which is passed on to the final customer. Just a couple of miles away is a "mom and pop" hardware store run by John and Jane. With only a single location in a small town, they buy small quantities of the things they believe will sell, so their prices are a little higher. They stay in business because they offer exceptional customer service.

John and Jane do not just sell drill bits and light switches; they help you understand which drill bit is the best for your application and which light switch will work best for you. If you ask, John will tell you how to wire a three-way switch. They sell a quality service that one does not typically find in one of the chain megastores. They sell solutions to your household problems. Jane does not ask you what you want to buy; instead she asks you what you are working on.

Many times what you went after is not what you really need. Customers may ask for a certain items but actually need help understanding what they really need.

Now That Is Meeting the Customer's Expectations!

The story is told in various management books of the overzealous American purchasing agent who thought he would be tough on his supplier by setting a completely unreasonable failure specification of 3 parts per 10,000 units. When this company's first order arrived from the Japanese supplier, there was a small package in the top of the box with a note that read, "Dear sirs, enclosed please find your first order of 10,000 widgets. We do not know why you desired 3 bad widgets, but at your instructions we have included them. For your convenience we have packaged them separately so you would not mistake them for good widgets."

Managing expectations is important. Let us say you buy a lightbulb that guarantees it will last for 5 years. You write the date of installation on the bulb when you put it in and find out later that it lasted for only $4\frac{1}{2}$ years. You are not happy because you expected this bulb to last for 5 years. Had you bought the same bulb at the same price but been told that it would last for 4 years, you would have been happy to find out that it lasted for $4\frac{1}{2}$ years because your expectations would have been exceeded. "Underpromise, over-deliver." This rule of thumb will make customers happy. Exceed customer expectations and they will keep coming back for more.

Keep in mind that expectations change with time. In the 1960s and 1970s, many kids received transistor radios for Christmas. They were not expected to last, but it made Christmas shopping easy for parents every year. Today we expect more.

Fifty years ago a car that lasted 100,000 miles was a good car. Today we expect every car to last longer than that. Times change and so do expectations. By meeting your customer's expectations today, you continually raise the bar on what the customer will expect tomorrow. As your customers' expectations change, so will your quality plans and quality programs. Twenty-five years ago, no one in the process industry was talking about Six Sigma. Twenty-five years from now, there will be something else to talk about. What SPC, Six Sigma, ISO 9000, and all the other components of a total quality system have in common is their focus on and commitment to customer satisfaction.

Effective Customer Service Interaction

Most of the time the discussions of expectations, requirements, and specifications take place outside of the control room and involve the technical staff more than they do the process technician. However, customers come to visit their suppliers. Customers often want to tour and visit the control rooms where the action takes place. Every single person they talk to is a customer service representative for the duration of that visit. Think about a purchasing experience you found to be "top notch." Describe that experience. You might include such things as the following:

1. Comfortable temperature
2. Well lit
3. Uncluttered, neat, and tidy
4. Clean
5. Friendly, helpful staff
6. Spacious parking

The next time a customer is coming to visit your plant, create that environment for him or her. Make the customer feel special and he or she will love coming back. An example of this occurred at a chemical plant back in the early 1990s. After several years of improving this customer visit process and actively managing customer expectations during their visits, we found that most of our customers were commenting about how they looked forward to returning to our plant. Because of this attitude, they began relaxing the scope of their audits and surveys. They came to expect a pleasant visit and excellent performance.

To be sure, you also have to have a good-quality product and a competitive price to stay in business, but managing the less tangible aspects of customer quality plays a key role in convincing customers to keep coming back. Our process technicians played a big part in setting the stage for success during these visits. We spent time coaching individuals on proper etiquette during these visits. We cleaned up the operating areas and then kept them clean between visits. In addition to the physical appearance of your plant, here are some other tangible ways to make a positive impression on your visiting customers:

1. Be professional.
2. Be courteous.
3. Be responsive.
4. Be polite.
5. Be friendly, yet businesslike.
6. Avoid profanity and inappropriate remarks.
7. Avoid disparaging remarks about your company.
8. Avoid negative attitudes.

Remember that you are now the seller. You are being judged, collectively and individually, by the customer. Your goal is to demonstrate to the customer that your company is worthy of the customer's business. If in doubt, apply the golden rule. It works in life and it works in customer service. Treat your customers as you would like to be treated in the same circumstances.

Your customer service interaction may be direct through a customer visit. It may be indirect by conducting a root cause investigation into a customer complaint. It may be that you are the one putting the paperwork in the appropriate envelope. Whatever your role in the customer contact process, make it a pleasant experience for the customer.

> **"Treat the customers as you would like to be treated"**
>
> *~The golden rule of customer service*

Take a lesson from Disney. This corporation does not have employees at its theme parks; it has cast members. The cast members do not work in an office or at a plant; they are on stage. The personnel department is referred to as "casting." Instead of customers, Disney has guests. Once a year all the Disney executives trade their three-piece suits for costumes, and they run rides and sell hot dogs for a full week.

When a customer calls on or comes to visit your plant, you too are on stage. If you need to adopt a stage personality for the visit, then do so. How do actors prepare themselves for a role? They study the role and they memorize their lines. You should do the same thing. A customer is coming to visit next week. What does this customer buy? What does it make with your product? How is your product packaged and delivered? What is the purpose of the visit? Is the customer auditing you to a company standard or the ISO 9000 standards? What kinds of questions will be asked? What kinds of preparations do you need to make in anticipation of the customer's arrival? Preparing yourself and your location for the customer visit in advance will allow you to relax and host your customer at ease.

The closest thing we have seen to this type of customer attitude in the process industry is a company in the northeast that used to designate one specific carrier to its most critical customers. This carrier would literally have the truck arrive at the customer destination the day before delivery was due, so the driver could have the truck and trailer cleaned. The driver would put on a fresh change of clothes, including a tie, and drive up to the customer's plant an hour before the scheduled delivery time and park on the side of the road. At 10 minutes to scheduled delivery time, the driver would brush his teeth and pull the rig up to the customer's gate, arriving exactly on time, immaculately dressed, and in a truck that was ready for the showroom. There was no doubt in the minds of the customers that the company cared about their impression of it.

Summary

As you have now learned, all the statistical improvement tools in the world are of little value if you fail to satisfy the customer along the way. You cannot stay in business without customers. They have money and you want it. Your jobs are to manufacture and sell a product that will make them desire to give you the money.

The statistical tools, quality management systems, and other improvement tools discussed in this book are all focused on improving your processes. If the output from these processes does not meet the needs of the customer, you will lose in the long run.

You should now be able to describe your process in terms of a flowchart whereby you receive inputs from your suppliers, add value through your work processes, and then sell this value-added product to your customers. You should also understand the difference between an internal customer and an external customer. Hopefully you also understand that only by satisfying the external customer can your business succeed.

Covered at some length in this chapter was the difference between requirements (performance characteristics important to a customer) versus specifications (acceptance criteria for your products) versus expectations (things that may not prevent the product from working but will cause customer dissatisfaction if not met). Finally, what makes an effective customer service interaction was discussed, and some practical tips were provided on how to play the role of customer service representative effectively for your company.

Checking Your Knowledge

1. Define the following key terms:
   a. Expectations
   b. Requirements
   c. Specifications
2. Which of the following is important to consider in customer contacts?
   a. Cleanliness of the facility
   b. Availability of technical information
   c. Demeanor of the people
   d. All of these plus more

3. The relationship of your function to the customer can be captured in flowchart form with which acronym?
   a. QFD
   b. SPC
   c. RCI
   d. BOC
4. The recipient of your process output is your:
   a. Accountant
   b. Supplier
   c. Customer
   d. Process
5. If the person or organization that receives your process output is part of your same company, you call it an:
   a. Internal customer
   b. Internal supplier
   c. External customer
   d. External supplier
6. If the person or organization that receives your process output pays green dollars to purchase that ouput, we call it an:
   a. Internal customer
   b. Internal supplier
   c. External customer
   d. External supplier
7. Which of these would you NOT do in preparing for an upcoming customer visit?
   a. Clean up the area
   b. Find out what the customer is interested in so you can be prepared to answer any questions
   c. Coach the staff on effective customer relation interactions
   d. Take the day off
8. When it comes to quality, how good is good enough?
   a. 90%
   b. 95%
   c. 99%
   d. 100% is always your target
9. (*True or False*) As long as your product quality is good, that is all a process technician needs to worry about. Service quality is other people's job.
10. (*True or False*) If you focus on quality (product quality and service quality), you expect that profits will naturally follow.
11. (*True or False*) Customer complaints provide your business with a great opportunity to demonstrate your commitment to quality to your customers.
12. (*True or False*) Internal customers are just as important as external customers and should be treated the same.
13. (*True or False*) The golden rule is an excellent judge of how to behave when dealing with customers.
14. (*True or False*) Process technicians can reasonably expect to have direct contact with customers at some point in their career.
15. (*True or False*) Service quality and customer service interactions are not as important as product quality.

Activity

1. During the course of the next week, pay attention to customer service interactions at the grocery store, the department store, the hardware store, and so forth.
   a. Write a brief summary of an example of the worst customer service interaction that you experienced. What made this experience unpleasant? What can you learn from this experience and apply to your job as a process technician?
   b. Write a brief summary of an example of the best customer service interaction that you experienced. What made this experience pleasant? What can you learn from this experience and apply to your job as a process technician?
   c. One of the best ways to learn how to treat your customers with respect is to put yourself in their shoes. List at least five tangible ways that you can be a better customer when you go to the store.

CHAPTER 10

Six Sigma

"Take away the causes and the effects cease."

~CERVANTES

Objectives

Upon completion of this chapter you will be able to:

- Explain what Six Sigma means in terms of the number of defects per million opportunities.
- Describe and define the stages of a Six Sigma project (define, measure, analyze, improve, and control).
- Identify key improvement tools used at each stage of a Six Sigma project.
- Explain how Six Sigma tools could be applied to new designs as well as existing processes.

Key Terms

Black belt—team leader on a Six Sigma team that has training in the basic and some advanced quality improvement tools plus training in project management.
Champion—a manager in the Six Sigma world responsible for the overall effectiveness of the Six Sigma implementation at a business or functional level.
Design for Six Sigma (DFSS)—the application of appropriate Six Sigma tools to new process designs rather than to the improvement of existing processes.
Green belt—team member on a Six Sigma team that has training in the basic quality improvement tools.
Master black belt—Six Sigma expert with training in many advanced quality improvement tools. Is responsible for coaching black belts and providing training to green belts and black belts.

Introduction

Six Sigma. It sounds statistical. You learned in Chapter 2 that the lowercase Greek letter sigma (σ) is used in statistics to represent the standard deviation. This chapter is not about statistics; it is about an improvement strategy. It just so happens that this improvement strategy has a name that sounds statistical and, in truth, uses statistics to accomplish improvement, but it is much more than just a collection of statistical tools.

In this chapter you will learn about the various stages of the Six Sigma improvement strategy and what tools are employed during each stage. The discussion explains the way Six Sigma is organized. Six Sigma can be applied to manufacturing processes as well as functional work processes. It can be applied to existing processes as well as new processes. It can even be applied in the design phase before a product or process is launched. Let us start with a little history of Six Sigma.

Brief History of Six Sigma

In Chapter 1 you were introduced to some of the great masters who helped shape the field of quality as it is known today. The term *total quality management* (TQM) was also defined as the consolidating of many different components of quality into a single system. Six Sigma is total quality management. It is the consolidating of many, many elements of quality into a single management system.

Motorola gets the credit for having "invented" Six Sigma in the mid-1980s. General Electric, Honeywell, Allied Signal, and Texas Instruments were some of the first companies to leverage the success of Motorola. One thing you might notice is that all of the early adopters of Six Sigma were electronics manufacturers. These companies make thousands and thousands of discrete parts, in contrast with what the process industries produce—buckets of matter. Because of this difference, it took the process industry a little longer to get on the bandwagon, but get on the bandwagon it did.

Chemical giants Dow and DuPont led the Six Sigma charge in the late 1990s. In the case of Motorola, it attributed its success in winning the Malcolm Baldrige National Quality Award in 1988 and again in 2002 to its implementation of Six Sigma throughout the organization at all levels and in all functions. You will learn more about the Malcolm Baldrige award in Chapter 12.

Again, thinking back to Chapter 1, recall that Dr. Deming advocated statistical expertise at the master's level. Dr. Juran advocated improvements with a project-by-project approach using a breakthrough strategy. Phillip Crosby taught an attitude of zero defects and measurement of the cost of quality. Tom Peters spoke about the need for high-level quality champions to drive improvement down through the organization.

In that chapter it was proposed that rather than picking which of the quality gurus was the "best," you should adopt the best of each of the gurus. This is exactly what Six Sigma has done as an improvement strategy.

IN A NUTSHELL

Six Sigma

A formal, disciplined, top-driven approach to continuous improvement using the best practices of the historically recognized experts in the field of quality.

Six Sigma drives improvement from the top by starting with management as the champion for the effort. It concentrates on using the language of management to quantify the value of the improvement. Six Sigma improvements are made by applying the appropriate statistical and graphical improvement tools in a team-based approach. Many of the tools employed in the Six Sigma process have already been discussed in previous chapters. Six Sigma is made up of the best practices from historically recognized experts in the quality field. It is a formal, disciplined, top-driven approach to continuous improvement.

Six Sigma Defect Definition

Six Sigma has grown and evolved from a metric to a methodology to a management system over the last 20 years. Recall from Chapter 3 how process variation relates to specifications by using the process capability index (Cpk) and that ±2 sigma is equal to 95% of the total variation of a normal distribution. If 95% of your process is inside these limits, then it stands to reason that 5% is outside the limits. If your specifications are set at these limits, then your Cpk would be 0.66.

Reading Table 10-1 you would say that a 2σ process has a Cpk of 0.66 and a defect rate of 5%. As you recall, your customers typically want a less variable process than that, often desiring a 4σ process, which equates to a Cpk of 1.33 and a defect rate of 63 parts per million (ppm). From Table 10-1 you can then see that a 6σ process would have a Cpk of 2.0 and a defect rate of 2 parts per billion (ppb).

If you read about Six Sigma on the Internet, you may see information about the Six Sigma process yielding a defect rate of 3.4 ppm and wonder why this differs from the calculations. The reason is that these calculations assume that the process is perfectly stable; thus the mean of the process never changes.

The Six Sigma world has taken into account that the process mean drifts around a little bit. It allows for a process drift of 1.5σ. This may sound a little confusing at first; but the gist of the matter is that a 6σ process is one wherein a 1.5σ drift in the mean is allowed and the process still has a capability of 4.5σ, which means a Cpk of 1.5 or only slightly better than the targets historically employed in the process industry.

When you make hundreds of thousands or even millions of discrete parts, even a relatively small defect rate can add up to a large problem. Let us say you make wood screws. You make 50 million wood screws per year. A 4σ process (Cpk = 1.33) would mean a defect rate of 63 ppm or a total of over 3,000 wood screws that were not sellable every year. On the other hand, because you work in the process industry making, let us say, 350 batches of product per year, a 4σ process means it would take 45 years to make a failing batch.

Let us put this in terms that are less industrial. You play 100 rounds of golf every year and average 2 putts for each of the 18 holes in each round. That gives you 3,600 opportunities to putt per year, on average. A 2σ process is a 5% defect rate. Missing 5% of your putts means you miss 180 putts per year or about 2 putts per round. Improving your process to 3σ means you miss only 10 putts per year. If you could achieve the process industry target of 4σ (Cpk = 1.33), you would miss a putt only once every 4.5 years.

TABLE 10-1 Six Sigma Defect Rate

Cpk	*Sigma*	*Percent Passing*	*Defect Rate*
0.66	2σ	95%	5%
1.00	3σ	99.73%	2,700 ppm
1.33	4σ	99.9937%	63 ppm
1.50	4.5σ	99.99966%	3.4 ppm
1.67	5σ	99.9999427%	0.5 ppm
2.00	6σ	99.9999998%	2 ppb

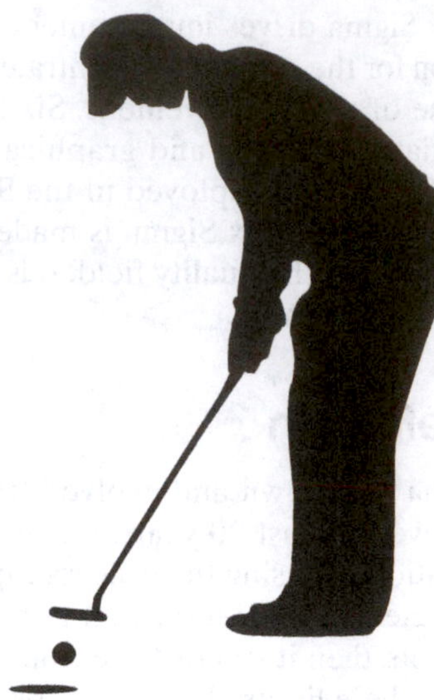

The purpose of this discussion is to put defect rates into perspective and to illustrate why the electronics industry sets goals so much higher than does the process industry. A defect rate of 5% is not too bad if you make only 10 batches per year, but it is completely unacceptable if you make 10 batches per hour. Most golfers would be happy to establish a 3σ process (Cpk = 1.0) and miss only 1 putt for every 10 rounds of golf they play.

Stages of Six Sigma

Six Sigma is a process for continuous improvement. As with most processes, it is made up of multiple steps or stages. Each of these stages has a purpose to fulfill for which there is a variety of tools available. Some tools can be applied in more than one stage; others are fairly specific in their application. As has been said several times throughout this textbook, the goal is to use the right tool for the right job in order to improve your process. To select the right tool you must understand the tools from which you can choose. Before you study a more detailed description of the stages of Six Sigma, you need to understand a few basic beliefs:

- Top-down deployment is an absolute must. Six Sigma cannot be successful if driven from the bottom of the organization. You probably will not get to decide whether Six Sigma should be implemented. But if you go to work for a company that has implemented Six Sigma, then get on board and do everything you can to participate and help make the program successful, because the tools and methodologies in Six Sigma are time-tested winners.

- Upper management must champion the cause—not just support it. For Six Sigma to work, upper management must be fully committed. Commitment is often gauged by the level of active participation. At Motorola, vice presidents were actively participating in team activities.

- Key measures are best quantified in dollars. Chapter 15 examines more about money, but that is why the company is in business. In most organizations, there are more opportunities to improve the process than there are resources and manpower to do so. Defining the opportunity in terms of dollars helps management ensure that it is working on the things that will help the most.

- The voice of the customer is key to success in the planning (define) stage. The preceding item states the importance of making Six Sigma important to managers by defining the opportunity in terms of dollars. The same concept applies here. If the improvement is important to your customer, then the improvement is important. If the customer will not recognize the improvement, then you have to challenge why you are working on the issue.

- $Y = f(x)$. Do not let the mathematical formula scare you. This is read as Y is a function of x (see Chapter 7: Y is your output, and x is your collection of inputs). Thus you can read this statement as "Our output is a function of our collective inputs." One premise of Six Sigma is that you focus on the inputs in order to drive the desired output.

Now, on to the stages of Six Sigma.

DEFINE OR CHARTER STAGE

Some publications include define or charter as the first stage of Six Sigma; some do not. Even if it is not included in the official process, somebody has to define what the project is before any work gets done. Management has to make a decision about where it would like to deploy its resources and charter a team with responsibility to accomplish improvement. Typically a process technician would not be involved in the define/charter step, but it might be valuable for you to know a little bit about how much work goes on behind the scenes.

Project Charter

What is wrong?
(Define the Defect)

What needs to be done about it?
(Goals and Objectives)

Where does it need to be done?
(Project Scope)

Who are the right people to get it done?
(Team Membership)

How long should it take?
(Timeline)

How will we know when they are finished?
(Deliverables)

The main output from the define stage is a project charter. The charter answers some basic questions about the improvement effort. This is where management defines the defect, establishes the strategic importance, assigns resources to work on the project, defines the scope of work, establishes a desired timeline, and documents the required deliverables. Successful accomplishment of the team's goals will require planning up front to ensure that the right people with the right skills and training are there, given the right amount of time to solve the problem. Planning for success is just as important as being a technical expert in the field of quality improvement.

Consider a project team that has been handed a charter with the following defect definition: Energy costs in the dibutyl futile unit are too high. The goal for the team is to deliver improvements to the process that will reduce energy costs by 50%. That sounds okay. But consider this: The consumption of utilities has remained fairly flat over time, but the price of the same utilities has risen by over 300% in the last 4 years.

This project is assigned to a production department team including salaried and hourly people. The plant engineers and process technicians may be able to make significant gains in consumption rates and energy efficiency, but they probably do not have any control over utility costs. In this case management is absolutely right in seeing a need to reduce energy costs, as these costs relate directly to the bottom line. The problem is that the goals and objectives of the team are outside management's span of control.

In this example the project team could implement some really elegant solutions such as reusing heat from one step of the process to preheat another step, shortening cycle times, or installing solar-panel-powered controller equipment. This team might be able to reduce energy consumption by 50% but still not meet its goals due to factors outside its control.

One of the first things a project leader should do when assigned a new project is review the charter with management to ensure that the team membership is capable of delivering results that meet the goals and objectives of the project within the timeline provided. Although arguing with management is not recommended, tactfully challenging the premises of the charter to ensure success is definitely in the best interests of both management and the team.

MEASURE STAGE

The main objectives of the measure stage are the following:

1. Finalize the project charter.
2. Document the current process.
3. Define the key input variables.
4. Establish the measurement system capability.
5. Establish baseline measurement of the defect.

The importance of a good project charter has been documented, so reviewing the charter with the team to gain a consensus understanding of what is going to be done, who is going to do it, and how they are going to do it is a reasonable way to kick off the project team. In Chapter 11 there will be a lot more coverage of team dynamics and various types of teams you may be called to serve on.

The next activity is to document the current process. In Chapter 6 flowcharts were discussed as a mechanism to document a process graphically. This can be an extremely valuable tool, as it establishes a common level of understanding among the participants. If everybody on the team has a different view of what the process looks like, the team will never come to an agreement on how to fix the problem assigned to it. In some cases, the act of mapping out the current process showed the team not only what the problem was but also how to fix it. Use whatever flowcharting, procedure review, or process mapping you desire, but never skip this step. One excellent technique for accomplishing this step is the Rummler-Brache approach to process mapping. It is not possible to condense its entire methodology to a couple of pages here, but it would be excellent reading for anyone interested in going a step or two beyond the basics.

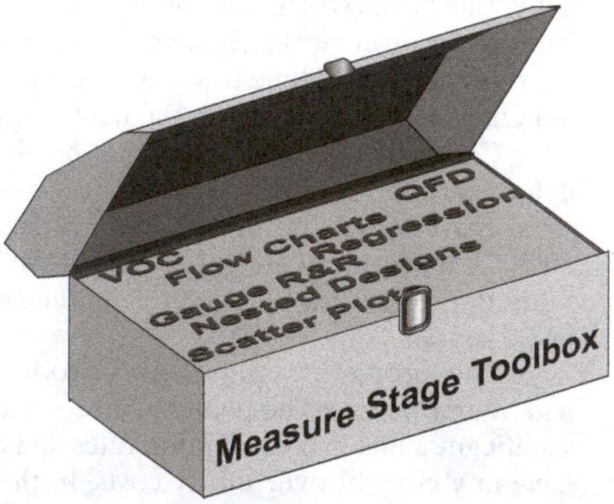

Management has given the Y in the equation $Y = f(x)$. It has specified which output is in need of improvement. The job of the team is now to define which process inputs are the ones that are generating that output. The team has to define the *x*s. The answer may be fairly obvious and require no more effort than reviewing the process flowchart and writing down the answer. In the energy consumption example just talked about, it should be straightforward to define where the company uses steam, where it uses electricity, and so on. But consider a project in which the Y is customer satisfaction.

Knowing exactly why customers are not satisfied is not typically quite as obvious. Sometimes even the customer cannot really put a finger on the problem; the customer just does not like doing business with you. Chapter 9 is set aside to talk about customers. The team might be able to make some guesses about why customers are unhappy but can work on the right *x*s only if it collects data to measure customer satisfaction.

The tools most often used to collect customer data are called voice of customer (VOC), quality function deployment (QFD), and the quality design plan portion of quality reliability planning, which will be discussed in depth in Chapter 13. On a more technical note, let us say the defect is related to product quality. The team must measure the amount of poor-quality material being produced and determine what process parameters are leading to poor quality. This is absolutely nonobvious and will require the use of statistical tools such as scatter plots, regression analysis, and designed experiments, all of which have been discussed in preceding chapters.

Now that you know the output (Y) and the inputs (*x*s), you need to understand your ability to measure them. Everything varies. If you do not think it varies, it is just because you cannot measure the variation. If you cannot obtain a "good" measurement of your inputs and outputs, there is no way to improve upon your current performance. You learned about this concept in Chapter 3 and how to use nested designs to quantify the measurement system variation.

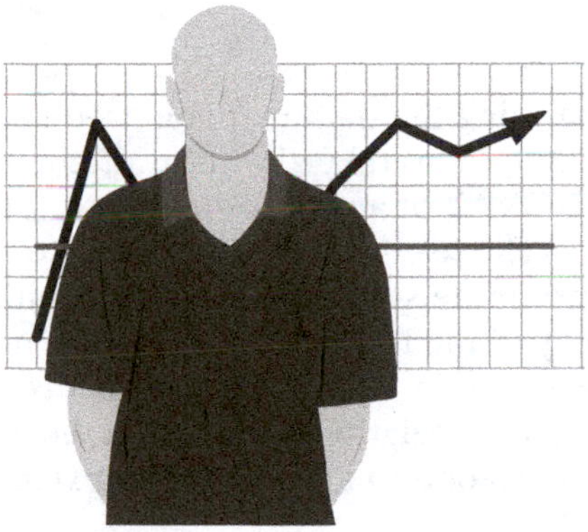

Another tool that accomplishes this goal is called the gauge reproducibility and repeatability (Gauge R&R) study. If the team finds itself in the awkward position of being charged to make an improvement in an area that cannot be adequately measured, it may have to spend some time in this stage improving the measurement system or even putting a new measurement system in place. You cannot improve it if you cannot measure it.

The final piece of work for the measure stage is the establishment of baseline performance using the accepted measurement practices established. Refer back to some of the statements made earlier about project charters: "Reduce energy consumption by 50%." The team has to establish some basis against which to measure this improvement. Is that a 50% reduction from last month's performance or from last year's average performance? Once you have established your baseline performance data, you are ready to move on to the next stage of the Six Sigma improvement strategy.

Occasionally a project dies right here. It is possible for a team to establish the measurement system capability and collect baseline data only to discover that the magnitude of the problem was not what had been expected. In that case, it is absolutely appropriate to thank the team for its efforts and disband it to reassign the resources to more pressing matters. This is not a failure; this is good management, using the data to drive decision making.

ANALYZE STAGE

Now that you have data, you can analyze those data. The main objective of this stage is simple: to determine the root cause(s) of the problem. You accomplish this by doing the following:

- Analyzing the data
- Brainstorming potential causes
- Validating causes
- Selecting root cause(s)

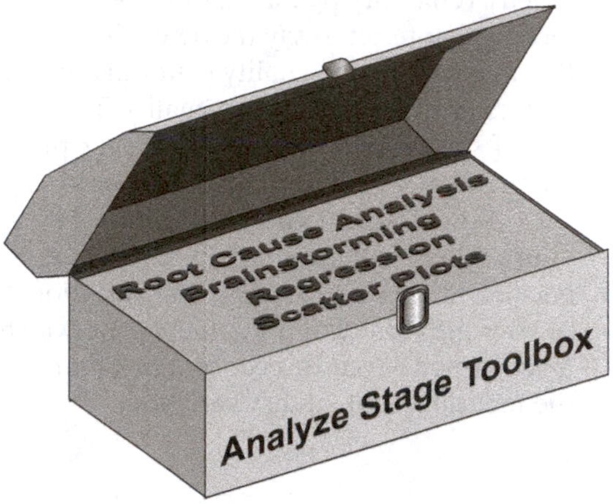

Analyzing the data that you have been collecting is the obvious starting point. This analysis may take the form of control charting, histograms, analysis of variation (ANOVA), or a myriad of other statistical techniques.

The team will typically brainstorm a list of potential root causes and then use the appropriate statistical analyses to either validate or invalidate each potential cause. Many of the tools mentioned in the measure stage are employed in this stage as well. For example, designed experiments and regression analysis are commonly used to establish a relationship between the inputs and the output. In addition, the root cause analysis tools that were covered extensively in Chapter 8 are commonly employed:

- Cause-and-effect matrix
- Five Whys
- Kepner-Tregoe root cause analysis
- Apollo root cause analysis

If all goes well, you now have a list of process inputs that correlate to the output that is the topic of concern for your project. This correlation validates which process parameters are root causes of the output and which ones are not. Oftentimes the analysis of the data takes place so closely tied to the collection of the data that the analyze stage is completed quickly after the measure stage.

Now that you know what may have caused your problem, you can progress to fixing the problem. There is no requirement to use each and every tool mentioned above. The requirement is to document the valid root cause(s). The team should feel free to employ whatever tools help it get there. A mistake made by many teams is to jump directly into fixing the things that "everybody knows" to be the problem without going through the disciplined root cause analysis. This typically leads to fixing a known problem but not fixing the problem assigned to the team.

IMPROVE STAGE

The objectives of the improve stage are all focused on eliminating or reducing the root cause(s) determined in the preceding stage. You accomplish this by doing the following:

- Brainstorming potential solutions
- Evaluating potential solutions
- Selecting optimum solution(s)
- Validating optimum solution(s)

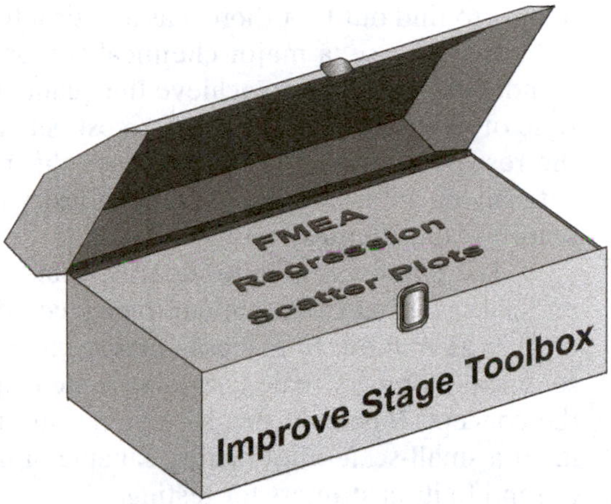

Sometimes the team identifies a single root cause that can be completely eliminated, driving the defect down to zero. More often the team identifies multiple contributing causes, each of which contributes to the defect. Sometimes these can be completely eliminated, but sometimes they can only be reduced.

Take, for example, a root cause of raw material contamination. If you have the ability to test the raw material before it is used in your production process, you may be able to eliminate the contamination completely. If the contaminant is unknown or you do not have the technology to detect the contamination, you may not be able to eliminate this cause completely. Improvements at your supplier's location may help reduce the problem, but they are outside your scope of control and may not be completely dependable.

The first step is the generating of ideas about how to fix the problem. Once the root cause or causes are known, thinking of ways to fix the problem is usually pretty straightforward. The goal of Six Sigma at this stage is to challenge the team to think of the possible solutions that may be less obvious. Just as there may be more than one cause, there may be more than one solution. Sometimes the best solution is made up of several different ideas that build off of each other.

With a wide range of options identified, the team will then need to evaluate the potential solutions. A matrix can be created that the team can use to rate each solution against selection criteria that it agrees to in advance. In discussions with the process owner, the team determines that the selection criteria include the following issues:

- Cost to implement
- Cost to manage after implementation
- Time to implement
- Impact on the project objectives (low, mid, high)
- Complexity
- Probability of success

The team can set limits around what constitutes low cost versus high cost; define short time to implement versus long time to implement; and so on. With these criteria defined, each potential solution can be evaluated to see which option seems to offer the most bang for the buck. If management places a higher importance on cost to implement than it does on time to implement, the team can even use a weighted average to give more weight to those issues considered the most important. The end result is a short list of solutions that are expected to achieve the most benefit for the least amount of money in the shortest period of time. With management's help, the team selects the solutions to implement.

Most of the time, a full-scale implementation of a huge project will not gain upper management approval without testing the solution on a smaller scale. This is called a pilot program. The company may be willing to spend a few tens of thousands of dollars to make sure the project will work in order to reduce the risk of spending millions of dollars to find out that there was a critical flaw in the logic somewhere.

In one case, a major chemical company was led to believe that a new style of blender was required to achieve the quality targets for a certain product line. This new style of blender was estimated to cost between $15 million and $20 million. Based on the research conducted by the team, this particular style of blender had been used extensively in the food processing industry in the making of bulk chocolate, but never before in the chemical industry.

The team was able to identify a small-scale blender that would mimic the commercial scale, but even that equipment could be obtained and installed only for a price of over $1 million. In this case management chose to spend approximately $10,000 to set up a laboratory scale test using rented equipment and hired technicians to validate the concept. Based on the positive results, it agreed to spend the $1 million to implement a small-scale pilot facility capable of making large enough quantities of product to send to its customers for testing.

When the customer feedback was positive, management then authorized the implementation of the $20 million commercial-scale solution. The team "knew" the answer up front but had to take steps to collect data that would convince management that the benefits were worth the risks.

Sometimes the right answer is to rent the solution. Sometimes the right answer is to buy a smaller-scale model first. Sometimes the team may be able to run the existing equipment using different procedures or different conditions. In each case the team must not only evaluate the beneficial impact of the solution but also identify any possible negative trade-offs. It would not do to implement a new procedure that achieves significant reductions in energy usage only to find out that you are now making more off-spec product than before. In other words, you have to make sure you are not trading one problem for another one that is potentially worse.

If the team is able to use the existing equipment, then the validation may be as simple as running a few experiments to prove the concept. One excellent tool for evaluating potential failure modes is the failure mode and effects analysis (FMEA), which will be covered in some depth in Chapter 13.

The end result of the improve stage is a list of validated solutions that will fix the problem assigned to the team.

CONTROL STAGE

The final stage of the Six Sigma process is the control stage, the objective of which is to ensure that the desired improvements are maintained. More than once many have heard seasoned veterans in the industry recall a problem that had to be fixed again and again because the organization implemented what is called a "reversible" fix instead of an "irreversible" fix.

Consider a project team that determined that running the reactor at 100 degrees C would provide a 10% quality improvement over the current operating condition of 90 degrees C. How should the team implement this change? Administrative controls work well, but there is no way to guarantee that they will stay in place.

- Instructing four shifts of process technicians to run at 100 degrees instead of 90 degrees is good.
- Updating the procedure manual to state that the desired operating temperature is 100 degrees is better.

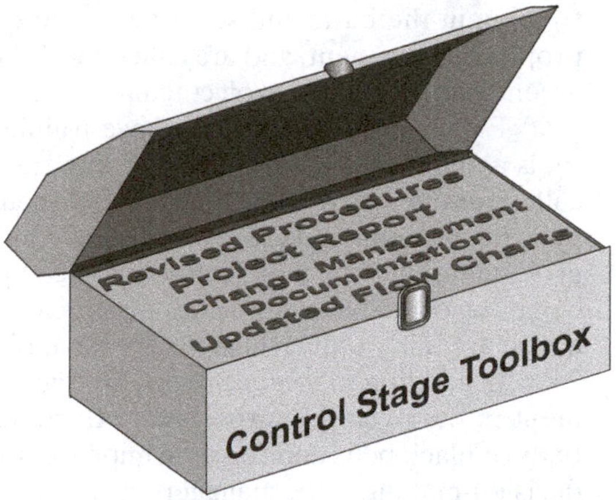

- Changing the set point in the process computer so that the change happens automatically is better still.
- Changing the set point in the process computer, updating the procedure manual with the new requirement and an explanation of why the change was made, and training all four shifts of process technicians on the nature of the change is the best yet.

One thing that gets overlooked far too often is the "why" of the change. If you do not explain why a change is made, then the organization soon forgets and may revert to the old conditions when a new team made up of people who either did not participate or cannot remember the reason for the change is chartered to work on a different problem.

The tools employed at this stage of the project are things such as revised procedures, updated flowcharts, a formal project report, and some type of change management documentation. If your company has a control plan for each product, then that control plan should be updated and a new FMEA constructed as well. The quality control plan and FMEA are both discussed in more detail in Chapter 13.

The last action that the team needs to take is to transfer ownership of the project back to the process owner. The team can draft the updated procedures, revise the flowcharts, and fill out all kinds of project documentation; but if the owner of the process does not take ownership in the change, the benefits will not be maintained. This activity is often accomplished by holding a formal transfer meeting at which the process owner has an opportunity to review and accept the work of the improvement team formally.

Six Sigma Participant Structure

The Six Sigma structure is patterned after the martial arts. Team members have training in the basic quality improvement tools and are called **green belts.** Process technicians are often trained as green belts in the process industries. Team leaders have

training in the basic and some advanced quality improvement tools, plus training in project management, and are called **black belts.** A black belt might manage two to ten green belts on a given project team.

Expert practitioners who have training in many advanced quality improvement tools and are responsible for providing training to the green belts and black belts are called **master black belts.** Master black belts may oversee any number of black belts, providing training and assistance as needed. In Six Sigma, as in the martial arts, progression through the ranks is accomplished by training and demonstration of the required competencies. Some companies may include additional rankings, but those mentioned above are fairly universal in Six Sigma applications.

In addition to these rankings, Six Sigma practitioners recognize the role of management in the overall improvement effort. Sometimes managers are certified as green belts or black belts and serve on improvement teams as leaders or participants. In their decision-making roles, managers may also serve as champions. As stated earlier in this chapter, a top-down commitment is absolutely required for success.

To ensure this high-level commitment, many companies identify business or functional champions. **Champions** are responsible for the overall effectiveness of the Six Sigma implementation at a business or functional level. The role of the champion is to identify and select high-priority projects, ensure that training is available to the team members, and remove roadblocks that might prevent the team from being successful.

Design for Six Sigma

The Six Sigma approach outlined in this chapter is the standard garden-variety Six Sigma. It is used extensively throughout the process industries to accomplish the improvement of existing processes. Another type of Six Sigma that is gaining ground and acceptance in the industry is called **design for six sigma (DFSS),** and it is the application of appropriate Six Sigma tools to new process designs rather than to the improvement of existing processes. Unlike the garden-variety Six Sigma—which is universally accepted as the define, measure, analyze, improve, and control (DMAIC) methodology—DFSS comes in many different flavors. Rather than explore the many varieties of DFSS in existence, let us just talk about DFSS in general and fairly brief terms.

DFSS is used in the design phase of a new product or process to ensure that the process does what it is supposed to do from the beginning, negating the need for improvement after start-up. Does this make sense? If you design it right the first time, you do not have to redesign it later. Many of the tools employed are the same tools as are used in the DMAIC variety of Six Sigma. Flowcharts, brainstorming, cause-and-effect matrices, and so on are all employed. Because everyone is starting from scratch, there is a strong need for tools such as voice of the customer and FMEA in design for Six Sigma. The statistical tools, such as designed experiments, require the use of data that often do not exist in a new application; but models generated from existing processes are sometimes used to predict the operation of a new process, so even these tools find use in DFSS.

You may be thinking, "I am training to be a process technician, not a design engineer, so I will never be a part of a DFSS team." Wrong. Process technicians and engineers working in similar processes to the one being designed are in the absolutely best position to review the proposed design and point out potential trouble spots. A good design team will take advantage of that expertise if given the chance. You might even find it fun.

Summary

This chapter has covered the basic methodology behind Six Sigma and compared the tenets of Six Sigma to earlier quality efforts such as total quality management. Notice that Six Sigma represents the best practices of each of the historically recognized quality gurus, pulling these pieces together into a single improvement strategy.

You learned that Six Sigma, as an improvement strategy, allows for a 1.5 sigma shifting of the mean. Because of this allowance, a process that is said to be a Six Sigma process is one that will produce defects at a rate of 3.4 per million opportunities.

You learned that Six Sigma is made up of multiple stages defined as the define, measure, analyze, improve, and control (DMAIC) methodology. Within each of these stages, the discussion identified the objectives and outlined the types of improvement tools used as follows:

Stage	Objectives	Tools
Define	Project charter • Defect definition • Strategic importance • Goals and objectives • Project scope • Team membership • Timeline • Deliverables	Charter template
Measure	• Finalize the project charter • Document the current process • Define the key input variables • Establish the measurement system capability • Establish baseline measurement of the defect	Gauge R&D Nested designs Quality function deployment Voice of the customer Regression analysis Scatter plots Flowcharts
Analyze	Determine root cause by: • Analyzing the data • Brainstorming potential causes • Validating causes • Selecting root cause(s)	Root cause analysis Brainstorming Designed experiments Scatter plots Regression analysis
Improve	Eliminate root cause by: • Brainstorming potential solutions • Evaluating potential solutions • Selecting optimum solution(s) • Validating optimum solution(s)	Designed experiments Scatter plots Regression analysis Failure mode and effects analysis
Control	Ensure that the gains are maintained	Quality control plan Failure mode and effects analysis Flowcharts Revised operating procedures Change management documentation Project report

You learned that Six Sigma practitioners are named using a martial arts–type convention with green belts being team members, black belts being team leaders, and master black belts being the implementation experts and trainers. In addition, many companies identify the role of champions (managers who are responsible for the overall effectiveness of the Six Sigma program).

Finally, the traditional application of Six Sigma to improve existing processes was contrasted with a relatively newer application of some of the same tools to the design of new products and processes. The application of Six Sigma tools to new designs is called design for Six Sigma.

As a process technician, you can expect to participate in the Six Sigma process, most probably as a green belt. Some of the tools you have learned or will learn in this textbook. Other tools will be taught to you once you join the workforce.

Checking Your Knowledge

1. Define the following key terms:
   a. Black belt
   b. Champion
   c. DFSS
   d. Green belt
   e. Master black belt
2. Team members on a Six Sigma project are called:
   a. Green belts
   b. Black belts
   c. Master black belts
   d. Champions
3. Team leaders on a Six Sigma project are called:
   a. Green belts
   b. Black belts
   c. Master black belts
   d. Champions
4. Managers responsible for the overall effectiveness of the Six Sigma implementation are called:
   a. Green belts
   b. Black belts
   c. Master black belts
   d. Champions
5. Six Sigma experts responsible for providing training in quality improvement tools are called:
   a. Green belts
   b. Black belts
   c. Master black belts
   d. Champions
6. Six Sigma can best described as:
   a. A metric used by managers to judge conformance to requirements
   b. A formal, disciplined, top-driven approach to continuous improvement
   c. A project team
   d. A Greek letter that means standard deviation
7. A Six Sigma process yields a defect rate of:
   a. 2 defects per billion opportunities
   b. 10 putts per year
   c. 64 defects per million opportunities
   d. 3.4 defects per million opportunities
8. (*True or False*) Six Sigma is not well integrated into the process industries.
9. (*True or False*) Six Sigma uses the best practices of the historically recognized quality gurus.
10. Which of these is NOT a stage of Six Sigma?
    a. Analyze
    b. Measure
    c. Design
    d. Control
11. Determining the root cause is the main objective of which stage?
    a. Measure
    b. Analyze
    c. Improve
    d. Control
12. Eliminating the root cause is the main objective of which stage?
    a. Measure
    b. Analyze
    c. Improve
    d. Control
13. Ensuring that the organization maintains the gains made is the main objective of which stage?
    a. Measure
    b. Analyze
    c. Improve
    d. Control
14. (*True or False*) A good project charter is nice but not essential to the success of the team.
15. (*True or False*) Six Sigma improvement tools can be applied only to existing processes.

Activity

1. Suppose that your class is struggling with the concepts of quality. Draft a project charter for an improvement team.

Project Charter
Title:
Defect Definition:
Strategic Importance:
Goals and Objectives:
Project Scope:
Team Membership:
Timeline:
Deliverables:

CHAPTER 11

Teams

"For there are few things as useless—if not dangerous—as the right answer to the wrong question."

~DRUCKER

Objectives

Upon completion of this chapter you will be able to:

- Explain the purpose of teams.
- List examples of types of teams in which process technicians will be expected to participate.
- Define and describe the stages of team development.
- Explain the roles of various functions within a team.
- Explain the purpose of quality tools employed by teams such as RACI charts, Gantt charts, and agenda.
- Describe key components of good team etiquette.
- Describe constructive methods to resolve team conflicts.

Key Terms

Quality circle—a group of people, typically organized by logical structure within a company, who work together for a common goal.

Synergy—the phenomenon in which the total effect of a whole is greater than the sum of its individual parts.

Team—a small group of people, with complementary skills, committed to a common set of goals and tasks.

Introduction

Throughout this textbook note has been made of how many quality improvement projects are accomplished through teams. In this chapter a purpose for working in teams is provided and the stages of team building discussed. These stages are generally accepted to be forming, storming, norming, and performing. You will gain some insight about how to deal with each of these stages in the life of a team.

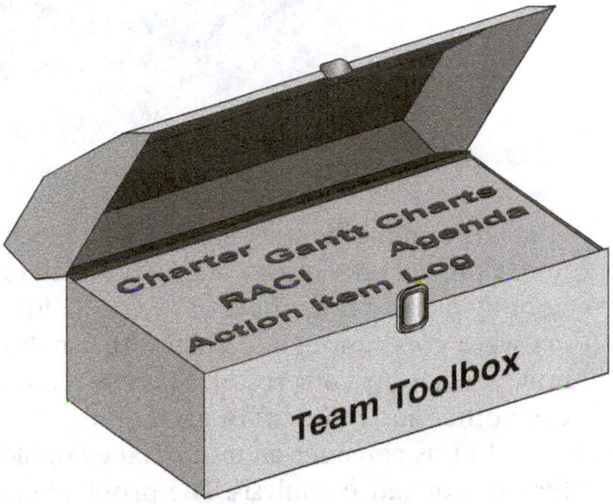

Also discussed in this chapter are the various roles required to keep a team on track such as leader, facilitator, team member, and recorder.

Several quality tools can be applied at different stages in the life of a team, such as the team charter, an agenda, RACI charts, Gantt charts, and an action item log. This chapter provides explanation and examples of each of these tools to help prepare you for your role as a team member in the process industries.

Regardless of their role on the team, all members should follow some basic etiquette points to help ensure the team's success. Some of these components of good team etiquette (active listening, focus, etc.) are discussed briefly.

Purpose of Teams

Research has shown time and again that employees will take more pride and interest in their jobs, and show more ownership in the success of the business, if they are allowed to make meaningful contributions to that success by participating in the decision-making processes. The contrary is also true; when management uses teams to justify a decision it has already made, the team concept will fail miserably.

Teams are not a cure-all for quality and productivity problems. Certain issues can be effectively handled by individuals working alone, but oftentimes a team approach is best. Working together in a team allows the organization to achieve results that could never be attained by a collection of individuals. Two heads really are better than one.

One college experiment involved four roommates pulling on a rope attached to a tension-measuring device. After each individual was given the chance to pull as hard as he or she could, the roommates were put into groups of two, then three, then all four

DICTIONARY

Synergy
The phenomenon in which the total effect of a whole is greater than the sum of its individual parts.

$1 + 1 + 1 > 3$

pulling at one time. The experimenter found that their working together could pull much more than the sum of each one pulling independently. That is synergy, and that is why teams are such an important part of the quality process.

A **team** is defined as a small group of people, with complementary skills, committed to a common set of goals and tasks. To be effective, it is important that all three of the components of this definition be met.

1. Small group: We can argue back and forth about what constitutes a small group of people, but all would agree that a group of 5,000 people can never operate effectively as a team.

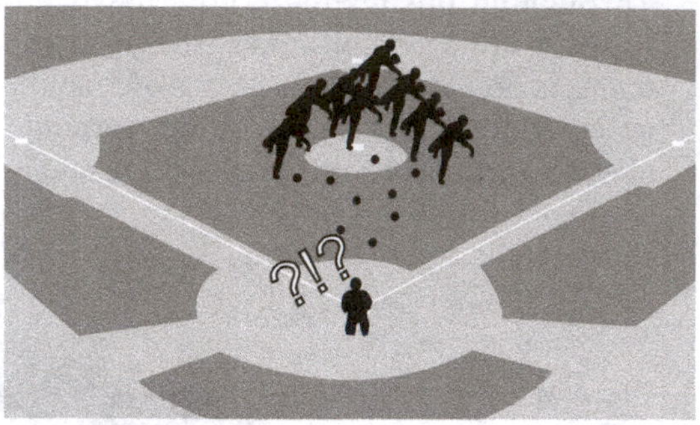

2. Complementary skills: The concept of complementary skills is equally important. Can you imagine a winning baseball team that had eight pitchers, one catcher, and nobody playing the infield or the outfield? Theoretically this might work if your pitchers were good enough to ensure that nobody would ever get a hit, but this just is not how you set up your roster. Success requires that the team members be able to assist each other and build off of each others' strengths.

Let us consider an industrial example using a case study mentioned in the chapter on root cause analysis. The problem encountered was technical, having to do with the molecular structure of your product and how your product was transformed during its intended use. The event that triggered the need for a root cause investigation was a third-party laboratory analysis performed on the spent product. A team full of chemists would know a lot about the technical structure of the product but might not have a lot of experience in routine lab analysis.

A team made up solely of hands-on lab technicians might miss the subtleties of the product chemistry. Without including the workers who actually use the product to help the team understand the conditions under which your product was being transformed, the team would not have all the information it needed to conduct a thorough analysis. A good team of people with complementary skills was assembled including the following:

- Two lab technicians from different labs, both experienced in this particular method
- Two process technicians–including the one whose job including taking the samples
- Two R&D scientists–
  - One with expertise in this chemistry
  - One from your business in order to provide an unbiased perspective
- The business manager
- An RCA facilitator who was familiar with many of these activities but not an expert in any of them

During the course of this particular root cause investigation, many other players were brought in for short durations to help with the process or to answer specific questions.

3. Common goals: Finally, there is the importance of commitment. Each and every person on a team must be committed to the same goal. Sadly, this is not always the case. Too often some team members have their own ideas about what is important.

On a professional baseball team, there may be a player or two who are more interested in padding their own stats than they are in winning the game.

Consider a typical business unit within the process industry. The safety, health, and environmental director is interested in safety and spill statistics. The manufacturing individuals are interested in making pounds of product. The quality members are working on process improvements. The financial representative is interested only in controlling costs. To function effectively as a team, these players would all have to subject their personal and functional goals to the goals of the team.

Everybody sees work from a different perspective. Let us be honest, most of us would not bother working if we did not have to do so. Few of us would continue to work without the benefit of the paycheck. The vast majority of us are in it for the money, but we still have different attitudes about work. Knowing how other people see the concept of work helps you understand what roles might be best for them in a team environment.

- Some see it as a compulsory activity engaged in solely for the purpose of making a living. These individuals are hard to engage in participatory team activities. They have little motivation to do anything more than the bare minimum required to get by.
- Some see work as a career path. What they do today is a stepping-stone to what they want to do sometime in the future. For these individuals, growth is important because it gives them an opportunity to expand their skills and knowledge. These individuals will be excellent contributors in a team environment.
- Others view their work as a status symbol. Their definition of who they are is wrapped up in what they do. For them, work is equated to personal worth and perhaps even to power. These individuals are often natural leaders as well as contributors in a team environment.

Understanding the skill set and level of commitment of each team member is important when forming a team.

Role of Process Technician on Various Types of Teams

Many different types of teams exist in the process industries. The quality function uses teams to accomplish diverse goals. Here are some examples of teams that you may reasonably expect to participate in as a process technician:

- Process improvement teams
- Root cause analysis teams
- Six Sigma teams
- Management system audit teams

These examples of teams are all project teams. They are assigned specific tasks to accomplish and are often given a specified time frame in which to complete these tasks. The expectation is that these project teams will disband when the tasks are complete.

There are also standing teams used in the quality function. A standing team, or committee, is a group of people chartered to provide oversight in an ongoing manner. In the quality field these are often referred to as quality circles. Quality circles are often attributed to Dr. Ishikawa. He formed quality control circles under the sponsorship of the Japanese Union of Scientists and Engineers while an engineering professor at the University of Tokyo in the early 1960s.

The concept of quality circles was first used in the United States at Lockheed in the 1970s. A **quality circle** is a group of people, typically organized by logical structure within a company, who work together for a common goal. This group may be made up of all the process technicians of a certain production line. Their job is the never-ending task of monitoring the quality of production and identifying opportunities for improvement. When an opportunity is found, the right answer may be to charter a project team to address the opportunity.

Another style of team is the ad hoc team. These short-term teams come together with little or no notice to address an immediate need. Here is an example:

The plant has an oversight committee (a standing team) that is charged with monitoring the environmental performance of the plant. When an incident such as a spill occurs, the supervisor on duty may form an ad hoc team to mitigate the impact of the spill as quickly as possible. This team may include representatives from operations, supply chain, maintenance, and even the safety, health, and environment (SHE) organization. When the immediate danger has passed, management may charter a root cause investigation team to review the incident to determine why the problem occurred and what improvements may be needed to prevent recurrence. Process technicians would certainly be needed on the ad hoc team and the RCI team. In some companies, process technicians also serve on standing teams such as the environmental oversight committee in this example.

Stages of Team Development

In the first section of this chapter, the composition of teams in terms of the skill sets needed to be effective was discussed. In the preceding section, the techniques employed by teams to achieve their assigned goals were covered. The techniques used depend on the specific purpose of the team: audit techniques for audit teams, RCI techniques for RCI teams, and the like.

Now let us talk about the process for teams or how the teams operate. Regardless of what type of team you may serve on, how the team works together is important to success, and many of these team processes are similar. Teams do not just spring to life fully formed and effective; instead, they must be cultivated, taught, and given time to mature before they can be effective. Nearly all teams go through a series of developmental stages that are commonly referred to as forming, storming, norming, and performing. At the conclusion of a team's existence, there may also be a final stage, called adjourning, so for the sake of completeness that option is discussed here as well.

FORMING

Characteristics of the first stage of team development include tentativeness and insecurity. A lot of individuals might be spending as much time trying to figure out what is going on and who everybody else is as they are working on the assigned tasks.

If you do not know the other team members, there may be a lack of trust. Individuals might be overly cautious about what they say and how they say it for fear of offending. Anxiety about "fitting in" or being able to contribute might make some team members uncomfortable in this stage. Different members of the team may have different opinions about the goals and tasks assigned to the team, about the makeup of the team, or about the approach being used to achieve success.

So how does the team deal with the forming stage? The team sponsor may want to schedule a kickoff meeting that is designed more to help the team get to know each other than to accomplish any tasks. Icebreakers are perceived by many individuals to be silly and even a waste of time, but in reality they can serve a valuable purpose in helping the team get acquainted. Here are a few examples of icebreakers that you may choose to use in your teams.

- Paired introductions—each person is to find somebody whom he or she does not know and obtain enough information about this person to introduce that person to the rest of the group.
- Scavenger hunt—in larger group settings you can issue a list of traits that each person is supposed to find in the other team members.

STORMING

The storming stage of team development is when people are starting to figure out what is required of them and the other team members. During this stage, it is normal for conflicts to develop. Some individuals will disagree with the assigned task or perhaps with the techniques the team is choosing to employ to achieve success. For example, because the team is made up of people from different functions or areas of the company, some may be trained in the Five Whys methodology of root cause analysis and others in the Apollo methodology.

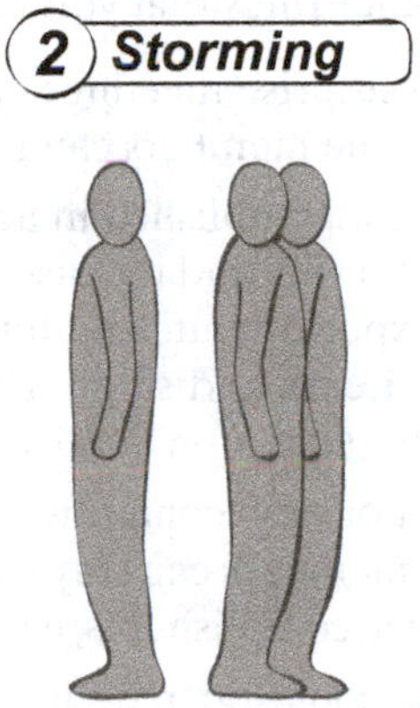

It is natural to be uncomfortable with employing a technique with which you are not familiar. There may also be some jockeying for position in the team. There is almost always an assigned leader. In addition to that person, most teams have a natural leader that others just tend to turn to for advice and counsel.

If you have more than one natural leader on the team, expect a little jousting to occur as everybody seeks to find his or her place. In the case of a team that is self-directed, the team may be responsible for electing a leader.

A couple of quality tools may help the team deal with the storming stage. The first is a team charter. A charter is a document that formally outlines the team membership, goals, responsibilities, timelines, and the like, clarifying the expectations of the team. This helps put everybody on an even playing field and removes some of the questions about who should be doing what.

Another useful tool for helping a team through the storming phase may be the use of a RACI chart to document the roles of the various characters on the team as well as outside the team. A RACI chart is a matrix showing who is responsible, accountable, consulted, and informed about team tasks and progress. In the section on quality tools later in this chapter, more information about these tools and an example of each is provided.

NORMING

When the storms start to pass, the team begins to settle down and establish operating rules that everybody can live with. At this stage the team composition may have changed a little because of the storms, or maybe the same players are there but they have all discovered their place on the team. This is the first stage in which the team members are quieting down and getting comfortable with the team and their place on the team.

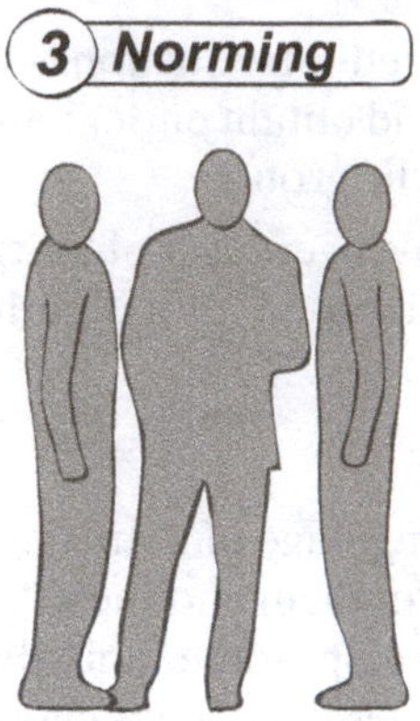

Some teams develop team ground rules that help formalize their agreement to work together. Here are some ground rules that we have seen employed in the past. Your team may want to adopt some of these as is or modify them for your own use. You may find other rules that you need to add that will help your team succeed.

- No surprise meetings—all meetings should be scheduled with at least 48 hours of notice so the members can be prepared.
- Agenda published in advance of the meeting—the agenda should include a list of topics, who will lead the discussion for each topic, how much time is allotted for each topic, the expected outcome for each topic, review of old action items and capturing of new action items, and so on. Examples of agenda format and action item logs can be found in the section on quality tools.
- Honor time commitments—stay on schedule. It is not reasonable to expect that most team members can stay over if the team does not accomplish its goals in the time allotted. To accomplish this, each team member must do the following:
    - Come prepared—if you are assigned an action item or a discussion topic in a meeting, you should come to the meeting with your ducks all in a row.
    - Stay focused—if you have an agenda as described above, then stick to it. Spending a few minutes chitchatting at the beginning of a meeting is okay, as long as it does not interfere with getting the work done.
- Everybody contributes—it is not fair to have some team members doing all the work while others just attend meetings. Everybody on the team has a role to play and should shoulder a fair share of the workload. In meetings, this also means that everybody should be paying attention and voicing opinions. Many a good idea fails to get implemented because a timid team member was afraid to speak.
- Respect each other—team members can have differing opinions. It is perfectly natural. What you cannot allow to happen is divisiveness in the team that prevents the team from working together. All team members must avoid personal behaviors that could prevent the team from working together.

PERFORMING

Finally, you reach the team development stage that you desire, which is performing. The goal of the team leader is to move the team through the various stages of team development as quickly as possible to get it to this spot. At this stage the team is working

together to accomplish their assigned tasks. Each team member knows the part he or she has to play and is ready to play that part. There is a strong commitment to the success of the team by each member.

The challenge at this phase of a team's life is to sustain the momentum. Problems and issues will still arise, but by this time the team has a mutually agreed-upon set of rules by which to operate and is in agreement regarding who should do what and by when. To keep the team on track and ensure progress toward the desired goal, many teams employ some type of time chart such as a Gantt chart.

A Gantt chart can be a huge, complicated mechanism for tracking every aspect of a team's performance or something as relatively simple as a timeline for when major milestones must be reached in order for the team to accomplish its goals in the desired time frame. As with the other tools mentioned, this will be covered in a little more depth in the section on the quality tools.

ADJOURNING

Teams come to a close for several reasons. Sometimes the team is unable to complete its assigned task and is disbanded. Sometimes the project is postponed for reasons completely unrelated to the team itself. Hopefully most of the teams on which you serve will come to a conclusion because the team successfully completed the assigned task and achieved its goal. Whether or not a team has been successful, it is appropriate for the members to be recognized for their contributions. The type and extent of the recognition will depend on how successful the team has been.

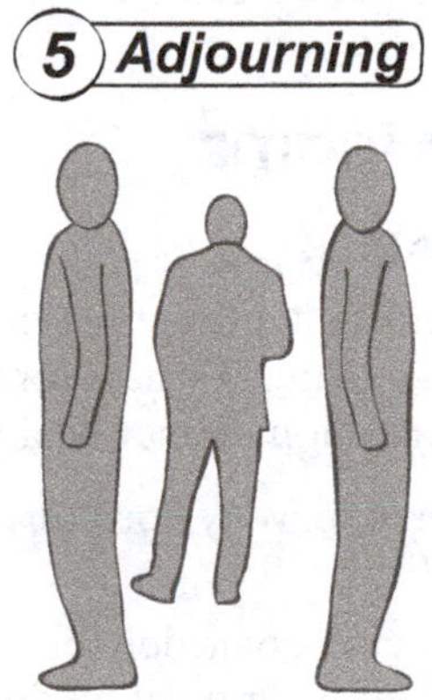

Some people are project oriented. For these people, the successful completion of the team is a reward in and of itself. They look forward to closing out a team effort because it means they have finished a task, and this is important to them. Other members may be more people oriented, and they will miss the team interactions. Some of these individuals may actually mourn the adjourning of the team.

Just as a team is kicked off with a special meeting where team members can get to know each other, the team may also close out with a wrap-up meeting where team members can celebrate their success and draw closure on the effort.

Roles of Various Team Members

Regardless of what type of team you are on, several roles will exist, either officially or unofficially. The key roles follow:

- The team leader—schedules the meetings; provides progress reports to the sponsor; ensures that the team is making progress according to the schedule; ensures team members are trained to perform within the team (RCI training, audit training, statistics training, etc.); and provides conflict resolution as needed.
- The facilitator—typically a process expert in the team methodology being employed but not necessarily in the technology involved. For example, the facilitator for a root cause analysis team may be an expert in the Kepner-Tregoe methodology but may know next to nothing about the manufacture of dibutyl futile. The facilitator can provide training to the team members and ask questions of the team without immersing team members in the details of the issue at hand. The facilitator's job is to make sure the team stays on track. In some standing teams, the role of facilitator rotates among the members from meeting to meeting.
- The recorder—somebody has to take notes. Minutes from the meetings provide an organizational memory for the team. In addition to summary notes of what was discussed, the recorder should record decisions made, actions taken, and action items to be completed. Issuing the minutes from today's meeting provides everybody with not only a record of what they accomplished in the meeting but also a list of what they need to work on in preparation for the next meeting. In many standing teams, the role of recorder rotates among the members.
- The team members—these individuals do the work. They should follow the team's ground rules as well as basic team etiquette: being respectful of their teammates, coming to meetings prepared, arriving on time, actively listening to others as well as participating themselves, and the like.

Depending on the size and makeup of the team, the facilitator may play the role of timekeeper, or somebody else on the team may play this role. In addition, these roles may not always be filled by the same person. In some teams the roles of recorder and facilitator are rotated throughout the team membership from meeting to meeting so that no one person always has to play the same role.

Quality Tools Used by Teams

TEAM CHARTER

The charter of a team is the document that spells out the who, what, where, when, how, and why of a team. Formats for team charters vary from company to company, but here are some common elements of nearly all team charters:

- Purpose: Why does this team exist? In broad terms, what is this team supposed to accomplish?
- Specific goals/deliverables: Create in specific and measurable terms and provide details of what must be done in order for this team to be declared successful. Is it a 10% reduction in quality problems? Then define 10% of what baseline and specify what quality problem is being reduced. The goal and baseline should be so specific as to make determining success at the end of the project a no brainer—either the team met its goals or it did not—but there is no room for interpretation.
- Timeline: When is the team supposed to complete the work? Are there milestone dates along the way? Some projects have certain milestone dates to meet along the

way. For example, in a typical Six Sigma team there are expected completion dates assigned for each phase (measure, analyze, improve, and control) of the project.

• Team membership: Who is on the team and why? When you see a baseball team roster, you see the names of the players and what position they play. Your team charter should include the names of the players and what roles they play on the team as well.

• Sponsorship: Include the name of the person or organization that is sponsoring the team effort. This provides the team with an escalation route should team members run into questions about the purpose or goals of the team or have issues with the team membership.

• Boundaries: It may be helpful for the charter to include boundary conditions for the team. For example, the quality problems that you are working on in the dibutyl futile unit may exclude the new experimental isobutyl futile product line. This helps focus the team on the issue. The team's efforts may be limited to the dibutyl futile production line at plant A but not include work at plants B and C just because of the travel costs involved. In this case management may desire that the team work on the issue locally at first and then attempt to leverage the learnings out to the other plants.

Usually the sponsor will provide a draft charter to the team. Reviewing the charter is a common action for the first couple of team meetings. This gives the team members a chance not only to ensure that they understand the charter but also to challenge portions of the charter that may seem unrealistic to them. More than one team has looked at the timeline proposed and told management that what it was being asked to do could not be done within that time frame with the resources provided.

RACI CHART

Another commonly used tool to help teams get organized is the RACI chart. This acronym stands for responsible, accountable, consulted, and informed. If a problem requires a team to address it, then it is safe to assume that the problem is not a small, isolated issue. The problem assigned to the team may involve quite a cast of characters, some of whom serve on the team and some of whom do not.

The purpose of the RACI chart is to help the team get a handle on who is affected by the project. Some people (or groups) may need to be consulted along the way. Some individuals just need to be notified or informed of the team's progress. Somebody has to take responsibility for the work getting done, and somebody is going to be held accountable for everything coming together. All of these individuals are collectively referred to as stakeholders. They all have a stake in the success of the team, but they all see the team's success from a different perspective.

The odds are that you, in the role of a process technician, will not be asked to complete a RACI chart on your own, but you should know what these charts look like and what purpose they serve. Figure 11-1 is an attempt to identify the roles of the key stakeholders for a family reunion BBQ for each of the tasks involved in that reunion.

TEAM GROUND RULES

The concept of establishing ground rules for the team has already been covered, but for the sake of simplicity, and to make them easier to find later, here they are restated in bullet form:

- Have no surprise meetings.
- Publish agenda in advance of the meeting.
- Honor time commitments—arrive on time and stay on schedule.
- Come prepared.
- Stay focused.
- Everyone contributes.
- Respect each other.

FIGURE 11-1 Example of a RACI Chart

RACI Chart for Family Reunion BBQ

Tasks (Parties →)	Dad	Mom	Children	Fire Dept.	Guests
Develop invitation list	C	AR	C		
Send out Invitations	I	A	R		
Notify authorities that Dad will be BBQing - again	R	A		I	
Provide the food for the party	A		R		
Cook the meat	R	A			
Cook everything else		AR			
Attend					AR
Put out the fire	I	I	I	AR	I
Clean up the mess		A	R		

Legend
R = Responsible
A = Accountable
C = Consulted
I = Informed

Remember that these are just a few examples. You will need to work with your team to establish ground rules that make sense for your team.

AGENDA

An agenda is a simple, but extremely valuable, tool. For each meeting, the team leader should insist that an agenda be published in advance. This way every team member knows what is going to be discussed, why it is going to be discussed, and what preparations are needed to ensure that the discussion is value added.

A well-formatted agenda not only helps ensure that all the players are prepared to play their roles during the meeting but also helps the facilitator keep the team on track with the timing of the meeting. You may see many different formats used. Figure 11-2 shows an example of how you may design an agenda for your meetings.

Here are the main things that your agenda should include:

- When and where of the meeting
- What topics are to be covered
- Time allotted for each topic
- Desired outcome for each topic
- Who is responsible for leading each topic

If so desired, a spreadsheet can be easily formatted to perform the time calculations so that the user just inputs the duration for each topic and lets the computer calculate the timing for the next topic. When used in this manner, it is easy to have ad hoc or guest members join the team at a certain time for a specific topic of interest.

ACTION ITEM LOG

An action item log is a simple listing of what the team members agreed to go do, when they agreed to have it done by, who would be responsible for getting it done, and then of tracking their performance in getting it all done. Figure 11-3 is a portion of an action item log that highlights actions that are not completed and highlights actions that took longer than expected.

FIGURE 11-2 Example of an Agenda

Quality Project Team Meeting

Date:	13-Jun-09
Call-in number:	1.800.567.5349
Facilitator:	Jim
Recorder:	Kim

Timing (CST)		Agenda Item	Person(s) Responsible	Desired Outcome	Comments
Start	Duration				
8:30AM	0:15	Time out for safety	(All)	Share knowledge	
8:45AM	0:05	Agenda review	Jim	Review meeting timeline and objectives	Identify participant time conflicts identify facilitator / action recorder for next meeting
8:50AM	0:20	Action item update	Angelica	Update action items from previous meeting	
9:10AM	0:30	Review financial performance	Scott	Identify new actions to meet teams goals	
9:40AM	0:30	Review project timeline status (Gantt chart)	Jack	Ensure team on track to meet the timing goals set down by management	
10:10AM	0:15	Update on engineering cost reduction	Cindy	Share project status with team	
10:25AM	0:10	Break	All	Relief	
10:35AM	0:45	Review customer data	Chris	Need a decision from the team regarding how to use the customer survey data. Three options to be presented.	
11:20AM	0:40	Brainstorming exercise	Charlie	We need more ideas about how to improve the logistical performance of our operation	
12:00PM		Adjourn	Jim		

GANTT CHART

The last tool to include in your team toolbox is the Gantt chart. A Gantt chart shows all the various tasks that a team needs to complete and when they need to be completed. Usually this is done in a graphical format that makes it easy to see the progress of the team at a glance. Do not expect to be asked to put together a Gantt chart on your own.

FIGURE 11-3 Example of an Action Item Log

Dibutyl Futile Quality Improvement Team Action Log

Assigned	Description	Lead	Due Date	Status/Action Taken	Complete Date
1/5/09	Set up a control chart for butyl rubber results on dibutyl futile production line	Charlie	3/1//09	Complete.	2/28//09
1/5/09	Contact 5 customers (at random) to conduct mini-survey regarding recent shipment troubles	Angelica	1/13/09	Contacted Ultimate Futility Inc; BooBooButyls; ACME Chemicals-R-Us; I-Glow Chem, and Mega Futile Futures. Results tabulated for presentation at next team meeting	1/14/09
1/5/09	Finalize 2009 financial goals and publish to team members	Scott	2/1/09	Complete.	2/1/09
2/10/09	Determine if we should attend quality convention in Orlando	Jack	2/15/09	We think we should - the boss thinks otherwise!	2/15/09
2/10/09	Prioritize customers for full-blown customer survey in last half of year	Chris	3/1/09		
2/10/09	Set up meeting with engineering department regarding improvement projects	Cindy	2/25/09		

FIGURE 11-4 Example of a Gantt Chart

Project Title: Quality Team

Today is November 3, 2008

Description of the task(s) to be completed:	Team Leader	Start Date	End Date	Days	Confirmed/ Tenative	Status
Tasks A	Angelica	1/1/09	2/10/09	40	C	
Tasks B	Jim	1/1/09	4/1/09	90	C	
Tasks C	Jack	2/1/09	3/1/09	28	C	
Tasks D	Cindy	2/15/09	4/1/09	45	C	
Tasks E	Chris	3/1/09	4/1/09	31	C	
Tasks F	Charlie	3/1/09	10/28/09	241	C	
Tasks G	Melissa	3/15/09	12/30/09	290	T	
Tasks H	Jerry	1/3/09	1/22/09	19	C	
Major Task I	Scott	2/1/09	3/15/09		C	
Mino Task I-1	Liz	2/1/09	2/20/09		C	
Mino Task I-2	Dylan	2/15/09	3/1/09		C	
Mino Task I-3	Josh	2/28/09	3/15/09		C	
Tasks J	Crystal	4/1/09	5/1/09	30	C	
Tasks K	Chloe	1/6/09	1/22/09	16	C	Cancelled
Tasks L	Adian	3/16/09	4/1/09	16	T	

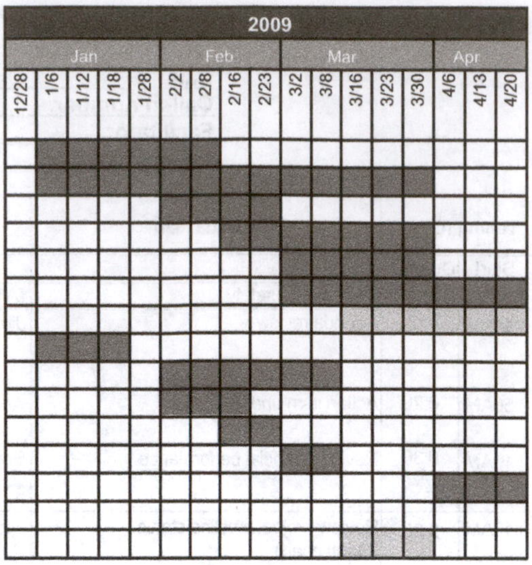

The purpose for including it in this text is so that you will recognize it when you see it and understand what purpose this tool plays in the life of a team.

In Figure 11-4, many of the tasks must be done in a sequential order because they rely on the results of previous tasks. Some of the tasks (Major Task I) have subtasks that must be completed in order to accomplish the major task. Some Gantt charts use color coding to make it easy to distinguish confirmed tasks versus tentative tasks, or maybe to highlight the different functions within the organization so managers can see at a glance who is responsible for what.

Did You Know?

Henry Gantt popularized this project timeline tool around 1910—making this tool nearly 100 years old.

Team Etiquette

Working together as a team does not always come easily. It can be made a little easier if everybody on the team practices a little team etiquette. Here are a few tips that will serve you well in your team efforts:

- Listen carefully—if you are not listening to others, you will not know what is going on and will not be prepared to participate in the discussion.
- Watch carefully—pay attention to nonverbal clues as well as to what people are saying. Have you ever noticed somebody sitting back in a team meeting with his or her arms crossed and eyebrows furrowed? Sometimes what people do *not* say speaks volumes. Pay attention to these nonverbal forms of communication as well. Get the issues out on the table so everybody can participate.

• Actively participate—teams succeed when every individual on the team plays his or her position. Do not be an absentee member even when you show up. Speak your mind and make your opinion known.

• Contribute clearly—sometimes your message will come across clearly; sometimes it will not. Each member on the team comes to the table with a different set of experiences and biases. How you say what you want to say is just as important as the words you use. Think about the message you are trying to communicate. Think about the best way to get that message across. Do you need a drawing or an example to illustrate your message? Then voice your thoughts clearly. If your message cannot be heard in the back of the room, it cannot be understood.

It is okay to disagree as long as you aren't disagreeable!

• Respect others—just as you would have them listen to you and value your opinion, you should listen to them and value their opinions. Avoid sarcasm, put-downs, name-calling, and so on. These communication techniques never build consensus. As Grandma used to say, "It is okay to disagree, as long as you just aren't disagreeable!"

• Focus on the job at hand—if possible, have all team members turn off their cell phones and spend team time working on team issues. Multitasking is almost always assumed to be a sign of a productive person, but in many cases attempting to work on too many things at one time really results in doing none of these tasks well. Take the time to focus on the team effort.

Conflict Resolution

Entire books are written on this subject, and you could easily fill an entire class with the topic; but for the purposes of this chapter about working in teams, here are just a few points about conflict and conflict resolution for you to think about:

• Conflicts will arise on your team. Get used to the idea. Do not let it get you down; just be prepared for it.

• If you can, keep the conflicts impersonal. Focus on the concepts, not the people. It is okay to disagree, as long as you are not disagreeable.

• Seek solutions that everybody can buy into. I think "A-42" is the right answer. You think "B-42" is the right answer. Neither of us can agree to the other's option. A good conflict resolution technique for the leader to employ would be to look for common components in our opinions. Because we both agree that the solution should include "-42," what compromise would take advantage of this common link and allow us to move forward? Perhaps there is an "AB-42" option that would make us both happy.

• Deal with the conflicts as they arise. Do not let them stack up and fester. They will only get worse with time.

• Recognize that sometimes the conflict you see in the team has its origins outside of the team. Sometimes you get individuals on your team who just do not get along. Do not try to make everybody be bosom buddies; just get them to agree to work together toward the team goal.

Summary

Teams are an important technique for getting work done in the process industries, and everybody can expect to serve on a team of some kind, in some role, eventually. Many of the quality improvement processes discussed in this book rely heavily on teamwork in order to achieve the desired goals. If you look back to Chapter 1, you will be reminded that Dr. Juran taught about the importance of using breakthrough teams to accomplish quality improvement.

By working together in effective teams, we can accomplish more as a group than the sum of what we can accomplish individually; this is the concept of synergy. In the realm of quality, we use the team approach for audits, Six Sigma projects, root cause investigations, and other objectives.

Regardless of what kind of team we are on, all teams go through a series of stages of development, which include forming, storming, norming, performing and adjourning. Recognizing that these stages are a natural evolution of the team helps us understand what the team is going through.

You can employ several team tools such as an agenda, ground rules, Gantt charts, action item logs, and RACI charts to help the team get through the more difficult stages of development and become effective more quickly.

There are many roles to be played on each team, and during different parts of your career you may well play them all at one time or another, including team leader, facilitator, recorder, and team member. No matter what role you play, you can always make the team a better place to be by following some simple team etiquette guidelines, most of which stem from the golden rule.

Checking Your Knowledge

1. Define the following key terms:
   a. Quality circle
   b. Synergy
   c. Team
2. Which of the following is an easy way to remember the definition of synergy?
   a. $1 + 1 + 1 = 3$
   b. $1 + 1 + 1 < 3$
   c. $1 + 1 + 1 \neq 3$
   d. $1 + 1 + 1 > 3$
3. Which of these answers is NOT a part of this chapter's definition of a team?
   a. Small group
   b. Highly trained participants
   c. Complementary skills
   d. Common goals
4. What types of quality teams might you serve on as a process technician?
   a. Audit teams
   b. RCI teams
   c. Six Sigma teams
   d. All of these and more
5. The stage of team development that is characterized by tentativeness and insecurity is:
   a. Forming
   b. Storming
   c. Norming
   d. Performing
6. The stage you want to get your team to as quickly as possible, because this is where the team members are getting the job done, is called:
   a. Forming
   b. Storming
   c. Norming
   d. Performing
7. In the early life of a team, there is often some jockeying for position and in-fighting among the members. This stage of team development is called:
   a. Forming
   b. Storming
   c. Norming
   d. Performing
8. Which of these roles might a process technician find him- or herself assigned to in a team?
   a. Team leader
   b. Facilitator
   c. Recorder
   d. Team member
   e. Any of the above
9. (*True or False*) A team charter is a tool that shows a graphical timeline of the activities that need to be accomplished.
10. (*True or False*) A RACI chart is used to document the interests and roles of the various stakeholders in a project.
11. (*True or False*) A really good worker can get a lot more done on his or her own than if slowed down by a team.

12. (*True or False*) Dr. Joseph Juran was a firm believer in the team approach to quality improvement.
13. (*True or False*) Setting ground rules for your team is an excellent way to establish order and help your team work together effectively.
14. (*True or False*) If your team leader is good at the job, there will not be any conflicts on the team.
15. (*True or False*) Treating others with respect—as you would have them treat you—helps make the team a more pleasant working environment.

Activity

1. Teams are not employed just at work. Many of us serve on committees outside of work. If you participate in any kind of a team, draft a team agenda for your next meeting. Did you find that the meeting went more smoothly when it was planned?

CHAPTER 12

Management Systems

"If the only tool you have is a hammer, everything starts to look like a nail to you."
~ABRAHAM MASLOW

Objectives

Upon completion of this chapter you will be able to:

- Explain the purpose of the ISO 9001 Quality Management System standard.
- Describe and explain the key elements of ISO 9001 that apply directly to process technicians (Control of Monitoring and Measuring Equipment, Monitoring and Measurement of Processes, Monitoring and Measurement of Product).
- Explain the purpose of the ISO 14001 Environmental Management System standard.
- Compare and contrast the requirements of ISO 9001 with ISO 14001.
- Define appropriate behaviors for an auditee to exhibit in order to maximize the effectiveness of an audit.
- Explain the purpose of the Malcolm Baldrige National Quality Award.

Key Terms

Audit—a review of the management system to see whether the management system is effectively implemented.

Auditee—a person being audited.

Auditor—a person conducting an audit.

Documents—instructions (written or electronic) that define how a task is to be accomplished.

International Organization for Standardization (ISO)—federation made up of over 100 member countries that have agreed to publish common standard methodologies for various aspects of conducting global business.

ISO 9001—the internationally recognized Quality Management System standard.

ISO 14001—the internationally recognized Environmental Management System standard.

ISO registered—the recognition granted to a company by an accredited registrar when the registrar has verified that the company has effectively implemented the management system.

Malcolm Baldrige National Quality Award (MBNQA)—a United States–based award established by Congress in 1987 that recognizes not only the effectiveness of the management system but also the company results achieved as a result of the management system.

Management system—a collection of activities, usually but not always documented, that are intended to ensure that products and services meet specified needs.

Nonconformance—evidence of deviation from documented requirements.

Records—evidence that a task has been accomplished.

Introduction

A **management system** is a collection of activities, usually but not always documented, that are intended to ensure that products and services meet specified needs. If you think back to Chapter 1, you will remember that the definition of quality was hard to pin down because quality is in the eye of the beholder. The "needs" that a management system is intended to fulfill are also somewhat hard to pin down. There are the needs of the customer, the needs of management, the needs of various governmental regulatory agencies, and many others.

In a typical processing industry plant, it is not a matter of choosing which needs to meet but of trying to meet all of the needs. As a process technician you will absolutely, positively be impacted by and involved in the management system of your company in some way or another. You will certainly be expected to follow the company's management system. You may be asked to participate as an auditor, verifying that other parts of the management system are effectively implemented. You may also be audited by an external party whose job it is to verify that your part of the management system is effectively implemented.

As our economy has become more globalized, it has become increasingly important that the systems and protocols among countries be standardized. An organization called the **International Organization for Standardization (ISO)** is a federation made up of over 100 member countries that have agreed to work together to publish common standard methodologies for various aspects of conducting global business.

The United States representative to the ISO is the American National Standards Institute, ANSI. ANSI, in conjunction with the American Society for Quality, ASQ, publishes the ISO standards in the United States. Some examples of the standards published include the ISO 9001 Quality Management Standard and the ISO 14001 Environmental Management Standard, both of which are commonly implemented in the process industry.

Discussion in this chapter covers the pertinent portions of ISO 9001 and ISO 14001, including a comparison of these two management systems. It will concentrate

Examples of ANSI / ISO Standards That You May Encounter

ANSI / ISO / ASQ Q9000-2000 Series:
Quality Management Standards

ANSI / ISO / ASQC - Q10011-1994 Series:
Guidelines for Auditing Quality Systems - E Standards

ANSI / ISO 14001-1996:
Environmental Management Systems Specifications with Guidance for Use

ANSI / ISO 14001-1996:
Guidelines for Environmental Auditing

ANSI / ISO / ASQ Q9000-2005:
Quality Management Systems - Fundamentals and Vocabulary

ANSI / ASQC B1-B3-1996:
Quality Control Chart Methodologies

specifically on the elements of these standards that apply to the day-to-day life of a process technician.

The role of a process technician as an auditor and as an auditee is also covered here. Finally, there is a brief discussion about the Malcolm Baldrige National Quality Award (MBNQA). Some companies are now moving toward the MBNQA criteria as a way of going beyond the "basic" requirements of ISO 9001.

What Is ISO 9001?

ISO 9001 is the internationally recognized Quality Management System standard. This management system has been implemented by thousands of companies around the world. Many customers require their suppliers to achieve and maintain ISO 9001 registration. ISO registration is a process whereby your company can hire an outside company called a registrar to **audit** (review of management system) your management system against the standard. To become **ISO registered,** the registrar has to verify that the company has effectively implemented the management system.

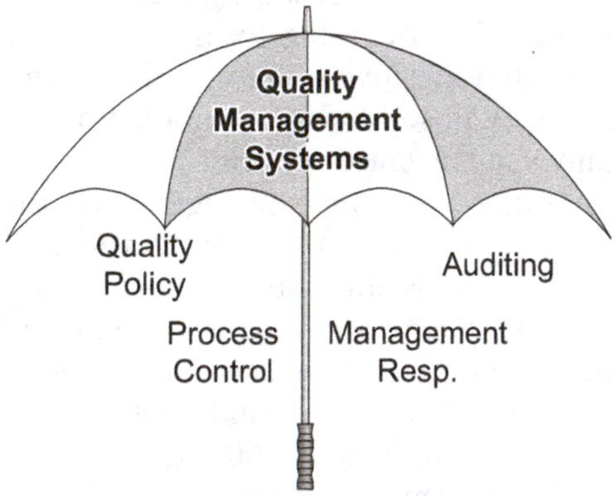

The registrar must be accredited by the ruling body of each country to grant registrations. In the United States, accreditation is managed by the Registrar Accreditation Board. If the registrar deems that your company has effectively implemented the entire management system, it will then grant you a certificate of registration.

The management system is made up of many elements (or clauses) and subelements (subclauses), each of which contains requirements that must be effectively implemented. In this sense the management system is like an umbrella covering many individual topics.

Did You Know?

When your company has been verified as meeting the requirements of the ISO 9001 Quality Management System standard by an accredited registrar, your company can be ISO registered—it cannot be ISO certified. Although you hear the phrase "ISO certified" frequently, this is a misnomer. The registrar does not certify your company to do anything.

The registrar audit to establish the initial registration will cover every element of the standard. Subsequent surveillance audits will then be performed to maintain the registration. In these surveillance audits, the registrar will have to cover all of the elements every 3 years but will not necessarily cover every element during every audit.

The first international standard for a quality management system was issued in 1987, published in the United States as ANSI/ASQC Q92-1987. The intent of the International Organization for Standardization is to review each standard every 5 years. Sometimes the schedule slips and the review starts in a 5-year time frame but takes a year or more to revise.

The first update to this initial standard was in 1994 and was called ANSI/ASQC Q9001-1994. The current version, issued in the year 2008, is called ANSI/ISO/ASQ(E) Q9001-2008. The first two versions were organized functionally, with each major element belonging to a certain function such as purchasing, quality, production, or management.

The current standard is organized more like a process flow, so you will see the customer requirements defined first, the manufacture of the product in the middle, and the checking and monitoring topics toward the end. Each time an updated standard is issued, every ISO registered company has to update its quality management system to meet the new requirements.

Here is an outline of all of the elements and subelements of the ISO 9001-2000 Quality Management System standard:

1. Scope
- **1.1.** General
- **1.2.** Application

2. Normative Reference

3. Terms and Definitions

4. Quality Management System
- **4.1.** General Requirements
- **4.2.** Documentation Requirements
  - **4.2.1.** General
  - **4.2.2.** Quality Manual
  - **4.2.3.** Control of Documents
  - **4.2.4.** Control of Records

5. Management Responsibility
- **5.1.** Management Commitment
- **5.2.** Customer Focus
- **5.3.** Quality Policy
- **5.4.** Planning
  - **5.4.1.** Quality Objectives
  - **5.4.2.** Quality Management System Planning
- **5.5.** Responsibility, Authority, and Communication
  - **5.5.1.** Responsibility and Authority
  - **5.5.2.** Management Representative
  - **5.5.3.** Internal Communication

- **5.6.** Management Review
  - **5.6.1.** General
  - **5.6.2.** Review Input
  - **5.6.3.** Review Output
- **6.** Resource Management
  - **6.1.** Provision of Resources
  - **6.2.** Human Resources
    - **6.2.1.** General
    - **6.2.2.** Competence, Training and Awareness
  - **6.3.** Infrastructure
  - **6.4.** Work Environment
- **7.** Product Realization
  - **7.1.** Planning of Product Realization
  - **7.2.** Customer Related Processes
    - **7.2.1.** Determination of Requirements Related to the Product
    - **7.2.2.** Review of Requirements Related to the Product
    - **7.2.3.** Customer Communication
  - **7.3.** Design and Development
    - **7.3.1.** Design and Development Planning
    - **7.3.2.** Design and Development Inputs
    - **7.3.3.** Design and Development Outputs
    - **7.3.4.** Design and Development Review
    - **7.3.5.** Design and Development Verification
    - **7.3.6.** Design and Development Validation
    - **7.3.7.** Control of Design and Development Changes
  - **7.4.** Purchasing
    - **7.4.1.** Purchasing Process
    - **7.4.2.** Purchasing Information
    - **7.4.3.** Verification of Purchased Product
  - **7.5.** Product and Service Provision
    - **7.5.1.** Control of Production and Service Provision
    - **7.5.2.** Validation of Processes for Production and Service Provision
    - **7.5.3.** Identification and Traceability
    - **7.5.4.** Customer Property
    - **7.5.5.** Preservation of Product
  - **7.6.** Control of Monitoring and Measuring Equipment
- **8.** Measurement, Analysis, and Improvement
  - **8.1.** General
  - **8.2.** Monitoring and Measurement
    - **8.2.1.** Customer Satisfaction
    - **8.2.2.** Internal Audit
    - **8.2.3.** Monitoring and Measurement of Processes
    - **8.2.4.** Monitoring and Measurement of Product
  - **8.3.** Control of Nonconforming Product
  - **8.4.** Analysis of Data
  - **8.5.** Improvement
    - **8.5.1.** Continual Improvement
    - **8.5.2.** Corrective Action
    - **8.5.3.** Preventive Action

As you can see, there are 23 main elements organized into the 5 main topics of Management System; Management Responsibility; Resource Management; Product Realization; and Measurement, Analysis, and Improvement. Because you live in the land of opportunity, there may come a day when you are in management and ultimately responsible for the effective operation of the entire quality system. In your role as a process technician, either many of these elements will not impact you at all, or the impact will be negligible. For the purposes of this book, the discussion concentrates only on those elements that are directly related to your role as a process technician.

Key Process Quality Elements of ISO 9001

Some of the elements of ISO 9001 have a minimal impact on the role of a process technician; other elements are the responsibility of the process technician. In order to provide an overview of the entire standard while focusing on the material that is important to you, here is a brief description of each of the clauses, with detailed coverage of only those that apply to you.

4.1. General Requirements

This section just outlines that the organization (your company) is to develop and implement a management system that includes all the various elements that are further defined in upcoming elements.

4.2. Documentation Requirements

This section requires that the company define its quality policy and have procedures in place to control its documents and its records. **Documents** are instructions (written or electronic) that define how a task is to be accomplished.

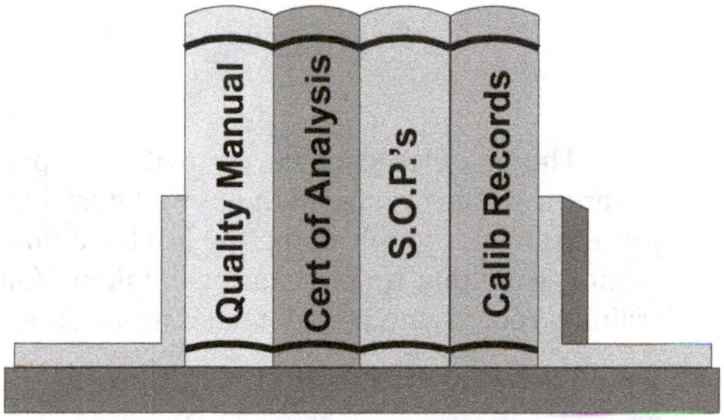

Controlling documents means ensuring that the current version of instructions is available to the people who need them and the instructions are protected. The specific portion of this that is applicable to process technicians is that a controlled document cannot be "changed" without appropriate authorization.

For example, let us say a member of the day staff comes to the control room and tells you to run at 100 degrees C instead of 105 degrees C, as is documented in the procedure. You cannot just scratch out 105 and handwrite 100 on the official procedure. There are usually multiple copies of the procedure, and by making a handwritten change on one copy you have not ensured that all of the copies are changed or that the change is truly "authorized."

A quality auditor would consider finding handwritten changes to one copy of an official procedure as a nonconformance. **Nonconformance** is evidence of deviation from documented requirements. **Records** are the evidence that a task has been accomplished. These might include completed checklists, e-mail notifications, invoices, and so forth.

Auditors look at records to determine how your management system is functioning. You will be filling out checklists, lab reports, calibration records, and other types of records that will be reviewed by auditors. It is important to fill out each one properly and file it in the appointed place. Otherwise, you could end up with a good management system but a lack of evidence because the records cannot be located. A checklist included in a procedure as a work instruction is a document. Once that checklist is filled out, it becomes a record.

5. Management Responsibility

All six elements within section 5 define various responsibilities of management. Management has to document its commitment, ensure that the customer is taken care of, define a quality policy, plan the quality objectives and management system, define responsibilities for implementing the quality management system, and then review the effectiveness of the management system. None of these activities typically includes the process technician.

6. Resource Management

Most of the four elements within section 6 will also be outside the responsibility of the process technician. The requirements are mostly broad statements about ensuring the right people are available to perform the work required by the management system.

The one element that you will be a part of is 6.2.2, Competence, Training and Awareness. The organization has to determine what kind of training is needed for each job, ensure the people doing the job have that training, and, of course, keep records to demonstrate that training has been taken. Your part of this will be to ensure that your training records are complete and up-to-date.

7.1. Planning of Product Realization

The organization must define how it intends to make good product, including specifications, validation procedures, and the like. These activities do not typically include process technicians.

7.2. Customer Related Processes

These requirements all deal with defining the specifications for your product (both customer and regulatory specifications), defining the contract terms with your customers, reviewing the specifications and contract terms, and establishing communication channels with your customers.

One example of a customer communication is customer complaints. Although you may be asked to participate in an analysis of a customer complaint, the defining of the customer-related processes is not typically the responsibility of a process technician.

7.3. Design and Development

This fairly lengthy section has to do with controlling the various aspects of the design and development of new products. These activities are not typically the responsibility of a process technician.

7.4. Purchasing

This element requires that the organization define how it will ensure that purchased products (raw materials) meet the needs of the management system. The ISO 9001 standard requires that a company manage the input to ensure a quality output. These activities are not typically the responsibility of a process technician.

7.5. Production and Service Provision

This element is broken down into several subelements. Because some of these will affect you directly and others will not, here is a further breakdown of this element:

> **7.5.1.** Control of production and service provision requires that the company provide that the workers (you) have the necessary information, work instructions, equipment, and so on to ensure that they can do their job.

7.5.2. Validation of processes for production and service provision requires that your production processes be validated to ensure they can achieve their planned results.

7.5.3. Identification and traceability is a subelement that impacts you as a process technician. This subelement requires that, *where appropriate*, you can identify your product throughout the various phases of production and that you can trace the product as it moves through the phases of production. The way that this requirement is accomplished is by having process technicians fill out production logs, transfer logs, and inventory sheets.

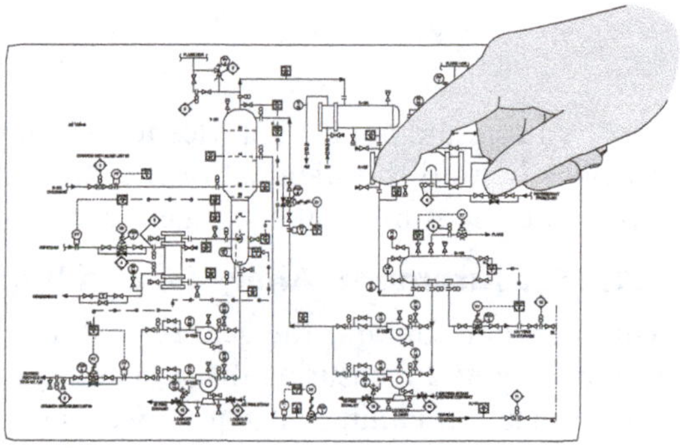

If you fail to fill out the appropriate logs, evidence will not be available to demonstrate the effective operation of your quality system. More importantly, if you do not keep a record that you have filled up a certain tank, you might try to fill it again, causing it to overflow. Production logs, transfer logs, inventory sheets, and other similar records absolutely must be kept accurately.

It is much better to document the truth, even if the truth is not what you were instructed to do, than it is to document what you were instructed to do, even though you were unable to complete the instructions. Accurate records are an absolute must in the process industry, and you will be on the front lines when it comes to process record keeping.

7.5.4. Customer property is a short subclause that requires you to take good care of anything that is in your possession but belongs to the customer, such as special packaging it provided, or perhaps a secret ingredient it ships you for incorporation into the final product.

7.5.5. Preservation of product is exactly what it sounds like. How do you take care of your product after it has been produced in order to ensure that the quality stays good? Some obvious examples in the process industry include nitrogen blankets on storage tanks to keep out atmospheric moisture, the addition of inhibitors in some products, and the packaging of the product under pressure. In each of these cases, the role of the process technician is to follow the documented instructions to the letter. Defining what kind of preservation technique is needed is usually done for you.

7.6. Control of Monitoring and Measuring Equipment

This element relates to any equipment that is used to monitor or measure your product or your process. This would include electronic measurements of pressures and temperatures as well as laboratory measurements such as purity and moisture content.

The reason this is part of the quality management system is because all of these measurement devices or meters have to be calibrated and validated as suitable to make the desired measurement. In order for you as a process technician to know the device is usable, it has to be marked with some kind of calibration sticker, and, of course, calibration records have to be maintained to demonstrate to management and the auditors that all is well.

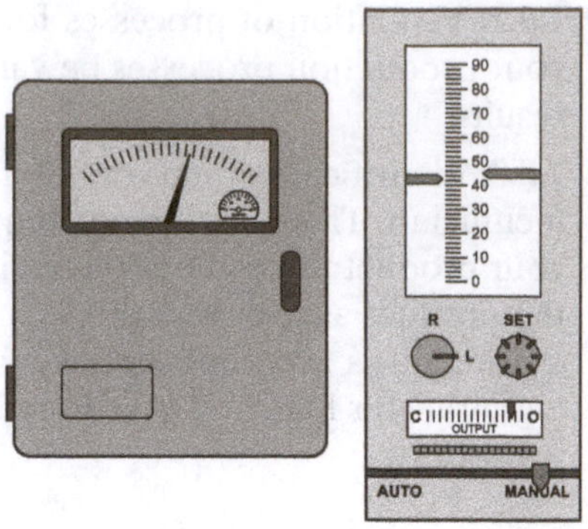

You should not use a device for which you cannot trust the results. The underlying purpose behind this element of the ISO 9001 standard is to make sure the readings from your measurement equipment are trustworthy.

8.1. Measurement, Analysis, and Improvement, General

This element contains the generic statements that the organization must provide for monitoring and measuring the process and the products to make sure that everything is okay. The remaining elements cover requirements that impact your role as a process technician.

8.2. Monitoring and Measurement

One subelement has to do with measuring customer satisfaction; another deals with measuring the overall performance of the management system via quality audits. You may be a member of an audit team, and if your company is ISO registered, you will almost certainly get audited eventually. More information about your roles as an **auditor** (person conducting the audit) and **auditee** (person being audited) will be discussed later in the chapter.

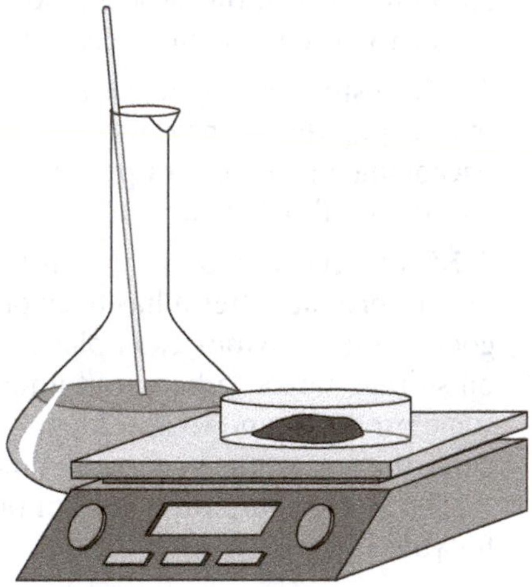

The other two subelements (8.2.3 and 8.2.4) establish requirements for the monitoring and measuring of your process and of your final product. Monitoring your process is the essence of a process technician's role. Instructions and procedures will be provided regarding how hot, how long, and so on, to run your equipment. Your job will be to follow those instructions to the letter and maintain legible, accurate records of what happened on your shift.

Some process technicians will also be expected to run laboratory analyses to check on the output from the process. Other process technicians will send samples to a central lab for analysis. Either way, monitoring the quality of the product you are making will be expected. Keeping good records for both your company and the auditors is a must.

IN A NUTSHELL

The Key Elements of the ISO 9001 Quality Management Standard for Process Technicians:

4.2 Documentation Requirements
6.2.2 Competence, Training and Awareness
7.5.3 Identification and Traceability
7.6 Control of Monitoring and Measuring Equipment
8.2.3 Monitoring and Measurement of Processes
8.2.4 Monitoring and Measurement of Product

8.3. Control of Nonconforming Product

Despite having the best quality management system and the best process technicians in the world, once in a great while you might make a batch of product that just does not meet the needs of your customers. This element of the ISO 9001 standard simply requires that the organization define how to handle that situation.

8.4. Analysis of Data

This element requires your organization to define what it does with all the data that you, the process technicians, record and file away.

8.5.1. Continual Improvement

The final element is that of continual improvement. The concept of continual improvement is so ingrained in the ISO 9001 standard that the plan-do-check-act cycle discussed in Chapter 6 is included in the forward to the standard. The other topics also covered within the element of 8.5. Improvement are Corrective Action and Preventive Action. Corrective action is fixing something that has broken in order to prevent recurrence. Preventive action is fixing something before it breaks to prevent occurrence. Although you may not be asked to analyze data to look for potential problems, it is quite natural for process technicians to notice opportunities for improvement and initiate preventive actions.

What Is ISO 14001?

Like ISO 9001, ISO 14001 is a management system standard. The difference is that **ISO 14001** is intended to provide process guidance for environmental management systems. The current version of ISO 14001, issued in 2004, is organized as follows:

1. Scope
2. Normative References
3. Definitions
4. Environmental Management System Requirements
  4.1. General Requirements
  4.2. Environmental Policy
  4.3. Planning
    4.3.1. Environmental Aspects
    4.3.2. Legal and Other Requirements
    4.3.3. Objectives, Targets and Programs
  4.4. Implementation and Operation
    4.4.1. Roles, Resources, Responsibility and Authority
    4.4.2. Competence, Training and Awareness
    4.4.3. Communication
    4.4.4. Documentation
    4.4.5. Control of Documents
    4.4.6. Operational Control
    4.4.7. Emergency Preparedness and Response
  4.5. Checking
    4.5.1. Monitoring and Measurement
    4.5.2. Evaluation of Compliance
    4.5.3. Nonconformity, Corrective Action and Preventive Action
    4.5.4. Control of Records
    4.5.5. Internal Audit
  4.6. Management Review

Comparison of ISO 9001 with ISO 14001

Did you notice that the standards are organized in much the same way, with planning up front, then implementation, and finally checking? Does it sound a lot like the PDCA cycle? You should also have noticed that many of the topics are similar

TABLE 12-1 Comparison of Key Elements of ISO 9001 and ISO 14001 for Process Technicians

Key Element of ISO 9001	*Matching Element in ISO 14001*
4.2 Documentation Requirements	4.4.4 Documentation 4.4.5 Control of Documents
6.2.2 Competence, Training and Awareness	4.4.2 Competence, Awareness and Training
7.5.3 Identification and Traceability	4.4.6 Operational Control (a broad topic)
7.6 Control of Monitoring and Measuring Equipment	4.5.1 Monitoring and Measurement
8.2.3 Monitoring and Measurement of Processes	4.5.1 Monitoring and Measurement
8.2.4 Monitoring and Measurement of Product	4.5.1 Monitoring and Measurement

between these two management system standards. The key elements of ISO 9001 that were identified are listed in Table 12-1, along with the matching elements of ISO 14001.

If you were to look at other elements of the standards, you would find even more overlap. Both standards require corrective and preventive action processes, both require internal auditing, both require management responsibility, and so on.

There is so much overlap that many companies now implement an integrated management system that meets all of the requirements of both standards. This allows the company to pursue compliance to both standards without having to set up parallel structures. This practice has become so common that the ISO 10011 Guidelines for Auditing Quality Systems standard (originally published in 1991) was replaced in 2002 with the ISO 19011 Guidelines for Quality and/or Environmental Management Systems Auditing standard.

As the name of the new standard implies, there is now a single standard for conducting quality audits and environmental audits. Your company may employ an integrated or harmonized management system that is designed to meet the requirements of both ISO 9001 and ISO 14001. Depending on the company you work for and the types and uses of the chemicals it produces, you may see many other management systems as well. Examples of these other management systems are listed in Table 12-2.

As you can see, depending on which company in the processing industry you work for, and which business unit within the company, you may be required to follow a management system that is traceable to a number of different standards. In the case of quality, failing to follow the guidelines is bad because it leads to poor-quality product and potentially dissatisfied customers. In the case of drug products, following the management system is required to protect the lives of the people who will consume your chemicals.

TABLE 12-2 Examples of Other Management Systems

Application	*Management System*
Laboratories	ISO 17025 Testing and Calibration Laboratory standard
Drug chemicals (active ingredients)	Food and Drug Administration Good Manufacturing Practices
Chemical excipients (inactive ingredients)	Good Manufacturing Practices Guide for Bulk Pharmaceutical Excipients
Ethanol products	Bureau of Alcohol, Tobacco and Firearms standards
Kosher applications	Supervised by competent rabbi and certified as kosher with a hechsher (a symbol or statement of compliance)

The Process Technician as an Auditor

One requirement of nearly all management systems is that of internal audits. The old ISO 10011 quality audit standard defined an audit as "a systematic examination of the acts and decisions by people with respect to quality to independently verify or evaluate and report degree of compliance to the operational requirements of the quality program, or the specifications or contract requirements of the product or service." In other words, auditing is examining evidence to demonstrate conformance to requirements.

Notice that the purpose is not to find fault but to demonstrate conformance. In our experience, process technicians make great auditors.

Let us describe what an auditor does and then you can see how a process technician could fill this role. A management systems audit can be broken down into two main components:

1. Does the management system meet the requirements of the standard?
2. Is the management system effectively implemented?

The design and setup of the management system will probably already be accomplished, so the first part of the audit is relatively easy for you. The second part is where the rubber meets the road; essentially, are you doing what you said you would do? The auditor will look at your procedures, then look at what you do and see whether you are doing what you are supposed to be doing.

An auditor will interview the people working the floor and ask questions about what you do. He or she will look at the records of what happened yesterday, last week, and last month to see whether you can demonstrate that the process is being managed according to the requirements of the management system.

The kind of things an auditor looks at depends on the element of the standard that is being audited. An auditor will look at process logs, transfer sheets, inventory records, training records, calibration records, and lab analyses. He or she will check to see whether the document control procedures are being followed. Are there any unauthorized copies of the procedures lying around? Are there any unauthorized changes made?

Auditors will validate that the record retention rules are enforced according to the management system requirements. An auditor must be independent of the area being audited, so you cannot audit within your own department, but you can audit another manufacturing unit within your plant. By using process technicians from one area to audit other areas, companies are able to take advantage of the process technicians' detailed knowledge of how things are REALLY done to drive continuous improvement throughout the organization.

The Process Technician as an Auditee

Even if you do not serve as an auditor, you will almost certainly be audited at some time in your tenure as a process technician. Unfortunately, some people struggle with being audited. It makes them uncomfortable and in some cases can even bring on a serious panic attack. It should not be that way. One major advantage of becoming trained as an auditor is that it makes you a much better auditee, because you know exactly what the auditor is looking for and you can provide the answers in a timely and efficient fashion. Here are some common scenarios that you will encounter as an auditee and some tips on how to respond positively:

Audit Scenario	How to Respond Positively
Auditors ask questions and take notes.	Answer the question that is asked as completely but succinctly as possible. Never volunteer information that was not requested. Be polite, but do not become a nervous chatterbox.
Some auditors will discuss non-business-related topics as a way of making you feel more comfortable.	Enjoy it. If they have an hour to spend with you and you take 30 minutes to discuss horses or cars or boats, that just leaves less time for them to dig around in your business. Do not stall. It is okay for the auditor to divert the auditee's attention, but it is not okay for the auditee to stall and attempt to divert the auditor's attention.
The auditor is going to ask you about various parts of the management system.	You need to know your management system procedures inside out. This does not mean everything in the company, but everything that has a direct impact on your specific job.
The auditor may ask you for information that is outside of your area of responsibility.	Do not let this fluster you. The auditor may not have a clear understanding of your area and is just fishing for information. It is okay to tell the auditor that he or she will have to seek that information elsewhere. If you know where to send the auditor, then by all means be helpful and redirect the question to the right person.
The auditor asks you a question but you do not understand.	Ask the auditor to clarify. Do not attempt to respond if you do not know the answer. Make sure you understand the question before you attempt to answer.
The auditor asks you how to perform a certain task.	Tell the auditor that you follow the documented procedure and then show him or her the procedure.
Being a little pushy, the auditor asks you just to tell how you do the task without looking at the procedure.	Tell the auditor that in order to ensure you are following the most current version of the procedure, you would never just perform the task without first reviewing the procedure. Do not fall for this trick.
Upon answering a question, the auditor makes a few notes, then looks up at you with a questioning look in his or her eyes and waits expectantly.	Return the auditor's gaze and sit patiently, awaiting the next question. Auditors are trained to use silence as a tool to get you talking.
The auditor asks you whether he or she might see a certain task being performed.	If the task is going to be performed anyway, then allow the auditor to watch, but there is no requirement that you (or anybody in the company) perform unnecessary tasks just to satisfy an auditor's question.

Audit Scenario	How to Respond Positively
The auditor asks whether he or she might perform some part of the task.	Politely decline. An auditor is to look and listen but not touch. In fact, one of your roles during an audit is to keep the auditor safe. Remember that this is your unit, not the auditor's, and he or she may not recognize all the hazards that you are familiar with.
The auditor asks you a question that you do not know the answer to.	Answer truthfully that you do not know the answer. Never try to pull the wool over an auditor's eyes; the auditor will see through it every time.
The auditor discovers a nonconformance in your area.	Acknowledge the audit finding, and thank the auditor for pointing out an opportunity for improvement. Make a note of the issue so you can follow up with the right person after the audit. This response demonstrates an attitude of continuous improvement and will be received positively. Unless directed to do so by your supervision or unless you see an obvious safety hazard that requires immediate response, do not attempt to address the audit finding on the spot with the auditor watching. This might give the auditor the impression that you are fixing the problem without giving the matter serious thought about other implications. Just accept the finding and move on.
The auditor asks to go off into the unit by him- or herself to look at something.	Politely decline. For the auditor's own safety, he or she should never be unescorted.

Here is a wrap-up of this topic with an outline of audit considerations for an auditee:

1. Ethical Considerations—ethical behavior is almost certainly a corporate requirement. Never put yourself, the plant, or the company in a compromising situation.
   a. Do
      - Answer every question honestly.
      - Answer only the questions that are being asked.
      - Narrow down the request to specifics.
   b. Don't
      - Lie to an auditor.
      - Volunteer information or opinions.
      - Point the auditor to someone else's deficiencies.
      - Air dirty laundry.
      - Attempt to "bluff" the auditor.
      - Alter data.
2. Social Considerations—contrary to popular belief, auditors are people too. Their perception is 90% of their reality; consequently, how they perceive your performance will influence how they document their audit findings.
   a. Do
      - Be friendly.
      - Be careful about joking—it could backfire.
      - Be on your best behavior.
      - Treat auditor with respect.
      - Take notes—show you are interested.
      - Watch your body language.
   b. Don't
      - Joke around unnecessarily.
      - Leave auditor alone.
      - Fight with auditor.
      - Argue—clarify and escalate if needed.
      - Whine or complain.
      - Panic—it is okay to ask for help.

3. Technical Considerations—the auditor's perception of your area will initially be based on you personally, then on your technical knowledge, so technical accuracy is a must.
   a. Do
      - Project the following to an auditor during the review:
        1. Good system in place
        2. Well organized
        3. Knowledgeable personnel
        4. Safeguards against single point failures
      - Look it up.
      - Ask for clarification.
      - Know your area well.
      - Ask to see the requirement.
      - Use the resources around you.
      - Your homework—be familiar with procedures, records, and manuals.
      - Admit it when you do not understand or do not know the answer.
   b. Don't
      - Fumble around looking for things.
      - Offer documents not requested.
      - Give the auditor entire files.
      - Come to close-out meeting to debate issues.
      - Fix issues immediately unless instructed to do so by supervision.
4. Personnel Considerations—sometimes the best answer to a question is, "Let me get the right person to answer that question for you." You should know whom to go to.
   a. Do
      - Steer the auditor to the right people.
      - Treat the auditor with respect.
      - Recognize that the auditor is here to help you.
      - Listen to the auditor. It is a sign of respect and interest.
   b. Don't
      - Try to answer every question yourself.
      - Point blame at somebody else.
      - Demean the auditor's position.
      - Appear victimized.
      - Ask the auditor to explain an internal or regulatory requirement.
5. Location Considerations—where the audit is conducted is crucial to the effectiveness of the audit. A good auditor will want to go where the action is. A good auditee will want to stay in a nice safe conference room.
   a. Do
      - Meet in your office or a conference room.
      - Go get requested information and bring it in to the auditor.
      - Provide the entire file only if the auditor insists.
      - Know your area so you can take the auditor where he or she wants to go without wandering around.
   b. Don't
      - Meet with the auditor in your file room.
      - Wander around the unit just to see what you might see.
      - Send an auditor anywhere—take him or her there.

Malcolm Baldrige National Quality Award

The **Malcolm Baldrige National Quality Award (MBNQA)** was established by the U.S. Congress and signed into law by Ronald Reagan in 1987. Named for Malcolm Baldrige, who served as secretary of commerce from 1981 to 1987, this award was created to stimulate American companies to improve quality and productivity by recognizing not only the effectiveness of the management system but also the company results achieved as a result of the management system.

The ISO 9001 Quality Management System standard has almost become the baseline for most companies. To distinguish themselves in the marketplace, companies must continually push forward and try to outdo their competition. Implementing the criteria for the MBNQA is recognized as a lofty goal for the next level of quality and excellence.

The management system for the companies that some of you will go to work for may include requirements based on the MBNQA criteria as well as many others that we have already discussed. The biggest difference between the MBNQA criteria and those of the other management systems that have been reviewed is the focus on results. ISO 9001 and ISO 14001 both focus on having the right work processes in place to ensure effective implementation. The MBNQA criteria require a demonstration of excellent results. The criteria are organized into seven categories as follows:

1. Leadership
2. Strategic Planning
3. Customer and Market Focus
4. Measurement, Analysis, and Knowledge Management
5. Workforce Focus
6. Process Management
7. Results

So far the only major chemical company to win this award was Eastman Chemical Company in 1993. Perhaps with your help, your company can add its name to the list.

Did You Know?

Malcolm Baldrige was named Professional Rodeo Man of the Year in 1980 and was installed in the National Cowboy Hall of Fame in 1984. He died at age 64 as the result of a rodeo accident in 1987. He received the Presidential Medal of Freedom from President Ronald Reagan posthumously in 1988.

Summary

You will encounter management systems of various types in your role as a process technician. Some may be quality management systems such as ISO 9001. Others may be environmental management systems such as ISO 14001. An increasing percentage of management systems are integrated, meaning the company system covers more than one standard. These management systems document the processes that a company should have in place to drive continuous improvement into its operations. Both ISO 9001 and ISO 14001 are organized around the plan-do-check-act continuous improvement cycle. Process technicians have direct responsibility for maintaining portions of a quality system, which may include the following:

- Documentation Requirements
- Competence, Awareness and Training
- Identification and Traceability
- Control of Monitoring and Measuring Equipment
- Monitoring and Measurement of Processes
- Monitoring and Measurement of Product

Keeping the procedures up-to-date and following those procedures to the letter are vitally important to produce a quality product safely and responsibly. If the procedures are wrong, you should work within the system to get them fixed, but you should NEVER arbitrarily think you have a better way and try it out to see what happens.

Additionally, nearly all process technicians will be involved in the audit function at some point as an auditee and perhaps as an auditor. As an auditee is important to understand the role you play in presenting a positive impression to the audit team. Remaining calm, having answers at your fingertips, and being a positive person help the audit team find the evidence of conformance that it is looking for. Due to their grassroots familiarity with the process operations, process technicians make excellent additions to the audit teams. Conversely, audit training makes better auditees out of process technicians and day staff alike.

For those companies looking to go above and beyond the minimum requirements for a recognized quality management system, the Malcolm Baldrige National Quality Award provides criteria that not only drive excellence but also require measurement of the results in order to demonstrate excellence.

Checking Your Knowledge

1. Define the following key terms:
   a. Audit
   b. Auditee
   c. Auditor
   d. Documents
   e. ISO 9001
   f. ISO 14001
   g. Malcolm Baldrige National Quality Award
   h. Management system
   i. Nonconformance
   j. Records
2. List six elements of the ISO 9001 Quality Management System standard that apply directly to the role of a process technician:
   a. ______________ d. ______________
   b. ______________ e. ______________
   c. ______________ f. ______________
3. "Examining evidence to demonstrate conformance to requirements" is the definition of:
   a. A management system c. A government inspection
   b. An audit d. An investigation
4. Which two of the following answers are components of a management system audit?
   a. Ensuring the management system meets legal requirements
   b. Ensuring the management system meets the requirements of the standard
   c. Ensuring the management system is effectively implemented
   d. Ensuring the management system is certified to national standards
5. Which of the following would not be reviewed during a quality audit?
   a. Medical records c. Process records
   b. Transfer records d. Calibration records
6. When it is okay to provide false information to an auditor?
   a. When the truth will get you in trouble
   b. When your supervisor instructs you to do so
   c. When you are not sure of the correct answer
   d. Never
7. All of the following are things you should NOT do with an auditor except:
   a. Joke around unnecessarily c. Argue with the auditor
   b. Treat the auditor with respect d. Leave the auditor alone
8. Which of the following is undesirable during an audit?
   a. Asking for clarification
   b. Redirecting a question to the appropriate party for an answer
   c. Using the resources available to you during the audit
   d. Fumbling around looking for things
9. All of the following actions are acceptable during an audit except:
   a. Pointing the blame at someone else c. Treating auditors with respect
   b. Steering auditors to the right people d. Listening to the auditor

10. (*True or False*) If possible, it is better to meet the auditors in a conference room rather than out in the workplace.
11. (*True or False*) Wandering around the unit to see what you can see is a valid way of hosting an auditor.
12. The national award for quality excellence in the United States is called the:
    a. Malcolm Baldrige National Quality Award
    b. National Quality and Productivity Council Award
    c. Presidential Freedom Award
    d. Deming prize
13. (*True or False*) The ISO 14001 Environmental Management System standards are required by the EPA in all chemical plants in the United States.
14. (*True or False*) The ISO 9001 Quality Management System standard is voluntary for all process industry plants.
15. (*True or False*) Malcolm Baldrige was the last secretary of commerce.

Activity

1. In this chapter you were provided with a cross-reference of the six key elements of the ISO 9001 standard to the ISO 14001 standard from a process technician's point of view. Now complete the expanded cross-reference guide below:

Key Element of ISO 9001	Matching Element in ISO 14001
4.2 Documentation Requirements	4.4.4 Documentation 4.4.5 Control of Documents
5.1 Management Commitment	
5.3 Quality Policy	
5.4 Planning	
5.5 Responsibility, Authority and Communication	
5.6 Management Review	
6.1 Provision of Resources	
6.2.2 Competence, Awareness and Training	4.4.2 Competence, Awareness and Training
7.5.3 Identification and Traceability	4.4.6 Operational Control (a broad topic)
7.6 Control of Monitoring and Measuring Devices	4.5.1 Monitoring and Measurement
8.2.2 Internal Audit	
8.2.3 Monitoring and Measurement of Processes	4.5.1 Monitoring and Measurement
8.2.4 Monitoring and Measurement of Product	4.5.1 Monitoring and Measurement
8.5.2 Corrective Action	
8.5.3 Preventive Action	

CHAPTER 13

Quality Reliability Planning

"Facts do not cease to exist because they are ignored."

~ALDOUS HUXLEY

Objectives

Upon completion of this chapter you will be able to:

- Define the three components of a quality reliability plan (QRP) and explain the value of using a QRP for process improvement.
- Explain the concept and components of a quality design plan.
- Construct a quality design plan.
- Explain the concept and components of a quality control plan.
- Construct a quality control plan.
- Explain the concept and components of a failure mode and effects analysis (FMEA).
- Conduct a failure mode and effects analysis (FMEA).

Key Terms

Failure mode and effects analysis (FMEA)—the third and final step in the quality reliability planning process, in which you examine every possible failure mode for your process in order to understand the consequences of process failures from the customers' perspective.

Quality control plan (QCP)—the second step in the quality reliability planning process. The specific process control handles are specified to ensure that your process is capable of meeting the needs of the customer. Process capability is captured at this stage.

Quality design plan (QDP)—the first step in the quality reliability planning process. The needs of the customer are translated into specific requirements (specifications), and the measurement techniques are specified to ensure that those needs are met.

Quality reliability planning (QRP)—a procedure designed to ensure that when a new product is being introduced, proper communications occur between all of the functional groups and that a system is in place that will allow the product to meet the customers' expectations consistently. The procedure consists of the quality design plan, the quality control plan, and the failure mode and effects analysis.

Risk priority number (RPN)—a calculated number within the failure mode and effects analysis process that helps you to prioritize the actions that need to be taken to minimize the risks of process failures and consequences to the customer.

Introduction

Quality reliability planning (QRP) is a procedure designed to ensure that when a new product is being introduced, proper communications occur between all of the functional groups and that a system is in place that will allow the product to meet the customers' expectations consistently. QRP is not another new procedure but rather a melding of the best parts of several procedures and techniques that have been known for many years. These procedures and techniques include control plans, process analyses, sample plans, failure mode and effects analyses (FMEA), and quality function deployment. The benefit of using these tools has been documented by a variety of companies.

In the grand scheme of this textbook, quality reliability planning is another improvement strategy to employ. For your purposes, this chapter will review the entire procedure, but the discussion concentrates on the sections most applicable to process technicians.

Over the years, many businesses have adopted the use of these tools as a process improvement technique. The implementation of these tools in the process industry has added a new dimension to their utility. The process industries are accustomed to producing new products on existing equipment; therefore, a procedure that is machine or equipment oriented becomes difficult to apply. QRP has been designed to be specifically beneficial to the processing industries and incorporates changes made to the widget industry procedures as a result of experiences in applying these tools.

Through the use of QRP, many problems that occur due to lack of communication between the product designers and the product producers are overcome. The industry has found that QRP better enables companies to manufacture products that can consistently meet and exceed the customers' expectations and has resulted in better initial system designs, which have significantly reduced problems after a product goes into commercial production. QRP consists of three main components, each of which serves a useful and complementary purpose in the overall scheme of process improvement.

The foundation of the QRP is the quality design plan, which is used to document the customers' expectations, translate those expectations into product specifications and methods, and analyze those specs and methods to ensure that they are capable of indicating compliance with the customers' performance requirements. Building upon this foundation is the quality control plan, which documents the manufacturing procedures for handling nonconforming product, describes the process parameters that control the quality of the product, defines the acceptable variation in the process parameters, and outlines the action steps needed to correct a process that is exhibiting unacceptable variation.

Finally, the FMEA is used to predict potential failures of the product, document the effects of these failures on the customer, and brainstorm potential causes for each failure. The FMEA is then used to generate a **risk priority number (RPN),** a calculated number that helps you to prioritize the actions that need to be taken to minimize the risks of process failures and consequences to the customer. In those cases in which an unfavorable RPN is calculated, the opportunity is provided to recommend process improvements.

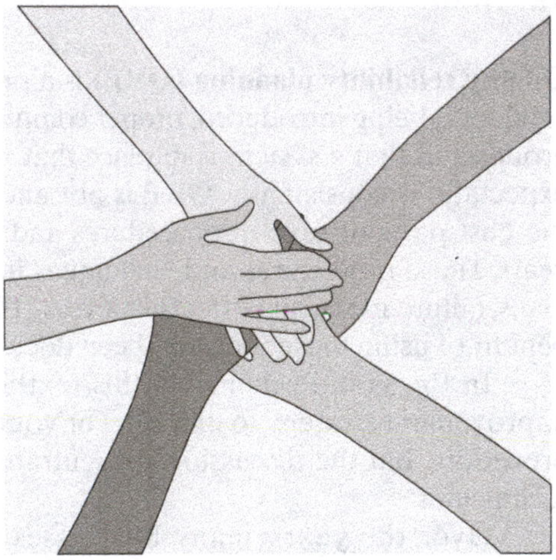

QRP is not intended to be an individual activity. It is much better applied as a team effort, although much of the initial work can be done in small groups to make better use of everyone's time. In addition to step-by-step instructions for applying this tool, charts will be provided outlining the sequence of events necessary to complete a QRP. You will notice on the charts that suggested primary leadership responsibilities have been identified. Please remember that any other interested party may be involved at any or all steps in the process. A suggested team membership roster would include representatives from the following:

Plant laboratory	LAB
Analytical measurements R&D	ANL
Applications R&D	APP
Process R&D	PROC
Production Department	PROD

A representative from the quality function, such as a quality engineer or quality manager, would also be included to serve as a facilitator of the QRP process and possibly act as a team scribe.

The rest of this chapter will be devoted to a column-by-column, form-by-form instruction guide to using QRP. Keep in mind that the underlying purpose of this tool is process improvement; the technical details of the forms are unimportant except as an aid to process improvement. Notice that blank forms for the quality design plan, the quality control plan, and the FMEA are included at the end of the chapter. On the forms you will see that each column is numbered. These column numbers will be used in the body of this guide as a reference for the instructions.

Quality Design Plan

The **quality design plan (QDP)** is the first step in the quality reliability planning process. The needs of the customer are translated to specific requirements (specifications), and the measurement techniques are specified to ensure that those needs are met. Its primary contribution to the overall process is that of building an information base, which greatly increases the efficiency of the rest of the process. Because the type of information captured in the design plan is usually provided by research and development (R&D), the role of the process technician in this phase of the QRP is to learn. Including operators and technicians in this step is an excellent training opportunity for you to learn not only what needs to be done but also why it needs to be done. A thorough understanding of what the customer expects and how to measure the quality of your product can only make you more aware of the impact of your actions in making the product.

An example of the header section of the QDP is shown in Figure 13-1. With the column numbers provided, here is an explanation of what is expected in each column.

Column Instructions

1–2. The first two columns are where you document what exactly the customer needs your product to do in terms of its performance characteristics. In some cases these may be the specifications for the product that your product is used to make. If possible, the customer should be asked to provide this information directly; otherwise, the representative from Applications R&D will need to take the lead in defining the customers' expectations. Examples of customer performance attributes might include the following:

Vent collapse	Force to crush
Cure rate	Rise time
Tensile strength	Solubility
Compression set	Volatility
Tear strength	Surface tension
Lubrication	Molecular weight distribution

3–4. In these columns, Process R&D has the leadership role, working with Applications R&D. The purpose here is to translate the customers' expectations into meaningful product specifications. Limits for your product specifications

FIGURE 13-1 Example of the Header Section of a Quality Design Plan

Quality Design Plan for:__________ Page ____of _____ Date: __________

						If Method Capability Is Less Than 2.50:			
Customer Performance Attribute	Acceptable Limits (If Any Apply)	The Quality Attributes That Control Performance	Proposed Limits for Quality Attribute	Detection Method for the Quality Attribute (Lab Method)	Method Capability, Cm	Recommended Actions to Improve Precision	Resp. Party	Actions Taken	Date Completed
1	2	3	4	5	6	7	8	9	10

should be set to ensure complete compliance with the customers' needs. Examples of typical quality attributes would be the following:

Hydroxyl number	Percent weight water
Total polymer	Alkalinity
Acid number	Purity
Viscosity	Color stability

5–6. Here the detection method for the quality attribute is identified. Process R&D, in conjunction with the plant laboratory, should be able to describe the method and produce historical data on this method or similar methods illustrating the expected precision of the method. A laboratory method should exhibit variation no greater than 40% of the specification range (this equates to a Cm of 2.5, which you learned about in Chapter 3) to ensure that your system is capable of meeting the specs. When this standard is not met, several options are available:

1. Improve the precision of the method.
2. Widen the specification band.
3. Devise a sample plan, as an integral part of the production system, to compensate for the excessive variation.

7–10. For those analytical methods that can be improved, columns 7–10 are available to document the recommendations and to record the improvements. The analytical measurements group and the plant laboratory people are the most likely candidates for initiating this type of activity.

Quality Control Plan

The **quality control plan (QCP)** is the second step in the quality reliability planning process. The specific process control handles are specified to ensure that your process is capable of meeting the needs of the customer. Process capability is captured at this stage. The control plan portion of the QRP process is designed to document the manufacturing reaction to nonconformance and the exact critical in-process parameters that control the quality attributes that affect the customers' expectations. This is a key piece in the puzzle of process understanding. The plant manufacturing department representative has the lead role in the control plan, working closely with Process R&D.

Process technicians can and should play a key role in this phase of the QRP, as they are the ones who control the process on a day-by-day basis. It is common during one of these exercises for the day staff to have a different understanding of how the process is controlled than do the process technicians. In this case, one of two things will happen. Either you will help to change the current control scheme because of your input, or you will be convinced to change your behaviors based on a better understanding of what is important to the customer.

An example of the header section of the quality control plan is shown in Figure 13-2. With the column numbers provided, here is an explanation of what is expected in each column.

FIGURE 13-2 Example of the Header Section of a Quality Control Plan

Quality Control Plan for:______________ Page ____of ____ Date: __________

	Sample Plan									
Quality Attributes	Sample Point	Freq.	No. Samples	No. Tests	Specification Limits for Quality Attribute	Corrective Actions for Non-conformance (Outside Limits)	Controlling Process Parameters	Acceptable Limits for Process Parameters	Process Parameter Control Scheme	Corrective Actions for Non-conformance (Outside Limits)
1	2	3	4	5	6	7	8	9	10	11

Column Instructions

1. In this column all of the quality attributes that were identified through the use of the quality design plan are reiterated, as well as any additional specifications that are included as double checks or any specifications that are indirect indications of quality.

2–5. These columns represent the results of any sample planning activities, as well as provide documentation of the current sample schemes. The sample plan is just a description of how many samples to take and from where to take them.

2. In this column enter the location(s) where samples will be taken. The list below provides examples of sample locations as well as shorthand forms for entering the information in the form.

ML—make line	PL—pipeline
MT—make tank	ST—storage tank
TT—tank truck	TC—tank car
DR—drum	CYL—cylinder

3. For periodic sample locations, such as line samples, input the frequency of sampling here. For example, this might be every 4 hours, daily, hourly, and so on. For those samples that are not periodic, such as tank samples, enter "each" or "lot" as the frequency to indicate that each container is sampled.

4. If replicate samples are taken, in accordance with an established sample plan, enter the number of replicate samples taken. For most of your products, this number will be a 1.

5. If replicate analyses are run on each sample, enter the number of replicates here. Many of your analytical methods routinely require duplicate analysis, to reduce the analytical variation, as seen by the manufacturing units.

6. Enter the current specification limits for the quality attributes either as a nominal value with the acceptable deviation, or as a specification range with minimum and maximum values. Be sure to include the unit of measure.

7. Document the current reaction scheme to nonconformance to the specifications. This piece of information is of utmost concern to your customers, who invariably bring up this topic in audits. Examples would include the following:

- Isolate batch and rework
- Destroy product
- Sell as distress

8–9. In these columns you document the parameters in your process that actually control the quality of your product and how much variability is allowed in each one. These would be the physical control parameters of your process, such as the following:

Reactor temperature	125–135 degrees C
Column pressure	40 psi max
Feed rate	90.5 pph
Transfer rate	1500 gpm, nominal
Mixing time	16 hours, minimum
Rate of addition	1 drum in 30 minutes

10. Now you must describe the system(s) used to control the process parameters. Process control schemes would include things such as the following:

Electronic controllers	Pneumatic controllers
Redundant RTDs	DP cells
Administrative controls	Standard operating procedures (SOPs)

11. The last column of the control plan is where you document the actions to be taken when the process control parameters deviate from their acceptable limits. In many of the types of process controls in the process industry, these actions are things such as repairing malfunctioning controllers.

Failure Mode and Effects Analysis

The **failure mode and effects analysis (FMEA)** is the third and final step in the quality reliability planning process, in which you examine every possible failure mode for your process in order to understand the consequences of process failures from the customers' perspective. FMEA is the portion of the QRP process that documents the effects of nonconformance on the customer and allows some freewheeling brainstorming of potential problems in your processes. This exercise is definitely a team effort.

All members of the team should have access to the quality design plan and quality control plan information for review prior to meeting to work on the FMEA. All team members should also have access to this instruction guide before attempting to complete the FMEA, in order to familiarize themselves with the ranking criteria listed in the following pages. The FMEA is applied here to quality failure modes. As a process technician, you may also find yourself part of a team that is analyzing potential failure modes for safety and environmental issues. The mechanism is the same, although the descriptions of the occurrence, severity, and detection sections may vary.

An example of the header section of an FMEA is shown in Figure 13-3. With the column numbers provided, here is an explanation of what is expected in each column.

Column Instructions

1. This column begins the part of the QRP exercise in which you analyze every potential failure of your product to meet specifications. Some failures will be two-sided failures, such as total polymer, which can fail high or low. Other failures will be one sided, such as water contamination, which will only fail high. In addition to specification failures, you should include any other pertinent, potential failure of your process that could impact the customer. Some examples of potential failures include the following:

   High hydroxyl
   Low hydroxyl
   Product not well mixed
   Contaminated when loaded
   Low total polymer
   High water
   Off color

2. For each failure mode documented in column 1, give a complete description of the effect(s) of that failure on the customer. Applications R&D will obviously play a large part in this portion of the exercise. For example, a low total polymer failure may cause low loads in the customers' foam; high hydroxyl number might cause vent collapse and shear collapse. This part of the QRP is your evidence that you have in-depth knowledge of how your product affects the customers' process.
3. Brainstorm all (within reason) possible causes for each of the failure modes listed in column 1. The generation of this list will involve the entire group. Consider historical information when generating the list. Typical causes in the past have been things such as the following:
   - Failure to follow operating procedures
   - DP cell failure
   - Pump failure

FIGURE 13-3 Example of the Header Section of a Failure Mode and Effect Analysis

FMEA for:________________ Page ____of _____ Date: ____________

			Existing Conditions						Resulting Conditions					
Give a Description of the Potential Failure	Effect on the Customer Due to This Potential Failure	List All Possible Causes for Each Potential Failure	O CC	S EV	D ET	R PN	Recommended Actions for RPNs >125	Resp. Party	Actions Taken	Date Completed	O CC	S EV	D ET	R PN
1	2	3	4	5	6	7	8	9	10	11	12	13	14	15

- Dirty tank truck
- Filter bag ruptured during loading
- Raw materials are off-specification
- Improper raw material ratio

4–7. These columns are used to analyze the system currently in place for its capability to achieve the desired outcome. There are three numerical rankings to assign: occurrence (OCC), severity (SEV), and detection (DET); and a risk priority number (RPN) to calculate. Each cause must have a separate set of rankings, because different failures may have different rankings depending on what caused the failure. An example would be the causes of a pump failure versus an airplane's falling out of the sky and landing on the equipment. Most people would agree that there is a higher probability that a given pump will fail than there is that an airplane will crash into the equipment. You must apply this thought process to each potential cause that was identified. Each column will be discussed separately.

4. Here you assign a numerical rank, on a scale of 1 to 10, which describes the probability of occurrence for each of the potential causes that have been documented. Because your ability to detect a failure is evaluated in a different column (6), you must narrow your thoughts strictly to the probability of occurrence. The following table gives a description of what each ranking means. A probability of 1/370 means that with the current system, out of every 370 batches produced, 1 will fail. The use of historical data, including SPC/SQC type data, will be helpful. The ranking assigned may be based on data but sometimes represents a consensus opinion of the participants. If the issue in question is one that you have been control charting, then data would exist to calculate the Cpk.

Criteria for Assigning the "OCC" Ranking

Rank	Criteria	Probability	Cpk
1	Remote probability of occurrence. Capability shows at least $x \pm 4\sigma$ within specifications.	1/16,000	1.33
2 3	Low probability of occurrence. Process in statistical control. Capability shows at least $x \pm 3\sigma$ within specifications.	1/2000 1/370	1.17 1.00
4 5	Moderate probability of occurrence. Generally associated with processes similar to previous processes that have experienced occasional failures, but not in major proportions. Process in statistical control. Capability shows more than $x \pm 2.5\sigma$ within specifications.	1/144 1/61	0.90 0.80
6 7	High probability of occurrence. Generally associated with processes similar to previous processes that have often failed. Process in statistical control, but capability shows $x \pm 2.5\sigma$ or less within specifications.	1/28 1/8	0.70 0.50
8 9	High probability of occurrence. In evaluator's view, failure is almost certain to occur.	1/4 1/2	0.40 0.20
10	System not in state of statistical control.	1/1	N/A

5. Into this column goes a numerical severity ranking, on a scale of 1 to 10, of the expected severity of each identified potential cause. Applications R&D or the commercial organization may be best suited to describe the severity of each cause. In some instances a particular failure mode may have the same severity ranking for each cause. In other cases it is possible that a particular failure will have more severe consequences for some types of causes than for others. The following table lists the severity criteria.

Criteria for Assigning the "SEV" Ranking	
Rank	**Criteria**
1	Unreasonable to expect that the minor nature of this failure will cause any noticeable effect on the product performance. Customer will probably be unable to detect failure.
2 3	Low severity ranking due to minor nature of failure causing only a slight customer annoyance. Customer will probably notice only minor system or product performance degradation. Defects that cause the customer to make minor adjustments to its process but have no other effects might be examples of low severity ranking defects.
4 5 6	Moderate failure that causes some customer dissatisfaction. Customer is made uncomfortable or is annoyed by the failure. For example, increased scrap rate or major process adjustments due to poor processing would be considered a moderate failure.
7	High degree of customer dissatisfaction due to the nature of the failure.
8	Customer will be unable to use the product. The problem results in a shutdown of the customer's processing lines but does not involve safety or environmental issues or noncompliance to federal or state law.
9 10	High severity ranking when a failure mode involves potential safety problems and/or nonconformance to federal or state regulations.

6. Into this column you enter a DET ranking that describes your ability to detect a failure, once it occurs, before the product reaches the customer. You may either assign a DET rank based on a perceived probability of shipping a defect (using customer complaint data is even better) or use the measurement system capability (Cm) such as described in Chapter 3.

7. The risk priority umber (RPN) is the product of multiplying the OCC ranking times the SEV ranking times the DET ranking. The RPN can therefore be any number between 1 and 1,000. By using the RPN, you can determine which potential causes should be addressed first. A potential cause that has a moderate probability of occurring (5), will cause a moderate severity situation with your customer (5), and has only a moderate probability of being detected before shipment (5) will yield an RPN value of 125.

8. A maximum RPN limit should be agreed upon by the team. Any potential causes having an RPN higher than the agreed-upon limit should have action plans to lower the RPN. A limit of 125 may work for most areas. The team may also want to set a maximum for the OCC, SEV, and DET categories separately. However, just because a potential cause has an RPN lower than the limit does not mean that it cannot be addressed if it is the opinion of the team that improvements can and should be made. This tool is meant as a process improvement device, and good engineering logic should never be omitted from the process.

9. All action items should have responsibilities assigned to ensure thorough follow-through. At this point the efforts of the original QRP team have been completed. The team scribe can tabulate the efforts of the team and submit the information to the quality manager.

10–11. When the action items have been completed, they should be documented as to what steps were actually taken and the dates that each was completed. The team should then be reconvened to complete columns 12–15.

12–15. New OCCs, SEVs, and DETs should be assigned, and RPNs should be recalculated, based on the new system(s) in place. These will use the same criteria as are used for columns 4–7, cited on the preceding pages. This then becomes documentation of the continuous process improvement effort.

Criteria for Assigning the "DET" Ranking

Rank	Criteria	Probability of Shipping	Method Cm
1 2	Remote likelihood that the product will be shipped containing the defect. The defect is a visually obvious characteristic (e.g., color).	1/16,000 1/1,000	3.25 3.00
3 4	Low likelihood that the product will be shipped containing the defect. The defect is easily detectable with routine analyses that contain a low degree of subjectivity and remote possibility for human error.	1/200 1/100	2.50 2.25
5 6	Moderate likelihood that the product will be shipped containing the defect. The defect is an easily identified characteristic, but the detection method contains some degree of subjectivity or a moderate potential for human error.	1/50 1/20	2.00 1.75
7 8	High likelihood that the product will be shipped containing the defect. The defect is a subtle characteristic that is detectable, but the detection method has a high degree of subjectivity or high potential for human error.	1/10 1/4	1.25 1.00
9 10	High likelihood that the product will be shipped containing the defect. The product is not checked or not checkable. Defect is latent and will not appear at manufacturing location (e.g., defect affects processing of product).	1/2 1/1	0.50 0.25

QRP Process Flow

Step	Description	Lead Function	Form	Columns
1.	Identify need.	Management		
2.	Charter a team and obtain membership.	QM		
3.	Identify the customers' requirements.	APP	QDP	1–2
4.	Develop product specification requirements.	APP PROC	QDP	3–4
5.	Identify detection (lab) methods. Document precision.	PROC LAB ANL	QDP	5–6
CHECKPOINT:	Is the method precision less than 30% of the specification range? If not, proceed to step 6. If so, proceed to step 7.			
6.	Develop action plan to reduce method variability, modify the spec limits, or devise a sample plan to accommodate excessive variation.	ANL LAB	QDP	7–10
7.	The quality design plan is now complete. Reiterate the quality attributes, the spec limits, and the sample plan on the quality control plan form to begin the next phase of the QRP process.	PROC PROD	QCP	1–6

8.	Document the corrective actions to be taken for nonconforming quality attributes.	PROC PROD	QCP	7
9.	Define the process parameters that control the quality attributes.	PROC PROD	QCP	8
10.	Define the normal operating limits (operating windows or envelopes) for the process parameters.	PROC PROD	QCP	9
11.	Describe the process parameter control scheme.	PROD	QCP	10
12.	Describe the corrective actions to be taken for nonconforming process parameters (how to react to out-of-control conditions).	PROD	QCP	11
13.	Describe every potential failure of the product (within reason).	TEAM	FMEA	1
14.	Describe the effect(s) of each failure from the customers' viewpoint.	APP	FMEA	2
15.	Brainstorm all potential causes for each failure mode.	TEAM	FMEA	3
16.	Using the information from the control plan, assign an OCC ranking for each documented possible cause.	PROD TEAM	FMEA	4
17.	Using the information in the FMEA, assign a SEV ranking for each documented possible cause.	APP TEAM	FMEA	5
18.	Using the information in the design plan, assign a DET ranking for each documented possible cause	PROC TEAM	FMEA	6
19.	For each documented possible cause, calculate the risk priority number, RPN. RPN = OCC × SEV × DET	TEAM	FMEA	7
CHECKPOINT:	Is the RPN greater than the limit set by the team? If not, proceed to next CHECKPOINT. If so, proceed to step 20.			
CHECKPOINT:	Are actions needed to reduce the RPN? If so, proceed to step 20. If not, proceed to step 22.			
20.	Develop action plans to reduce the risk priority numbers, and assign accountabilities and responsibilities.	TEAM	FMEA	8–9
21.	Document actions taken and then recalculate the OCC, SEV, DET, and RPN values for the new process per step 19.	TEAM	FMEA	10–15
22.	All potential failures have now achieved acceptable levels of risk. Document the final results of the team effort and publish it to the business team.	TEAM		

Legend	
ANL	Analytical measurements R&D
APP	Applications R&D
QCP	Quality control plan
FMEA	Failure mode and effects analysis
LAB	Plant laboratory
PROC	Process R&D
PROD	Production department
QDP	Quality design plan
QM	Quality manager
TEAM	The entire team

Dibutyl Futile Example

Figure 13-4, Figure 13-5, and Figure 13-6 show examples of what portions of the QRP might look like for the fictional product dibutyl futile. To make it easier to read, full-page examples can be found at the end of the chapter.

FIGURE 13-4 Portion of a Completed Quality Design Plan

Quality Design Plan for: Dibutyl Futile | Page 1 of 1 | Date: February 10, 2009

Customer Performance Attribute 1	Acceptable Limits (If Any Apply) 2	The Quality Attributes That Control Performance 3	Proposed Limits for Quality Attribute 4	Detection Method for the Quality Attribute (Lab Method) 5	Method Capability, Cm 6	If Method Capability Is Less Than 2.50: Recommended Actions to Improve Precision 7	Resp. Party 8	Actions Taken 9	Date Completed 10
Cold load	2.4 - 2.8 ft ^{2}	Total polymer	38% ± 4%	IR	4.1	NA			
		Hydroxyl	456 - 556	Wet chemical titration with phenyl whatsis	1.33	Investigate need for an automated method	DRF	Reliable instrument not available at this time	11/2/08
						Devise sample plan to compensate during the short term	JLL	Sample plan implemented at plant lab	2/08/09

FIGURE 13-5 Portion of a Completed Quality Control Plan

Quality Control Plan for: Dibutyl Futile | Page 1 of 1 | Date: February 10, 2009

Quality Attributes 1	Sample Plan: Sample Point 2	Freq. 3	No. Samples 4	No. Tests 5	Specification Limits for Quality Attribute 6	Corrective Actions for Non-conformance (Outside Limits) 7	Controlling Process Parameters 8	Acceptable Limits for Process Parameters 9	Process Parameter Control Scheme 10	Corrective Actions for Non-conformance (Outside Limits) 11
Total polymer	ST	Each	1	1	38% ± 4%	Isolate batch to repolymerize	Monomer feed rate	Target ± 5 gpm	Mass flow meters	Calibrate the meters
							Reactor temp	345°C ± 10°C	Controlling DMV redundant RTDs controlling the tempered water system	Calibrate the RTDs; check tempered water loop for plugs, etc.
Hydroxyl	ST	Each	2	2	456 - 556	Batch must be scrapped	Hydroxyl of incoming raw material	660 - 700	Vendor COA	Out of spec raw material rejected, sent back to vendor. purchasing notified of the problem
Acid content	ST	Each	1	1	3.5 - 4.0 pH	Isolate batch for acid adjustment with carbonic acid	Acidity of the incoming raw material	2.0-2.5 pH	Vendor COA	Reject, return; notify purchasing
							Reactor temp during cookout	350°C Max	Redundant RTDs controlling the tempered water system	Emergency cooling water added; instrument loop recalibrated

FIGURE 13-6 Portion of a Completed Failure Mode and Effects Analysis

FMEA for: Dibutyl Futile | Page 1 of 1 | Date: February 10, 2009

Give a Description of the Potential Failure 1	Effect on the Customer Due to This Potential Failure 2	List All Possible Causes for Each Potential Failure 3	Existing Conditions: OCC 4	SEV 5	DET 6	RPN 7	Recommended Actions for RPNs >125 8	Resp. Party 9	Resulting Conditions: Actions Taken 10	Date Completed 11	OCC 12	SEV 13	DET 14	RPN 15
High total polymer	Stiff foam, high cold load	Monomer/base feed radio incorrect;	3	4	2	24								
		Rx temp too high (>360°C)	5	4	2	40								
Low total polymer	Soft foam, low cold load	Monomer/base feed radio incorrect;	3	4	2	24								
		Rx temp too low (<340°C)	4	4	2	32								
Hi or lo hydroxyl	Affects cold load	Contaminated with a diluent;	2	8	2	32	Need to establish a better measuring for incoming raw materials	JJP	Customer supplying quarterly SQC data and plant lab is double checking raw material hydroxyl	2/9/09	4	8	2	64
		raw material off spec	4	8	5	160								

Summary

Quality reliability planning is an improvement strategy that brings the various functions within a business together to ensure continuity of purpose and understanding, so that you can make a product that consistently meets your customers' expectations. There are three components to the QRP.

The quality design plan builds an information base so that all functions of the organization have the same foundational customer and process knowledge, thus greatly increasing the efficiency of the rest of the process.

The quality control plan documents the manufacturing process control parameters needed to achieve the desired quality output and provides a path forward for handling nonconforming product should the process fail. Process technicians play a key role in ensuring that the procedures and practices out on the shop floor are properly documented and meet the needs of the business.

The final component is the failure mode and effects analysis. The FMEA is a tool employed by safety, environmental, and quality practitioners to evaluate the potential problems you might encounter in the manufacturing process and understand the impact of those problems on the customer.

Checking Your Knowledge

1. Define the following key terms:
   a. Failure mode and effects analysis (FMEA)
   b. Quality control plan (QCP)
   c. Quality design plan (QDP)
   d. Quality reliability planning (QRP)
   e. Risk priority number (RPN)
2. A procedure used to document the "baseline" knowledge of the organization in order to improve communication between the functions is the:
   a. Failure mode and effects analysis
   b. Quality control plan
   c. Quality reliability planning
   d. Quality design plan
3. A procedure used to document the process control schemes used to ensure the manufacture of a quality product is the:
   a. Failure mode and effects analysis
   b. Quality control plan
   c. Quality reliability planning
   d. Quality design plan

4. A tool used to evaluate potential problems so you can make improvements to prevent them before they happen is the:
   a. Failure mode and effects analysis
   b. Quality control plan
   c. Quality reliability planning
   d. Quality design plan
5. A procedure designed to ensure proper communications occur between all of the functional groups within a business and establish a system to ensure the manufacture of a product that consistently meets the customers' expectation is:
   a. Failure mode and effects analysis
   b. Quality control plan
   c. Quality reliability planning
   d. Quality design plan
6. Participants in a QRP exercise should include:
   a. Process technicians
   b. R&D scientists
   c. Production engineers
   d. All of the above
7. (*True or False*) Process technicians play a key role in the quality control plan.
8. (*True or False*) Quality reliability planning is a new technology that is just now being explored in the process industries.
9. (*True or False*) The information learned in early chapters on process capability does not help you complete the QRP.
10. (*True or False*) A process flowchart would be a good way of establishing a common understanding of your work process.

Activities

1. *Quality Design Plan.* You work in a small manufacturing plant. Your product is lemonade. You sell your lemonade in bulk (5-gallon containers) to several food services companies around town as well as directly to consumers by the glass. Using a blank quality design plan template, document customer critical performance attributes. What might be important to your customers? Remember, different customers have different needs and expectations. Establish reasonable spec limits for these attributes. Define some quality attributes that might be used to measure "goodness" relative to your customers' performance attributes. Establish limits for these quality attributes. How might you measure these quality attributes? (Answering these questions should complete columns 1–5 on the quality design plan.) To get you started: Customers are interested in color. They want the color between two certain shades, and they use paint samples to demonstrate goodness. You measure color in your lab with a colorimeter that provides a wavelength output. As long as the wavelength of your measured sample is okay, the color seen by the customer should be okay.
2. *Quality Control Plan.* Using three or four of the quality attributes you identified in column 3 of your quality design plan, continue this example by filling in relevant portions of a quality control plan template. In the case of the color that you identified in Activity 1, specification limits of ±10% from the target have been established. What will you do if you make a batch that is outside this range? (column 7) What is it in your process of making lemonade that controls the color of the final product? (column 8) Establish limits for the process control parameters in column 9. How are these parameters controlled? (column 10) What will you do if the process fails? (column 11)
3. *Failure Mode and Effects Analysis.* Finally, let us take each of the quality attributes from the previous activity and describe a way in which this attribute might fail, using a blank FMEA template. In your example, the color might be too dark, it might be too light, or it might be too cloudy. What might the impact of each of these failures be on the customer? (column 2) What might cause this failure to occur? (column 3) It will be completely subjective, but take a crack at assigning OCC, SEV, and DET rankings for each failure and calculate a risk priority number (RPN) for each using columns 4–7. If any of these potential failures ranks higher than 125, use column 8 to document some recommended corrective actions to improve your process.

Blank Templates and Examples

Quality Design Plan for:__________________ Page ____of _____ Date: ______________

Customer	Acceptable Limits	The Quality Attributes That Control Performance	Proposed Limits for Quality Attribute	Detection Method for the Quality Attribute (Lab Method)	Method Capability, Cm	If Method Capability Is Less Than 2.50:			
						Recommended Actions to Improve Precision	Resp. Party	Actions Taken	Date Completed
Performance Attribute 1	(If Any Apply) 2	3	4	5	6	7	8	9	10

Quality Control Plan for:____________________ Page _____of _____ Date: ______________

Quality Attributes	Sample Plan				Specification Limits for Quality Attribute	Corrective Actions for Non-conformance (Outside Limits)	Controlling Process Parameters	Acceptable Limits for Process Parameters	Process Parameter Control Scheme	Corrective Actions for Non-conformance (Outside Limits)
	Sample Point	Freq.	No. Samples	No. Tests						
1	2	3	4	5	6	7	8	9	10	11

FMEA for:__________________ Page ____of _____ Date: _____________

Give a Description of the Potential Failure	Effect on the Customer Due to This Potential Failure	List All Possible Causes for Each Potential Failure	Existing Conditions						Resulting Conditions						
			O CC	S EV	D ET	R PN	Recommended Actions for RPNs >125	Resp. Party	Actions Taken	Date Completed	O CC	S EV	D ET	R PN
1	2	3	4	5	6	7	8	9	10	11	12	13	14	15

Quality Design Plan for: Page 1 of 1 Date: February

						If Method Capability Is Less Than 2.50:			
Customer Performance Attribute 1	Acceptable Limits (If Any Apply) 2	The Quality Attributes That Control Performance 3	Proposed Limits for Quality Attribute 4	Detection Method for the Quality Attribute (Lab Method) 5	Method Capability, Cm 6	Recommended Actions to Improve Precision 7	Resp. Party 8	Actions Taken 9	Date Completed 10

Quality Control Plan for: Page 1 of 1 Date:

	Sample Plan									
Quality Attributes 1	Sample Point 2	Freq. 3	No. Samples 4	No. Tests 5	Specification Limits for Quality Attribute 6	Corrective Actions for Non-conformance (Outside Limits) 7	Controlling Process Parameters 8	Acceptable Limits for Process Parameters 9	Process Parameter Control Scheme 10	Corrective Actions for Non-conformance (Outside Limits) 11

FMEA for: Page 1 of 1 Date:

			Existing Conditions						Resulting Conditions					
Give a Description of the Potential Failure 1	Effect on the Customer Due to This Potential Failure 2	List All Possible Causes for Each Potential Failure 3	OCC 4	SEV 5	DET 6	RPN 7	Recommended Actions for RPNs >125 8	Resp. Party 9	Actions Taken 10	Date Completed 11	OCC 12	SEV 13	DET 14	RPN 15
		Monomer/ base feed radio incorrect; Rx temp too high (>360°C)												

CHAPTER 14

Lean

"People seldom hit what they do not aim for."
~HENRY DAVID THOREAU

Objectives

Upon completion of this chapter you will be able to:

- Define lean manufacturing as used by the process industries.
- Explain the growth in popularity of lean tools in industry in recent years.
- List the key process improvement methodologies included in the lean toolbox.
- Describe and explain the use of value stream maps.
- Describe the concept of continuous improvement (Kaizen).
- List and explain the 5 Ss.
- Describe the concept of just-in-time production (Kanban).

Key Terms

5 Ss—a floor-level improvement methodology taken from five Japanese words: Seiri, Seiton, Seiso, Seiketsu, and Shitsuke.

Just-in-time (JIT)—process in which the company creates a product for the customer when it needs it.

Kaizen—Japanese term for continuous improvement or incremental improvement.

Kanban—a signal card used to pull production.

Muda—Japanese term for traditional waste.

Mura—Japanese term for unevenness.

Muri—Japanese term for overburden.

Takt time—a German word that means beat. In quality it is the available production time divided by the customer demand.

Value stream map (VSM)—a high-level flowchart of a process. Included within each step of the process is the amount of time and what value it adds from the customer's perspective.

Introduction

It is time for the last improvement strategy topic. This one is called lean and is heavily influenced by Japanese improvement efforts. Because of the Japanese influence, lean is awash with lingo that makes lean harder to get started than it should be. In order to familiarize you with lean as it is implemented in industry, this chapter will introduce you to dozens of Japanese terms; but you will concentrate on the concepts and are not required to learn Japanese in order to succeed in this class or understand lean.

In this chapter you are introduced to the lean manufacturing practices used in the process industries. There is a brief discussion about the growth of popularity of lean tools in recent years. Different books on the subject concentrate on different aspects of lean, but some tools in the lean toolbox that are pretty commonly applied include value stream maps, Kaizen, 5 Ss, and just-in-time production. As an improvement strategy, you see lean implemented as a stand-alone effort and also as an add-on to an existing Six Sigma effort. Sometimes you even see it referred to as lean Six Sigma. However you see it packaged, it is still the same lean.

What Is Lean?

Think of your process as a lean, mean quality machine. Everything works smoothly. There is no wasted effort. No fat. Efficiency is at an all-time high. That is exactly the concept of lean, no waste. Waste costs money. If you can eliminate the waste, you get to keep the money. If your management sees waste, it sees money that could have been put on the profit line of the company's books put on the expense line instead.

Unfortunately, many times the managers are busy managing and do not know how to transform their operation into a waste-free, lean operation. They must rely on the workers on the floor to accomplish this. Of course, the workers on the floor do not have a chance if the effort is not supported by management, but lean really is a shop-floor level, or operating control room level, effort.

To build on the management support point a little further, lean really is a philosophy or a culture more than it is a collection of tools. The tools are implemented at all levels of the organization but must be implemented in a supportive environment that starts at the top. If you do not have a supportive environment, you will not have a lean culture and can look forward to only marginal success implementing the lean tools.

Recall from Chapter 1 that many of the tools in the field of quality were originally developed by Americans but really came into their own as they were implemented and popularized by the Japanese during their rebuilding period following World War II. Lean is an exception to this rule. Lean started out in the Toyota Motor Corporation over 50 years ago. Toyota called its quality improvement process the Toyota Production System, or TPS. The terms *TPS* and *lean* are synonymous in the quality industry today.

It was not until the 1980s that Dr. James Womack published a book about the TPS and called its quality operation "lean." The Toyota Production System was designed so that Toyota used fewer suppliers, had quicker turnaround times, produced product with fewer defects, carried less inventory, and in general had fewer wasted efforts than were common in traditional assembly-line operations.

In the Toyota Production System, and in lean today, waste is divided into three categories: Muda, Mura, and Muri.

Muda

Muda is the Japanese term for traditional waste, which is further broken down into seven subcategories:

a. Transport–in your production facility, if you finish operation A on one side of the plant and have to transport your intermediate product to the other side of the plant to continue with operation B, the time it takes to move your material from A to B is wasted transport time. No value is being added to your product during this time.

b. Waiting–when your product arrives at operation B, it has to sit and wait for somebody to pick it up and do something with it. There is no value being added to the product while it is sitting and waiting.

c. Overproduction–if the customer needs 1,000 pounds of product and you make 1,200 pounds in order to ensure that nothing goes wrong, you have incurred waste. You have individuals working on making product that is not destined for immediate sale and for which there is no immediate need. These individuals could have done something value added instead.

d. Defects–obviously, making a defective product is wasteful. A great portion of this book has dealt with various tools and techniques employed to analyze, detect, and reduce defects. The time, materials, and resources spent making the defective product will be wasted. Good product will then have to be made to replace the defective product, consuming even more time, materials, and resources. You will probably have to expend yet more effort to deal with the defective product. Even the act of throwing away the defective product costs time and money.

e. Inventory–inventory is necessary because it is what you sell. More inventory than you need is wasteful. You pay taxes on inventory that you carry. You pay for storage facilities to store the inventory. You pay for people to manage the storage facilities. You paid good money to make product just to sit on the shelf. In the Toyota Production System, you make only what you need and make it right when you need it.

f. Motion–excess motion is another form of waste. Let us say you work on an assembly line, and your job is to pick up headlights from a box on the floor and install them on a chassis moving along at chest level in front of you. On a good day

Motion may
be good for
your waist.
But in quality
motion is
waste!

you can install 250 headlights. This means that you are bending over and touching your toes 250 times a day.

g. Excess processing—this one is a little harder to grasp, but in general, any process step that is not adding value to the product that you sell is by definition waste. Some would argue that this would be too obvious and any excess processing would be eliminated, but you may be surprised to find operations in which what is done is what has always been done, even if nobody can tell you why it is done.

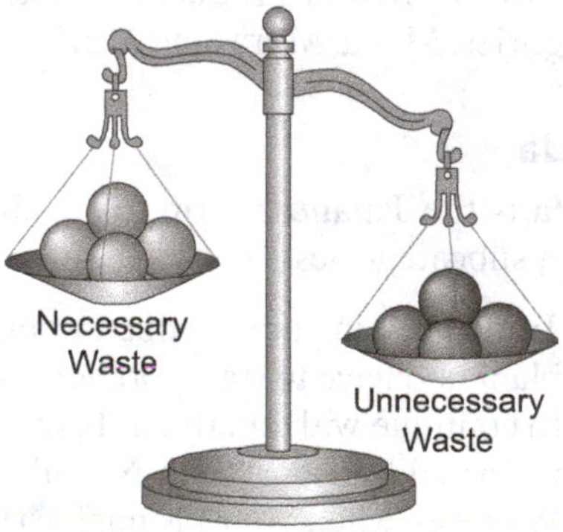
Necessary
Waste
Unnecessary
Waste

One more comment about Muda, or traditional waste: You may be able to think of some instances in which waste is necessary, such as inventory. You are absolutely right. Oftentimes in the process industry you make specialty products that take days or even weeks or months to produce. You cannot wait for the customer to demand the product and then tell the customer to hold on for a few weeks while you produce it. In the retail business you must have inventory sitting on the shelf or the customer has no reason to shop in your store.

Another example of necessary waste is in the excess processing area. Sometimes you have process steps that do not add value to the product that you sell but they help you ensure regulatory compliance or address some environmental or safety issue. You are perfectly happy to have those toxic off-gases treated so they are safe to vent, even if that part of the process does not add value to the customer. Even though all transport, motion, inventory, and so on are waste, companies allow that some waste is necessary. Your job is to determine which waste is unnecessary and get rid of it.

Mura

Mura is the Japanese term for unevenness. Americans would call this variation. A good portion of the early part of this book talked about variability and how to recognize it, quantify it, and reduce it. Variation is the enemy of quality. Variation is a classification of waste because due to this variation you must expend time and energy to measure the product to determine the amount of variation present and see whether your product can satisfy the customer.

If you made your product exactly the same every time, you would eliminate all these wasted steps. Some process technicians work in the laboratories or have small field laboratories associated with their processes. By the lean definition, all laboratory effort is a form of waste, made necessary by the variation in the processes and products. Eliminate the variation and you eliminate the wasteful laboratory work.

Muri

Muri is the Japanese term for overburden. This is the hardest one for Americans to grasp. It has to do with asking too much of your employees or your processes. Working overtime is often considered a good thing by employees looking to pick up a little extra cash, but working too much overtime can tire you out, cause family problems, and even increase the probability of safety incidents on the job.

Muri is the waste caused by asking too much of your employees. Consider the individual that had to pull a 36-hour shift during the start-up of a new process. He got the work done, but the work he was doing during the last 10 to 12 hours of that shift was not his best. Another potential form of Muri is ergonomic issues. Companies are paying more attention to ergonomics today than in years past, focusing on making it more comfortable for the employees to get their jobs done while reducing safety and health issues caused by repetitive motion, strain, and so forth.

Compared with many of the more sophisticated tools you have studied, such as designed experiments, some of the lean tools are much easier to work with. Some are just common sense and make you wonder why you need a separate name for them at all. Many of the tools that have already been discussed are designed to achieve a great step change in the performance of the operation, whereas lean focuses more on incremental improvements at the shop-floor level. The simplicity of the lean tools and the ability to implement the tools at the worker level have contributed to the growth in popularity of lean in the process industries in the last few years.

Key Lean Tools

You know that lean is the absence of waste, and you know about all the different kinds of waste that you can look to reduce. How do you go about making your organization lean? You accomplish this by applying the tools of the lean toolbox. Dozens of tools are available, and all are focused on continuous improvement and the elimination of waste. What is presented here are four of the most commonly applied tools or concepts associated with lean.

The tools value stream maps, Kaizen, the 5 Ss, and just-in-time production have been selected to focus on in this chapter. Keep in mind that lean is a culture of operating on the edge, so the tools you apply will depend on the operations in which you work. If you are in the supply or logistics area of the plant, then managing inventory to keep it at an absolute minimum will be your focus. If you are in the manufacturing area of the plant, then minimizing defects may be your focus. In the laboratories you might focus on the transport and motion types of waste in order to accomplish your necessary waste task of inspection with as little unnecessary waste as possible.

Value Stream Maps

The foundation of lean tools is the value stream map. A **value stream map (VSM)** is a high-level flowchart of a process that identifies within each step of the process the amount of time and what value it adds from the customer's perspective. Figure 14-1 is a simplistic way to show you what you can expect to see in the future.

A real VSM may include dozens of process steps and for each step may identify not only the amount of time that step consumes but also other measurements of interest such as manpower requirements, utility requirements, or inventory loading. Starting from the upper right-hand corner, you see that the customer wants product, so it places an order with you.

Your production control and scheduling center must order raw materials from the necessary suppliers and schedule your production facility to make the product. If you have the product on hand, then you might place an order to ship the product from the warehouse to the customer while you make product to replenish the warehouse.

Your process is broken up into several different sections; each one pulls material from storage, conducts some processing step, and then puts that material into inventory to have available for the next step in the process. When your product is completed, it is drummed and put into inventory to await the next customer order.

In this example it takes 48 hours from the time you contact the supplier until the time you have raw materials on hand. It takes you 12 hours to prepare your premix, which is a blend of the right amounts of the right materials for your product. This premix is put into storage for an average of 24 hours so the reaction group can use it when it needs it. Reacting your premix with the catalysts and other bulk ingredients is a 12-hour process.

FIGURE 14-1 Example of a Value Stream Map

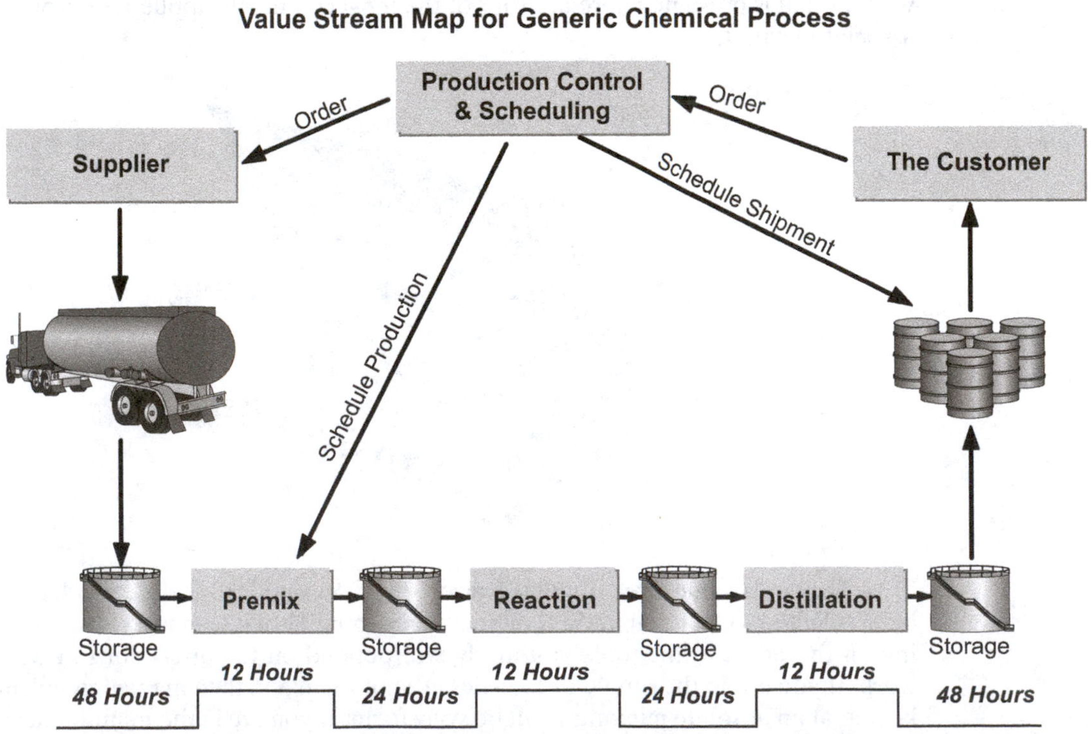

The product that you make in the reaction step is called your "raw" product. The product is made but it is still too dirty to sell. The raw product goes into storage to be ready for the next processing step. The average storage time is 24 hours. The raw product is then run through a distillation column, where the impurities are removed and the product is cleaned up for sale. This step takes approximately 12 hours.

The final product is then pumped to a large storage tank, where it awaits the drumming operation. It takes approximately 48 hours to get the product drummed, labeled, analyzed, and delivered to the warehouse, where it can finally be sold to customers.

From the time your customer wants product shipped to it to the time the product is delivered to the warehouse for shipment is 180 hours total—7½ days. Because your customers demands delivery within 48 hours of placing an order, you must maintain inventory on hand to meet their demands. Only 36 hours of the total 180 hours needed to make this product was value-added time. Eighty percent of the time was spent waiting on equipment to be ready or waiting for the product from the previous step of the process.

Let us say the warehousing costs for this operation are on the order of a million dollars per year. That is a million dollars that could be saved by eliminating the need to carry inventory. You could start by scheduling your production so that each step of the process is ready to take material from the previous step as soon as the previous step is finished. Your supplier may be willing to maintain a storage tank on your premises and manage the raw material inventory for you.

You can pull inventory whenever you need it and keep records to balance the books at the end of the month. You install an automated drumming machine at the end of the process that fills drums and palletizes them for shipment directly to the customer.

In a perfect world, you would not need a warehouse for this example. Because your world is not perfect, you may have to keep some inventory on hand, but you may be able to reduce that inventory from an 8-day inventory to a 2-day inventory.

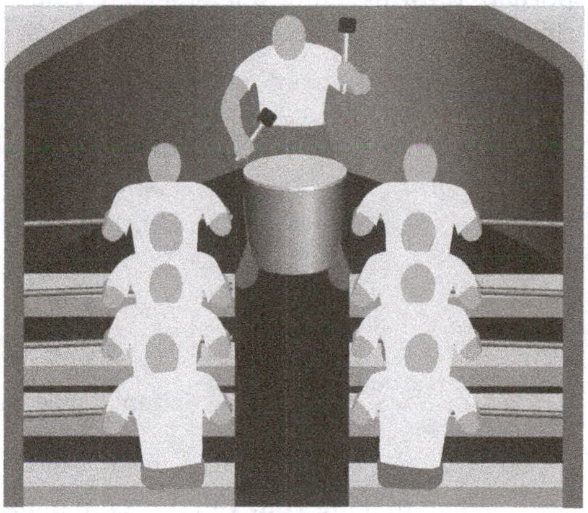

The idea is to match the time it takes to make product to the consumption rate of your customers so that the process is perfectly balanced. This rate of consumption to meet customers' need time is called **takt time,** a German word that means beat and is defined as the available production time divided by the customer demand. Think of takt time like this: If every part of the process is rowing to the same beat, then you have an efficient, lean, well-run process.

Let us say you have 12 hours of production available per day. The customer demand is 360 drums per day; your takt time is 1/30 hour, or 2 minutes per drum. You need to be kicking a pallet of 4 drums off the end of the process every 8 minutes in order to keep up with customer demand.

Here is a real-world example using the concept of takt time to pace a process. We were having a cookout at the church and planning to feed hamburgers to over 200 people. In order to get that much food ready by noon, the head cook thought we should start grilling hamburgers at 9:30 A.M. Our grill holds 20 patties, and it takes about 15 minutes to cook a batch of 20 patties; so in order to have 200 patties ready at noon, we could cook about 80 patties per hour and would need 2½ hours to prepare all 200 patties.

Of course, every single hamburger would be cold by the time we served dinner at noon, but at least we would be ready. We applied the takt time concept as follows: It takes approximately 1 minute for each person to go through the line to get his or her meat, bread, fixings, chips, drink, and dessert, so we would need to take a patty off the grill every 1 minute. It takes about 15 minutes to grill a burger, so we would need to start cooking the first patty 15 minutes before noon. The next patty would go on the grill at 14 minutes till noon and the next one at 13 minutes till noon.

Our grill can hold about 20 hamburger patties at a time, so we have plenty of production capacity to have at least 15 patties in various stages of doneness going at any point in time. We put a patty on every minute and we took a patty off every minute. Every single person in the line of 200 people was able to get a hot juicy hamburger patty without our doing anything other than pacing our process to match the customer need.

The other thing to consider at every single step of the process is whether this step is value added. To be value added, a process step must do the following:

- Add value from the perspective of the customer (the customer must be willing to pay for it)
- Transform the product in some way
- Be defect free (doing something over again is not value added)

All three conditions must be met; otherwise the step is considered some type of waste.

Of course, as with most of the improvement strategies and tools that have been covered in this book, the work process is to apply the tools to the current state, analyze the information, make improvements based on what you have learned, then reapply the tool to see where you are. In the case of VSM, it means making a current state map, perhaps drawing a future state map to guide the improvement process, and then drawing the resulting improved process map, which becomes the new current state map.

現地現物

Genchi-genbutsu: Go and see

改善

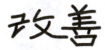

Kaizen: Continuous Improvement

When making your value stream maps, be sure to talk to the people who work in the parts of the process that you do not, and be sure to walk through the process yourself. It is amazing how many times problems get solved when individuals get out of the conference rooms and take a walk to see what is happening.

The Japanese word for "where the action is" is *gemba*. The Japanese phrase for the concept of going out where the action is to see for yourself is called *genchi genbutsu*. These are just a couple more terms you may hear individuals throw around in the lean world.

Now, a value stream map can show you where improvement opportunities lie, but the map does not tell you how to improve your process. That is why this tool and the concept of lean are classified as an improvement strategy. To employ this strategy, you also have to make use of capability studies, control charts, designed experiments, and Pareto charts. The tools covered in the preceding chapters fit together within the lean improvement strategy to accomplish the improvements that everyone seeks.

Kaizen

Kaizen translates from the Japanese as "continuous improvement" and is typically rendered as "incremental improvement" in the quality world. Many of the tools discussed in this book are focused on making a step change difference in the performance of your process. Kaizen is focused on the small, incremental changes that can be made quickly and cheaply on a day-by-day basis.

No matter what you do or where you go, leave the place in better shape than you found it and you will always be welcomed back!

Grandma used to teach us kids that we would always be welcomed back to anyplace that we left better than we had found it. Of course, what she was really telling us was that we should clean up our own messes and help with the dishes, but the concept also applies to the process industries.

Every single day you should look for every opportunity to clean up around you and make small, incremental improvements to your processes. Do not go home until you have made your work environment a better place to work. This is the concept of Kaizen.

One of the basic tenets of Kaizen is that you should employ the plan-do-check-act (PDCA) cycle, which was discussed in Chapter 6. Kaizen also requires that you operate to standards, as discussed in Chapter 12. If you do not have standards, set them, and then measure your process against the standard and continually improve the process, setting new standards. If small, incremental progress is not enough to get you where you need to be, then use Kaikaku (radical change) instead of Kaizen.

Kaizen projects can be carried out by individuals, by work group teams, or by management teams. Some improvements need to be made at a pretty high level in the company; and only management can accomplish these types of improvements. Other improvements can be accomplished by individuals or by small, self-directed work group teams.

Much attention is given in the lean world to the concept of Kaizen, but it all boils down to this: While working on the big breakthrough improvement projects, do not forget to work on making everything you touch just a little bit better every day. It is a culture, a way of life. It is about respecting the people around you, the environment, and the workplace.

In his book *Kaizen*, Masaaki Imai spends a lot of time talking about the differences between Western philosophy and Eastern philosophy. Western managers like big improvements. That is what gets rewarded in our society. No one gets a bonus for reorganizing the control room to make running the process easier and more efficient, yet doing so improves the process. It is difficult to transplant the culture from East to West, but transplant it we must in order to apply the philosophy of continuous improvement.

Five Ss

Here is a straightforward and simple concept that can be used by anybody at any level in the organization: the 5 Ss. The **5 Ss** is a floor-level improvement methodology taken from the following five Japanese words:

1. Seiri—organization
2. Seiton—neatness
3. Seiso—cleaning

4. Seiketsu—standardization
5. Shitsuke—discipline

You were promised that you would not have to learn Japanese to be successful in this course of study. Here is an attempt to make the Japanese 5 Ss into the English 5 Ss:

1. Sort or segregate
2. Straighten up
3. Sweep or scrub
4. Standardize (at least this one was easy)
5. Self-control

The concept of each S is almost self-explanatory, but here follows an explanation anyway.

Step 1: Sort or Segregate

Get organized. Everything in your office or on your workbench or in your closet can be segregated into categories:

- Stuff that you use all the time
- Stuff that you use occasionally
- Stuff that you never use
- Stuff that is not even usable

In order to apply the 5 Ss, you have to sort through your stuff and segregate it into these buckets.

Step 2: Straighten Up

Put stuff that is used all the time close by where it can be accessed easily. Put stuff that is used occasionally farther back. Put stuff that you hardly ever use all the way in the back. Get rid of stuff that is not used at all or is not usable.

Take a look at your computer keyboard. The home position (the place where your fingers are supposed to stay while typing) has your index fingers over the F and J. Many keyboards put tabs on the F and J keys so you can find them by touch without looking down. With your hands in the home position, the letters that are used the most often in the English language are under or around your index, middle, and ring fingers. Letters that are used the least often are farther from the center. Your keyboard is organized to put the stuff you use the most often in the most accessible place.

Another example is in the organization of tools. Many mechanics and hobbyists organize their tools on pegboards to make them easy to find and easy to put away. There should always be a place for everything, and everything should be in its place.

Step 3: Sweep or Scrub

Clean up the place. Most companies call this housekeeping. A clean place to work is a lean place to work. One look at the pegboard in the following picture tells you that there are a couple of tools missing. They are out of place. Find them and put them where they belong. Otherwise, when you need them, you will not know where they are.

Step 4: Standardize

You see this occurring over and over again in the field of quality. Chapter 12 was devoted to the discussion of standards. Many process technicians will work in companies that operate on rotating shifts, which means there will be three or four people working the

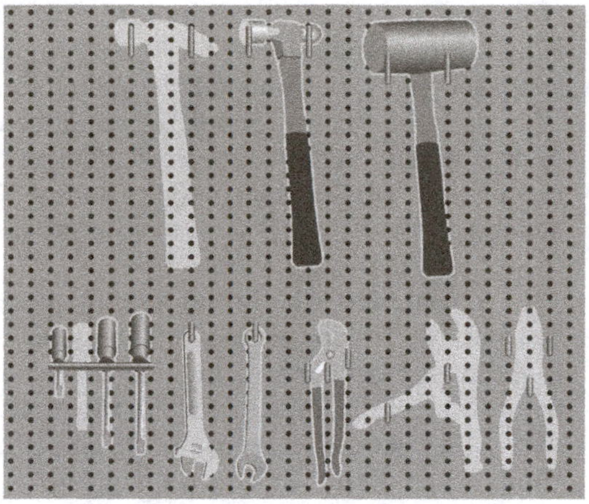

same job in the same location using the same facilities during different hours of the day. If everybody has a different way of doing things, everybody will be wasting time and energy looking for procedures, looking for equipment, or undoing something somebody else has done. If everybody does things the same way, then all of the procedures, tools, and equipment are always in the right place and easy to use.

Step 5: Self-Control

Discipline. The goal here is to maintain the gains. Having a pretty painted pegboard for your tools does not do any good if you never put the tools back on the board. If you have a team of four process technicians that work on rotating shifts and three of them are disciplined but one is not, the whole organization suffers. The person on the shift following the undisciplined technician will be frustrated that things are not where they belong and the place is a mess. You do not want to work with the undisciplined technician, so you should not want to be one.

Just-In-Time and Kanban

The traditional logistics process in America today is a push system. Teams of commercial and marketing experts forecast what they believe sales will be in the coming months, and production is scheduled to meet these forecasts. These forecasts are used to push the production schedule. In order to ensure that the company can meet this projected demand, product is made in advance of the orders and inventory is stockpiled.

As you already have learned, inventory is waste, and the whole idea of a lean organization is one without waste. In a lean organization you would wait for the customer actually to need the product and pull the product from you. You would make the product as needed, and there would be no stockpiled inventory. The product would be made "just-in-time" to meet the customers' needs. **Just-in-time (JIT)** is the process in which the company creates a product for the customer when it needs it.

The processing industries have a little trouble implementing JIT production because of the lead time often needed to produce the products, but it can be done. There is a major chemical manufacturer that makes a specialty catalyst product used around the world. A typical batch of product takes 4 to 6 weeks to manufacture, and another 2 weeks to blend, package, and test.

In order to manufacture this product, the company must order a custom-produced raw material from another state. This raw material takes 4 to 6 weeks to manufacture. To produce this raw material, the supplier must order base materials that can take 6 weeks to prepare. Add up all the lead time, and you see that from the time the customer tells the company it needs product, it will be 4 to 5 months before that product is ready to ship. Depending on where the customer is located, shipping time can be another month.

Amazingly, this particular organization does not carry any stock inventory. The customers and the producing plant have put in place a system of communications and reviews that ensures that orders are placed with at least 6 to 8 months of lead time. This is just-in-time manufacturing. The customers pull the product out of the organization when they need it, with minimal wasted time and inventory.

In some writings about JIT manufacturing, you actually see the concept referred to as *Kanban,* the Japanese word for card or sign. **Kanban** is a signal card used to pull production. The technical application of Kanban to JIT manufacturing is as follows: When you have several steps in your process, your product must flow from step to step to step within the processes.

Kanbans are the inventory management cards that are used to signal the previous step in the process that more product is needed. Literally, they are the signal used to pull inventory forward into the next step of the process. As outlined above, JIT has been successfully implemented in the process industry. There may be examples of Kanban cards' being used successfully in the process industries as well, but we have not come across them. We have seen examples of Kanban cards used in the process industries, but not in a way that added any value to the operation. Still, they are mentioned here because you may be part of an organization that uses them.

A Few More Japanese Words

In the introduction to this chapter, it was noted that the discussion would be outlining only a few of the many beliefs of lean manufacturing and that the study of Japanese vocabulary would not be the focus of this lesson. However, in order to ensure that you are fully prepared to take your place in a lean company, there are a few more Japanese words that you should at least see explained.

- Andon—a signal to alert people to a problem. A red light that flashes when you pull the fire alarm is an andon. The "check engine" light in your car is another example of an andon.

- Jidoka—making automated machines smart enough to stop the process when something is wrong. Many processes now have sensors that stop the process when the process is not flowing properly, thus preventing the process from making defects. Sometimes individuals that made the mistake of drinking and driving have to have a Breathalyzer installed in their car. This machine reads their blood alcohol content and does not allow the car to start unless they are under the limit. This is a jidoka device.

- Poka-yoke—a device or system that prevents defects. Essentially this means error proofing the process. You cannot accidentally introduce a diesel-dispensing nozzle into your unleaded gas tank because of the size difference. The process has been error proofed.

- Sensei—master or teacher. Just as the Six Sigma process refers to practitioners as green belts and black belts, lean leaders are called sensei.

Summary

Now you are familiar with lean. Lean means there is no fat or waste in your organization. You can categorize waste into three categories:

1. Traditional waste
2. Variability
3. Overburden

Traditional waste can be any of seven different types: transport, waiting, overproduction, defects, inventory, motion, or excess processing.

The simplicity of the lean tools and the ability to implement these tools at the worker level have contributed to the growth in popularity of lean in the process industries in the last few years.

Some of the key process improvement methodologies that you learned about include value stream maps, Kaizen, the 5 Ss, and just-in-time manufacturing.

The value stream map, or VSM, is a graphical technique used to analyze the steps of the process to show which steps add value and which steps do not. You can use VSM to show where improvements are possible within your process, then apply the appropriate improvement tool to the process.

Kaizen is the Japanese word for continual improvement. Many of the statistical tools that you have learned throughout this textbook drive toward a breakthrough or step change improvement. Kaizen is the philosophy of making small, incremental improvements every day.

The 5 Ss are a simple way to accomplish the desired small, incremental improvements of the Kaizen philosophy. The 5s stand for the following:

1. Sort or segregate
2. Straighten up
3. Sweep or scrub
4. Standardize
5. Self-control

You also learned about just-in-time manufacturing (JIT), a mechanism that can be employed to reduce inventory, which is one of the seven types of traditional waste. In JIT manufacturing, the customer PULLS product when needed instead off forecasts' being used to PUSH production schedules.

Finally, it was decided that it was okay to talk about waste instead of muda, that you could agree to go out to where the action is instead of genchi genbutsu to the gemba, and that error proofing your process is just as good a term as *poka-yoke*. In short, you discovered that lean is a philosophy of excellence and that you can learn this philosophy without getting hung up on all the Japanese terminology that has been published, which too often serves only to make the implementation more difficult.

Checking Your Knowledge

1. Define the following key terms:
   a. 5 Ss
   b. Just-in-time (JIT)
   c. Kaizen
   d. Takt time
   e. Value stream map (VSM)
2. Lean manufacturing can best be described as:
   a. A philosophy of working without waste
   b. A plant with only skinny workers
   c. A plant with the bare minimum number of workers
   d. A philosophy of working with the minimum amount of equipment
3. What we call lean today was actually started in which automotive company?
   a. Ford
   b. General Motors
   c. Toyota
   d. Honda
4. Which of these is NOT a type of waste as defined in lean?
   a. Traditional waste
   b. Variation
   c. Overburden
   d. Overtime
5. List and describe the seven types of traditional waste:
   a. ______________________________
   b. ______________________________
   c. ______________________________
   d. ______________________________
   e. ______________________________
   f. ______________________________
   g. ______________________________
6. A technique for graphically analyzing your process to look for opportunities to reduce wasted time is:
   a. Just-in-time
   b. Value stream maps
   c. Kaizen
   d. Flowcharting

7. Takt time is:
   a. The rate of consumption to meet the customers' needs
   b. The time it takes to takt
   c. The total process time divided by the number of customer orders per month
   d. The ratio of non-value-added time to value-added time
8. To be considered a value-added process step, which criteria must be met? (Select all that apply)
   a. Customer willing to pay for the service
   b. Ensures compliance with government regulations
   c. Transforms the product
   d. Is free from defects
9. (*True or False*) All waste is unnecessary.
10. List and describe the 5 Ss:
    a. ______________________________
    b. ______________________________
    c. ______________________________
    d. ______________________________
    e. ______________________________
11. (*True or False*) Kaizen is a philosophy of small, incremental, continuous improvement.
12. (*True or False*) In order to successfully implement lean, you must use only Japanese words to describe your efforts.
13. (*True or False*) Just-in-time production means making the product when it is needed instead of making the product to put into inventory.
14. (*True or False*) Just-in-time production cannot be used in the processing industries.
15. (*True or False*) Some of the lean tools are so easy to use that they can be applied by anybody at any level in the organization.

Activities

1. Think back to the registration process for this class. Construct a value stream map of that process. What percentage of the total registration time was value added?
2. Take a look at some area of your home such as your workshop, your garage, or you closet. Can you see opportunities to apply the 5 Ss there? Describe what you would do in this area for each of the Ss.

CHAPTER 15

Quality Costs

"Only good things happen when planned; bad things happen on their own."

~CROSBY

Objectives

Upon completion of this chapter you will be able to:

- Explain the importance of expressing improvement opportunities in the language of managers—money.
- List and explain the four types of quality costs (internal failure, external failure, appraisal, and prevention), and provide examples of each.
- Explain what is meant by the term *price of nonconformance* (PONC), and explain how nonconformance can financially impact the organization.
- Describe the concept of the "hidden factory."

Key Terms

Appraisal costs—the cost of checking performed to attempt to catch the mistakes before they get out the door.
Cost of poor quality (COPQ)—another term commonly used in industry to describe the cost of poor quality, of failing to meet the customers' needs.
External failure costs—the cost of mistakes that made it out the door to customers.
Hidden factory—the portion of the factory that produces non-value-added product.
Internal failure costs—the cost of mistakes made that were caught before they got out the door.
Prevention costs—the costs of improving processes so that they do not have to be checked to catch failures because everything was done right the first time.
Price of nonconformance (PONC)—a term coined by Philip Crosby used to describe the costs of poor performance and failing to meet the customers' needs.

Introduction

Macroeconomics, microeconomics, economies of scale, P/E ratio, return on investment, EBIT, EBITDA, cash flow, debt to equity ratio—the language of finance can be daunting, filled with acronyms and unfamiliar terms. You could take an entire class in economics or even obtain a 4-year college degree in finance. The breadth of finance will not be covered in this chapter. Instead, the discussion of costs is purely from a quality vantage point.

This chapter covers the language of management, which is money, so you will understand where financial people are coming from. It examines four different types of quality costs (internal failure, external failure, appraisal, and prevention) and how these costs affect you as a process technician. The term *price of nonconformance* (PONC) will be defined, and the negative impact of PONC on business will be explained. Finally, the concept of the hidden factory will be discussed.

Money: The Language of Management

Drive quality and the profits will follow. This theme is evident throughout the quality literature. Unfortunately, many managers have little quality training. Unless the customer complaints result in returned product or lost business, management may not equate quality problems to profits. In some ways, adopting a quality business attitude is a faith-based effort.

Rule Number 14:

The brutal truth is still the truth.

You do the right thing because it is the right thing to do and trust that the results will eventually justify your decision. Sometimes you see managers with this attitude. Sometimes you do not. Over the course of your career, you will undoubtedly work for some of each kind of manager.

There will be times when you know about an improvement opportunity that will make your process better, but management will not implement the project because there is no guaranteed return on its investment. This is reality. Sometimes that decision is tough to swallow, but as one grizzled old manager used to say, "The brutal truth is still the truth."

At least part of a quality effort has to be an honest evaluation of the tangible benefits an improvement might bring. Let us say your process runs 99% purity now. You and your coworkers have an idea that will increase the purity of your product to 99.5%. If this idea does not cost any money, then management will be pleased to support you. If it does cost money, then management can be expected to question the value of the improvement to the business.

Will this change in the process make your product worth more money to your customers, or are they perfectly content with the product as it is now? Will the change in the process result in a reduction of raw materials, reduced run time, less maintenance, or in some other way reduce costs and save the company hard dollars? Once the benefits are understood, you can estimate how long it will take to recoup the investment.

Let us say your improvement idea costs $5 million to implement. If you can show management how this project will save the company $10 million next year and every year after that, there is not a manager around who would not be supportive of the project. On the other hand, your idea may cost only $5,000, but if the time to recoup this investment is 10 years or more, do not expect management individuals to get too excited about it.

Companies do not operate with unlimited funds. A big part of management's job is prioritizing what to do with the resources at its disposal, and that includes spending the budget wisely. The improvement idea may be great, and from a purely philosophical viewpoint the company should do it because it is the right thing to do; but if you cannot express the value of the project to managers in their language (money) and show how this great idea will benefit the bottom line, the project is doomed never to get off the ground.

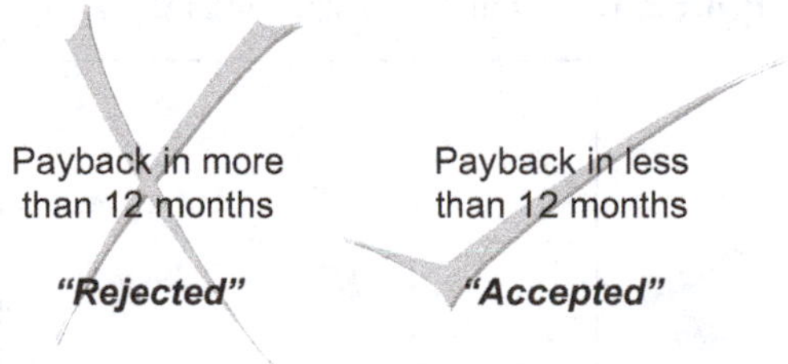

The hardest part of this concept is the element of strategic importance. If the idea is great, and it is the right thing to do, then why cannot management see that and just make it happen? Some managers are visionaries. They will see the long-term benefit and perhaps support the project even if there is no guaranteed return. Many are not so visionary and manage "by the numbers." In their case, if the numbers do not justify the project, the project will not get supported. No critical thinking skills are required. If your choices are that clear-cut, then managing is easy.

Approving funding for projects will not likely be part of your role as a process technician, but constantly working to improve your process is part of your role. Being able to express your ideas in language that managers understand can help you get your ideas implemented. Knowing how managers think will help you see the improvement opportunities from their perspective. To the extent that the improvements will save money, will make money, or can be accomplished without money, you will make a lot of progress. Be prepared to talk dollars when you talk about improvement opportunities.

Types of Quality Costs

Let us take a look now at the most traditional of views regarding quality costs. The categorization of quality costs into four distinct buckets began with Dr. Juran's *Quality Control Handbook* in 1951 and has been covered in quality literature by many of the quality greats through the years. In other words, this concept has been around for a long time.

You work in an imperfect company and mistakes happen. You have your laboratory checking your product before it goes out the door so most of the mistakes get caught, but every now and then a bad batch slips through. When a bad batch gets past you, you have to fix the problem with your customers. Knowing that your performance is not perfect, you work on making it better. All of the costs associated with making a quality product (or conversely failing to make a quality product) are referred to as quality costs. Quality costs are categorized as follows:

- **Internal failure costs**—the cost of mistakes made that were caught before they got out the door.
- **External failure costs**—the cost of mistakes that made it out the door to customers.
- **Appraisal costs**—the cost of checking performed to attempt to catch the mistakes before they get out the door.
- **Prevention costs**—the costs of improving your processes so that they do not have to be checked to catch failures because everything was done right the first time.

When you first launch your quality program, your quality costs might look like the graph shown in Figure 15-1.

In the beginning, you see a high rate of failures, both internal and external. Clearly, something must be done. But what? Costs are already out of control, so how can you fund an improvement effort? Your quality manager takes a stab at the problem by instituting a comprehensive system of checks and balances and double checks

FIGURE 15-1 Quality Costs—Initial View

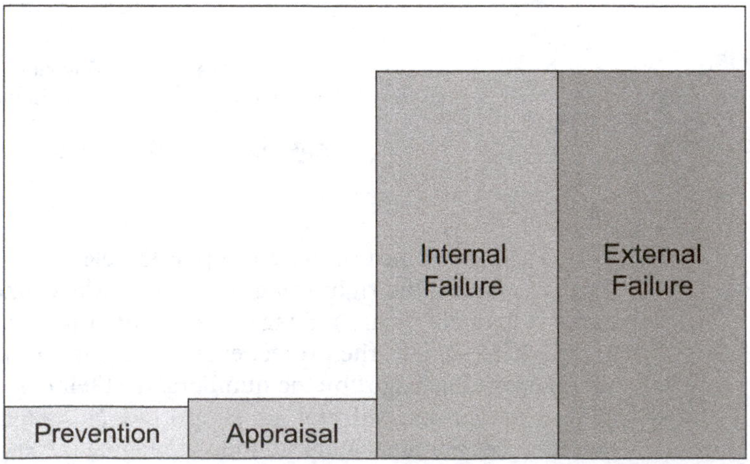

FIGURE 15-2 Quality Costs—Phase II

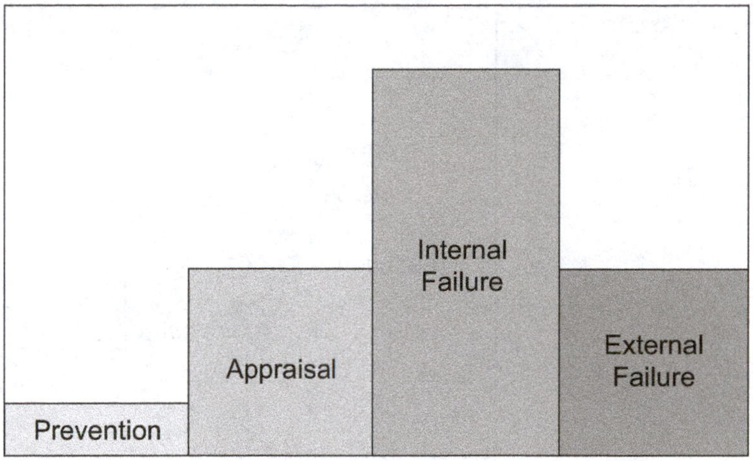

and more balances to make sure problems in the plant do not get out to the customer. In just a little while you see a new quality cost report that might look like the one shown in Figure 15-2.

You have succeeded in driving down the external failures by increasing your appraisal practices. Unfortunately, a closer look indicates that you have not lowered your overall quality costs; all you have done is to shift them from external failure to appraisal. It is in the direction of goodness, but still not where you want to be. Progress has been made. Now what do you do? Your quality manager has gone off to training and comes back with a lot of great ideas. With the support of management, and full participation from the technical staff and process technicians, your revitalized quality manager launches a vigorous quality improvement initiative that includes the following:

- Capability studies to understand the process variation and the significance of that variation on the needs of the customer
- Control charts throughout the process to help individuals understand when that variation is normal and when to react
- Designed experiments to optimize your now statistically stable processes
- Six Sigma teams to drive breakthrough improvements in the existing processes
- Standardized management systems based on the ISO 9001 standard to drive consistency into every aspect of your operation
- Quality reliability planning studies to ensure that newly developed products and processes are directly linked to the needs of the customer and all quality issues are understood and addressed in advance of implementation
- Root cause analysis tools to ensure that every problem is solved at the source and prevented from recurring
- Lean tools implemented throughout the organization to ensure that waste is minimized, costs are driven down, and everybody is engaged in the quality transformation
- Quality circles implemented throughout the manufacturing organization to ensure the individuals on the front lines are intimately involved in the quality process

The quality initiative costs a little bit to implement, but the results show that the money is well spent. Your prevention costs go up, but the total quality costs are down versus your baseline. Phase III (shown in Figure 15-3) is a huge success.

In fact, after just a few short years of effective implementation, your quality initiative results in reduced failure rates that lead you to cut back on appraisals altogether. Now you have figured out how to make it right the first time, so you do not have to spend so much time analyzing it after the fact. You know where to run your processes, and you know the expected result from running your process at the optimum levels. You

FIGURE 15-3 Quality Costs—Phase III

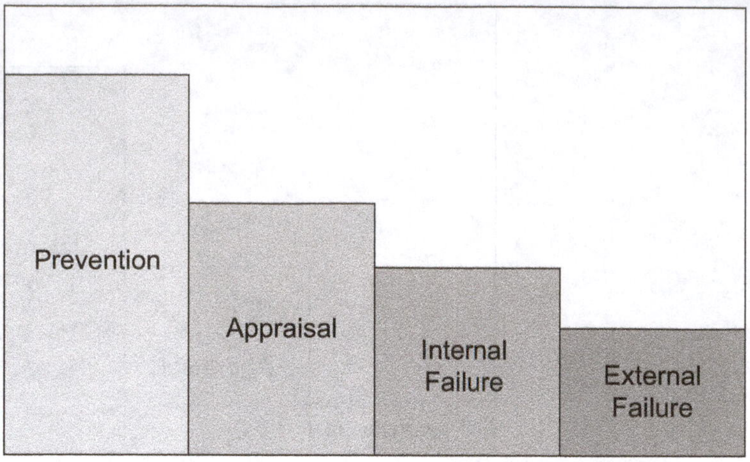

have tools in place to signal when the process is behaving in an abnormal fashion so you can fix the problem when it occurs. As a result, your quality costs continue to decrease (shown in Figure 15-4).

By now the concept of the various types of quality costs should be pretty obvious, but for the sake of completeness each type is defined here with examples.

Prevention costs:

- Quality planning
- Capability studies
- Quality improvement projects
- Quality training

Appraisal costs:

- Raw material testing
- Final product inspection and testing
- Audits
- Equipment calibration

Internal failure costs:

- Cost of scrapping the product
- Cost of reworking the product
- Cost of retesting the reworked product
- Delta cost of selling inferior grade product

FIGURE 15-4 Quality Costs—Phase IV

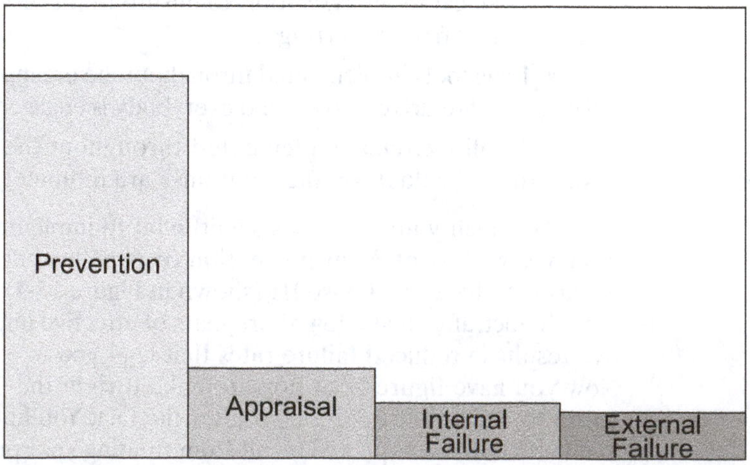

External failure costs:
- Customer claims
- Product returns
- Product recalls
- Costs of investigating customer complaints

Price of Nonconformance

If you were to add up all the various failure costs that have been listed, both internal and external failures, you would be able to calculate the total cost of not doing it right the first time. Together these costs are referred to as the **price of nonconformance (PONC),** a term coined by Philip Crosby used to describe the costs of poor performance and failing to meet the customers' needs. Depending on where you work and which quality guru your particular business unit follows, you may also hear this referred to as the **cost of poor quality (COPQ).**

In addition to the obvious costs included in this calculation, there may also be less tangible but no less real costs to the business in terms of lost business opportunity, lost customers, reduced customer trust, and so on.

It is difficult to calculate just how much damage can be done by sending your customer a bad batch of product. The cost of replacing the product is easy to calculate. The claim paid to a customer because your product shut down its system or produced bad product in its process is easy to calculate. The cost to the business relationship is much more difficult to quantify.

If you keep your customers happy by meeting and exceeding their expectations, they will have no reason to check out the competition. The ultimate price of nonconformance is a business that cannot stay in business. If you give your customers a reason to go elsewhere, they will.

The Hidden Factory

Internal failure costs are just as bad in the long run as external failure costs. The preceding section covered the real and obvious price paid when you send bad product out to your customers, resulting in hard dollar expenses as well as intangible business consequences. What about the costs of the internal failures?

Every minute you spend making bad product is a wasted minute. Every minute you spend blending, reworking, retesting, and trying to deal with the bad product you made is another wasted minute. The cost of the utilities needed to reprocess the bad product is wasted. The cost of the lab materials used to retest the bad product is all wasted. In fact, even the time spent tracking bad product is wasted time. Make it all correctly the first time, and none of these expenses have to be incurred. From a customer's perspective, all these wasted resources and all this wasted time add up to a more expensive product for the customer.

It is as though you have two different factories running, one within the other. You have one factory that makes good-quality, salable product. You have another, hopefully smaller, factory that makes junk. This can be called the **hidden factory,** the portion of the factory that produces non-value-added product. It costs money to keep your hidden factory running. Some of the money that your factory makes is diverted to cover the cost of your hidden factory.

It is not uncommon in some manufacturing plants to see hidden factory costs amounting to 20% of the cost of sales. Think about that. If you can reduce the costs of your hidden factory by reducing waste and reducing nonconforming product, you can make the same profit with a 20% reduction in your pricing structure.

If the market is extremely competitive, this means you can beat out your competition. If you have a great market position and offer an advantaged product, you might be able to keep the prices up and pocket the additional profits. Either way, the company wins. If the company wins, you get to keep your job and ultimately the employees win too.

Summary

So let us wrap up this chapter on quality costs. As much as it would be wonderful to exhort all businesses to work just on the quality issues and let the profits come, this concept is not practical in the real world. Managers speak their own language, the language of money. If you want them to listen to you, you have to speak their language, because they will not necessarily learn to speak yours.

When you can show management the value of your work in quality, it will support it without hesitation. To do this you need to be able to describe the value of your projects in terms of payback to the company.

Golden Rule of Management:
He who has the gold makes the rules!
$$$

You need to be able to quantify the benefits of reduced failures when you increase your efforts to prevent failures in the first place. When you can show management the true cost of poor quality or the price of nonconformance, and show how much potential profit is being wasted in support of your hidden factory, then you can get management's attention and, more importantly, its support.

Checking Your Knowledge

1. Define the following key terms:
   a. Appraisal costs
   b. Cost of poor quality (COPQ)
   c. External failure costs
   d. Hidden factory
   e. Internal failure costs
   f. Prevention costs
   g. Price of nonconformance (PONC)
2. The language of managers is:
   a. Productivity
   b. English
   c. Money
   d. Greek to most of us
3. To get quality projects funded, you have to:
   a. Fill out a project authorization form
   b. Be willing to lead the project
   c. Convince management that quality is a worthwhile goal
   d. Describe the value of the project to managers in terms they understand
4. Which of these costs is not a traditional quality cost?
   a. Cost of capital
   b. Internal failure costs
   c. External failure costs
   d. Appraisal costs
   e. Prevention costs
5. How might one work toward reducing total quality costs?
   a. Implement a standardized quality management system
   b. Implement Six Sigma
   c. Implement variability reduction programs (control charting, etc.)
   d. All of the above
6. Which of these graphs best represents where you want to be in terms of quality costs?

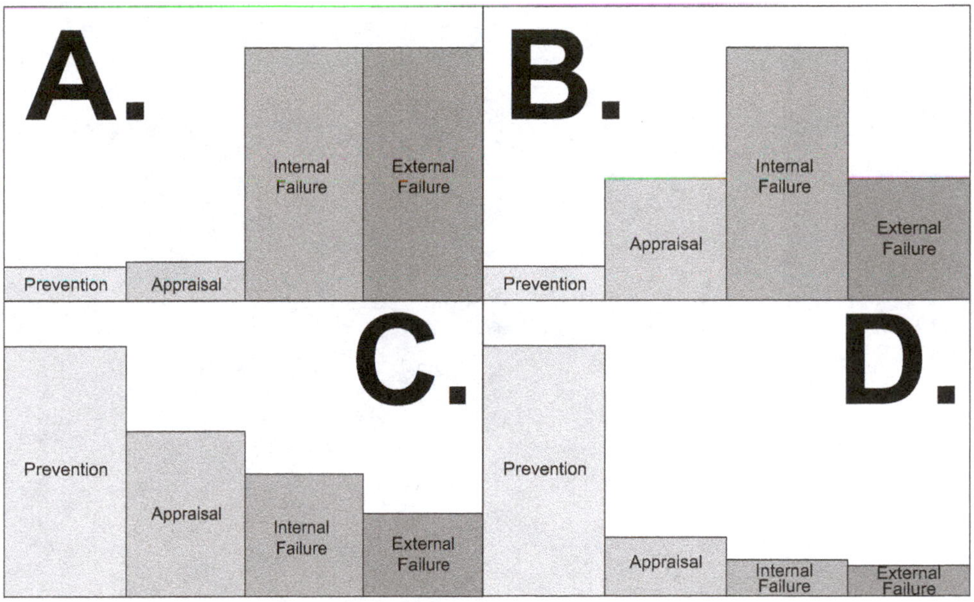

7. Prevention costs include which of the following?
   a. Costs of reworking bad product
   b. Costs of paying customer claims
   c. Costs of quality improvement projects
   d. Costs of testing product
8. The "hidden factory" is:
   a. The portion of your plant that makes poor-quality product
   b. The part of your plant that you choose not to show customers
   c. The portion of your plant that is not painted
   d. The part of your plant that is behind the fence from the public

9. (*True or False*) Companies will invariably implement quality programs based on the philosophical belief that quality is good and then just sit back, expecting the profits to follow.
10. (*True or False*) A basic principle of quality costs is that increasing spending in prevention will ultimately reduce total costs by reducing failure costs.
11. (*True or False*) Quality costs and profits are the responsibility of managers and do not apply to the role of the process technician.
12. (*True or False*) Failure costs are an issue only when the failure occurs at the customer's location.
13. (*True or False*) To work on quality costs effectively requires a degree in economics or finance.
14. (*True or False*) The ultimate price of nonconformance is the inability of a business to stay in business.
15. (*True or False*) Appraisal costs include the costs of conducting quality audits.

Activities

1. Why is it important to assess quality costs?
2. List and explain ways to reduce internal failure as described in the text.

CHAPTER 16

Putting the Puzzle Together

"The intelligence of the team exceeds the intelligence of the individuals in the team."
~SENGE

Putting the Puzzle Pieces Together

You have come a long way in your quality journey. When you started down this pathway, you did not know which way to go and were not sure how you would get there.

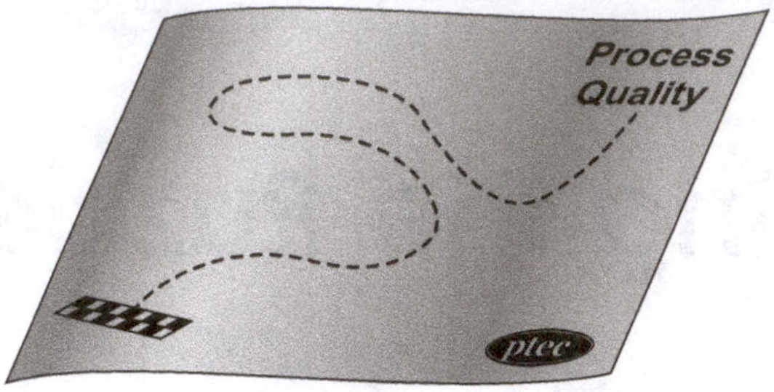

Now you have a firm foundation regarding the history and importance of the quality movement in the United States, as well as a working familiarity with the basic quality tools employed.

You now understand that the topics covered in the various chapters of this book all tie together neatly and point toward a single goal—to make a consistent quality product that meets or exceeds the needs of your customers so that they continue to purchase your products, keep your company in business, and keep you gainfully employed.

INTRODUCTION TO PROCESS QUALITY

In Chapter 1 you began your quality journey by learning a little of the history of the quality movement, starting out in Japan after World War II and gaining momentum in the United States in the 1970s and beyond. You gained an appreciation for how well established the quality concepts are today, with few, if any, industries that do not practice some form of quality improvement. The process industries are no exception.

Quality is well entrenched in the daily activities of the process industries, and process technicians play a vital role in the success of the quality effort. Indeed, without the participation of the process technicians, no quality effort can ever be truly successful.

Each of the recognized quality gurus—Deming, Juran, Crosby, Peters, and others—offered a different twist on how quality should be approached. By learning from each of them, you can implement a quality program that is above reproach.

Deming taught you the importance of applied statistics and the need to constantly improve everything you do. In his famous 14 points, he outlined many important leadership tenets that most companies would do well to revisit.

Juran showed you the importance of working in teams, even touting that all improvements come by way of teamwork and in no other way.

Crosby showed you the importance of having a quality attitude, an attitude of zero defects, accepting nothing short of perfection. We all know that being perfect is impossible, but by setting your goal at zero defects you drive the continuous improvement that both Deming and Juran told about.

Peters taught you the importance of championing quality and being passionate about quality and excellence. Having the right attitude about quality before you start means you are halfway there when you take your first step.

Wrapping up this introductory chapter, you learned that there are a myriad of quality tools to be employed and many ways to employ them. The goal of this textbook is to introduce you to the most commonly employed tools and help familiarize you with these tools so that when you see them again outside the classroom, you will be prepared to take your place in industry as a value-added member of the quality team.

VARIABILITY CONCEPTS

In Chapter 2 you took your first steps, laying the groundwork for much of what would come later. Some of the tools most commonly employed by process technicians in the process industry are the various forms of control charts. To understand control charts, you have to understand variation in your process and how to measure it. In this chapter how processes vary and why you should care were discussed in some detail.

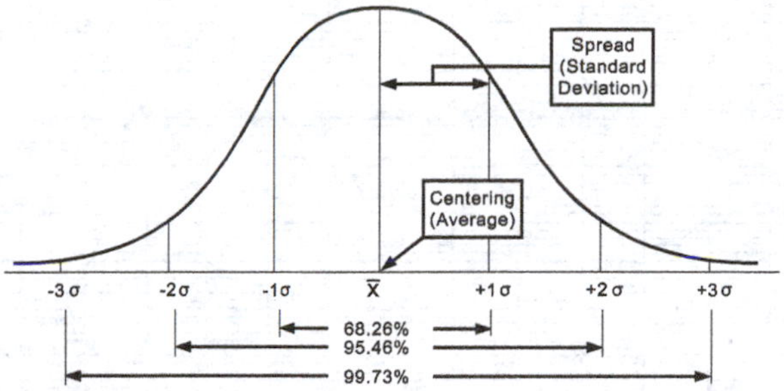

This chapter covered how to use the average (or mean) to measure the center of a process and the standard deviation (or sigma) to measure the spread of the process. It introduced the histogram as your first graphical tool to display the variation in your processes and began referring to this picture of your process as a distribution of data. You discovered that there are many types and shapes of distributions depending on the type of process that you have in place.

When your process is behaving normally, exhibiting only normal variation, you deem your process to be stable or in control. When your process is not behaving normally, but is exhibiting some kind of "extra" or abnormal variability, you deem your process to be instable or out of control; and the things causing this abnormal behavior are referred to as special causes.

PROCESS CAPABILITY

In Chapter 3 the concept of process variability was carried to the next level by introducing several process capability indexes such as Cpk, Cp, and Ppk, which provide a measure of relative goodness of your process compared with the specification limits established with your customers.

Process Capability Indexes
Rules of Thumb

<1.0 - Opportunity for Improvement
1.0-1.33 - Okay but borderline
>1.33 - Good - this is what we want

If your process variation is much less than what the customer desires, then your process is good. If your process variation is more than what the customer desires, you have an opportunity for improvement.

In general, you choose to grade your process variation using these indexes by saying any index below 1 is an opportunity for improvement, between 1 and 1.33 is borderline, and above 1.33 is good.

You learned about the natural variation that occurs in measurement processes as well, and took a quick look at nested designs, a mechanism for mathematically separating out the variation of your measurement process from the manufacturing process.

VARIABLES CONTROL CHARTS

In Chapter 4 you started building upon the concepts of variation with the variables control charts. These types of control charts are extremely common throughout the process industries, and you should expect to apply them in your jobs.

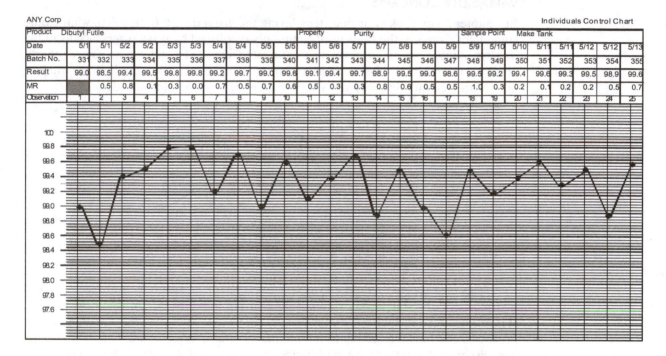

ANY Corp — Individuals Control Chart

Product: Dibutyl Futile — Property: Purity — Sample Point: Make Tank

Date	5/1	5/1	5/2	5/2	5/3	5/3	5/4	5/4	5/5	5/5	5/6	5/6
Batch No.	331	332	333	334	335	336	337	338	339	340	341	342
Result	99.0	98.5	99.4	99.5	99.8	99.8	99.2	99.7	99.0	99.6	99.1	99.4
MR		0.5	0.8	0.1	0.3	0.0	0.7	0.5	0.7	0.6	0.5	0.3
Observation	1	2	3	4	5	6	7	8	9	10	11	12

Date	5/7	5/7	5/8	5/8	5/9	5/9	5/10	5/10	5/11	5/11	5/12	5/12	5/13
Batch No.	343	344	345	346	347	348	349	350	351	352	353	354	355
Result	99.7	98.9	99.5	99.0	98.6	99.5	99.2	99.4	99.6	99.3	99.5	98.9	99.6
MR	0.3	0.8	0.6	0.5	0.5	1.0	0.3	0.2	0.1	0.2	0.2	0.5	0.7
Observation	13	14	15	16	17	18	19	20	21	22	23	24	25

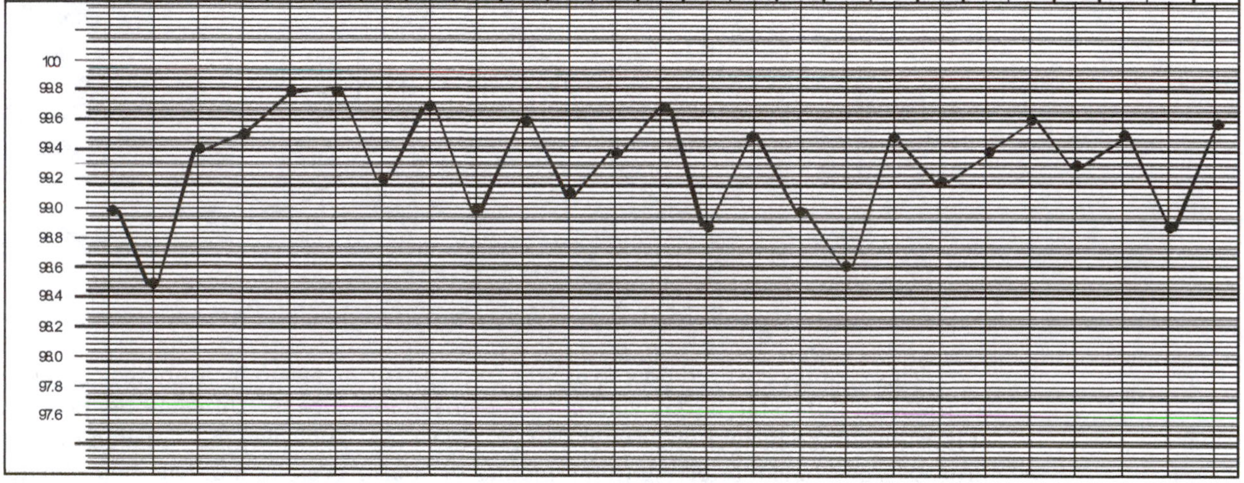

The two main control charts focused on in this category were the $\overline{X}R$ control charts and the individuals control charts, both of which are built upon the normal distribution of data. This chapter also introduced some of the other variables control charts that you may see in practice, such as zone charts, CUSUM charts, and EWMA charts, although the construction of these last two types was left for others to explain due to their complexity.

ATTRIBUTES CONTROL CHARTS

In Chapter 5 you continued the study of control charts by adding control charts from non-normal distributions such as the C and U charts, which are built on the Poisson distribution, and the P and NP charts, which are built on the binomial distribution. As you recall, the Poisson distribution is the distribution of counting data and is useful for counting of defects and is (or should be) employed extensively in the charting of safety and injury statistics. The binomial distribution is the distribution of defectives and is most commonly used when charting things with only two possible outcomes, such as pass/fail data.

Wrapping up this series of chapters on variation and how to measure, plot, and interpret that variation, you learned that regardless of what type of control chart you employ the interpretation rules are the same.

The three main rules are search for points outside the control limits, search for shifts up or down in the data, and search for trends up or down within the limits. Violations of the rules indicate that something abnormal is going on in your process, which gives you the opportunity to investigate and learn from the occurrence, ultimately improving your process.

ANY Corp

Attributes Charts (C) U P NP

Business Unit: Isopropyl soapyl (WorldwideOps) | Chart: Customer Complaints | Description: Minor & Major

Month/Yr	01-06	02-06	03-06	04-06	05-06	06-06	07-06	08-06	09-06	10-06	11-06	12-06	01-07
No. Defects	16	5	10	7	4	9	1	9	7	3	5	4	3
Sample Size													
No./Sample													
UCL	12.3	12.3	12.3	12.3	12.3	12.3	12.3	12.3	12.3	12.3	12.3	12.3	12.3
LCL	0	0	0	0	0	0	0	0	0	0	0	0	0
Observation	1	2	3	4	5	6	7	8	9	10	11	12	13

Month/Yr	02-07	03-07	04-07	05-07	06-07	07-07	08-07	09-07	10-07	11-07	12-07	01-08
No. Defects	6	4	4	9	2	2	2	5	2	4	4	5
Sample Size												
No./Sample												
UCL	12.3	12.3	12.3	12.3	12.3	12.3	12.3	12.3	12.3	12.3	12.3	12.3
LCL	0	0	0	0	0	0	0	0	0	0	0	0
Observation	14	15	16	17	18	19	20	21	22	23	24	25

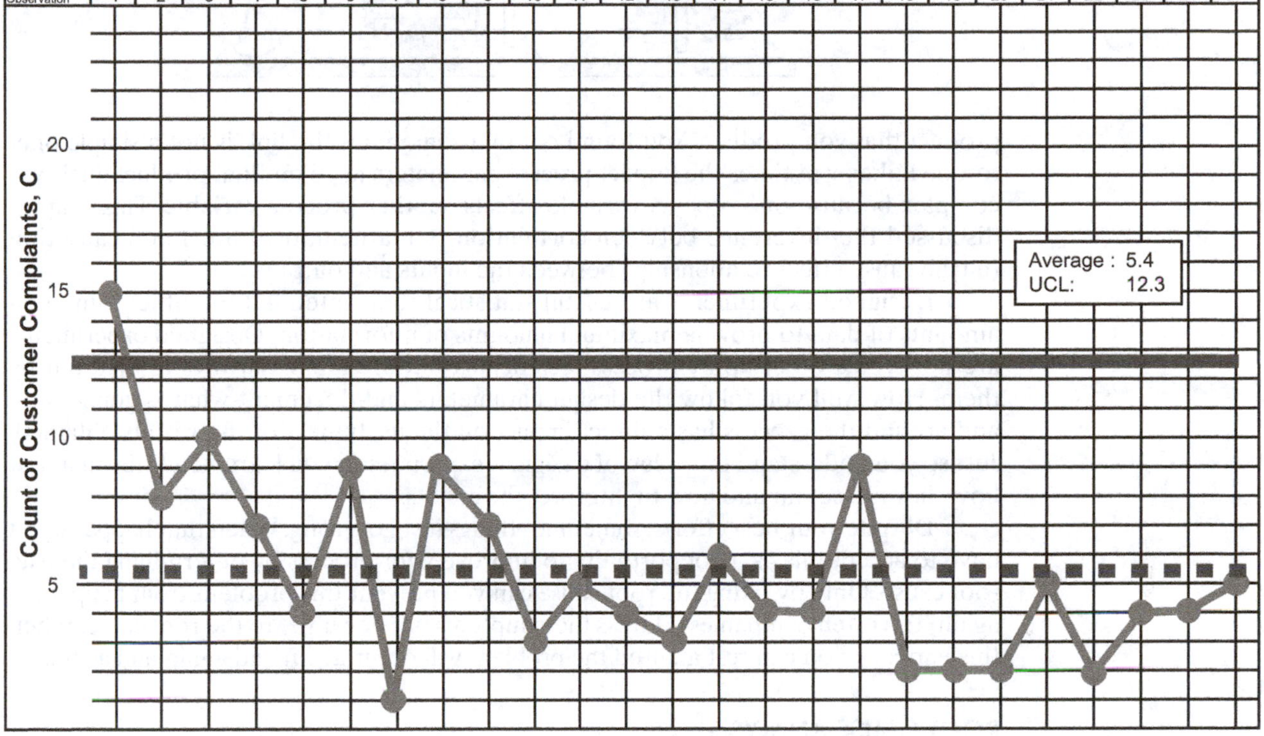

OTHER BASIC QUALITY TOOLS

These first few chapters form the foundation upon which much of the house of quality is built. In Chapter 6 the basic toolbox was rounded out with a collection of commonly applied quality tools such as the PDCA cycle, the fishbone diagram, flowcharts, and the Pareto chart, just to name a few.

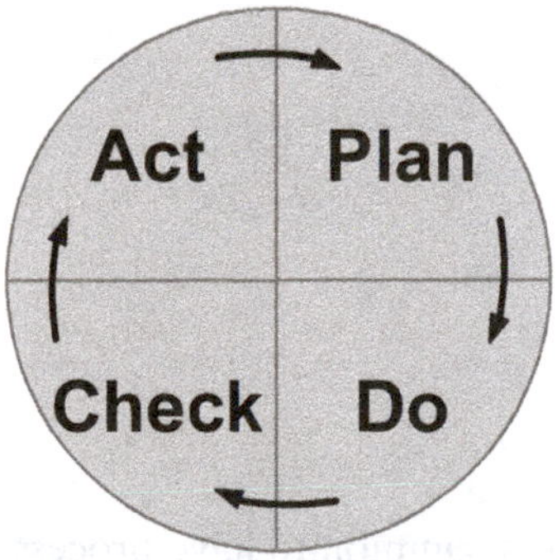

Some of these tools are meant to be applied independently, whereas others are used in a variety of quality improvement activities such as are outlined in other chapters. Each of them is a useful tool when applied in the appropriate way to the appropriate problem.

DESIGNED EXPERIMENTS

In Chapter 7 you jumped into the concept of modeling your process, establishing a mathematical link between the process parameters that you physically control and the

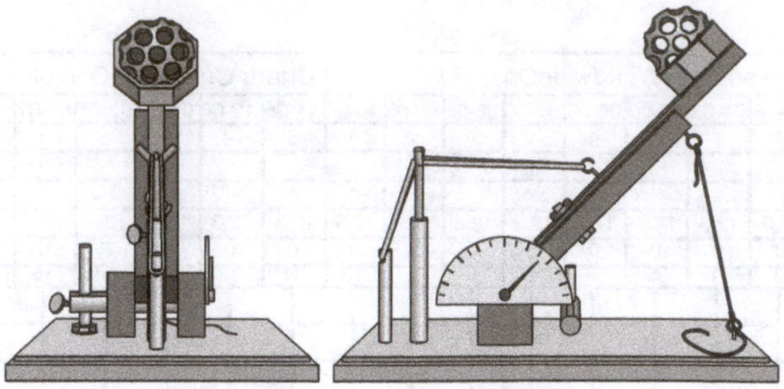

product that you produce. You found out that sometimes the link is not a simple one-to-one linkage; at times the impact process parameters have on final product quality is complex because one process variable affects another process variable. This chapter discussed the difference between correlation (a mathematical link) and causation (a true cause/effect relationship) between the inputs and outputs.

Designed experiments are useful statistical tools intended to gather minimum amounts of data to provide maximum amounts of information. Designed experiments are not typically set up by process technicians, but are almost always carried out by them. How well you follow the design parameters and document what is going on in and around the process has a direct impact on the usefulness of the results obtained. Just so you understand the value of designed experiments, you learned a little bit about how they are set up and how to interpret the results.

Despite your best efforts, sometimes things just go wrong. When that happens, you have to accept this as an opportunity to improve your process by determining the true root cause. Only by fixing the root cause can you prevent this problem from happening again. Too often companies address the symptoms but never get to the root cause. When this happens, you can rest assured the problem will occur again and again and again.

ROOT CAUSE ANALYSIS

In Chapter 8 several different styles of root cause analysis (RCA) were presented. All of these styles of root cause analysis are valid, and all are useful. Different companies use different styles, and it is important that you are able to support any type or hybrid of root cause analysis they throw at you.

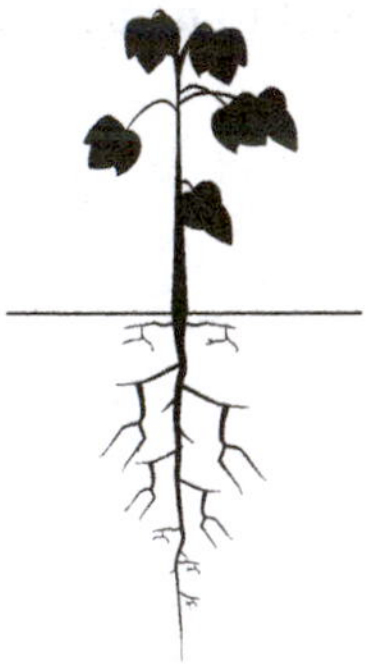

It is common to have process technicians participate on root cause investigation teams because they are the closest to the process. The three styles of root cause analysis most commonly used throughout the process industries include the Kepner-Tregoe, Five Whys, and Apollo methodologies, and this chapter briefly presented a couple of other RCA tools that you may see used.

CUSTOMER QUALITY

In Chapter 9 you were given a break from the technical side of quality to learn about a somewhat softer subject—customers. Too often process technicians are not included in

the customer service loop, even though they are the individuals with the hands-on experience with the product. Keeping the customer happy is paramount to the success of the business. Unhappy customers find another place to shop. Let that happen too many times and you have to find another place to work.

It is common for customers to conduct audits and surveys of their suppliers, often on site at your location. When this happens, you as a process technician become a market-facing employee. What you say, how you say it, and how you portray yourself and the company will have a direct impact on that customer's perception of your company. Prepare yourself for this important part of your job. It will not happen every day, but it will happen.

SIX SIGMA

In Chapter 10 what may well be the most commonly used approach to process improvement in the process industry today—Six Sigma—was described. Six Sigma really does pull together the vast majority of the various quality tools in your toolbox into a single improvement strategy. Six Sigma improvements are accomplished by the use of improvement teams.

These teams measure the extent of the problem to determine the value proposition. They then analyze the data using control charts, histograms, Pareto charts, and other tools to determine the root cause of the issue. These teams employ designed experiments and other techniques to improve upon the existing process and then develop control plans to ensure that these improvements last.

In many process industry plants, process technicians serve on Six Sigma teams, and some achieve certifications as green belts. Six Sigma is not an improvement tool; it is a formal, disciplined approach to continuous improvement that employs all of the quality tools in your toolbox. It is not a tool itself but an improvement strategy that can be applied to manufacturing processes as well as functional processes such as purchasing or shipping and logistics.

TEAMS

As you travel along your quality journey, note that you rarely travel alone. Much of the work of quality is accomplished in a team atmosphere. Quality circles are employed at the control room level to get the various shifts working together on quality issues. Root cause investigations are rarely conducted by a single person but rather are the result of a multifunctional team.

As just mentioned, Six Sigma efforts are almost always accomplished as a team activity. In Chapter 11 teams in general and several facets of working on teams were talked about. All teams go through growing pains. If you recognize the various stages of team building (forming, storming, norming, and performing), you can work through the stages and get the work done with a minimum of pain.

Several quality tools are available to help teams maintain their focus and function effectively. This chapter discussed the use of a charter, a meeting agenda, an action item register, Gantt charts, and RACI charts to get your teams focused on the task at hand and track your progress along the way. It even outlined some basic rules of good team etiquette that most of us could benefit from practicing on and off the job. Each team member has a role to play. If all team members know their roles and play them well, they can all succeed together.

MANAGEMENT SYSTEMS

In Chapter 12 an overview of the ISO 9001 Quality Management System standard was provided. This document describes the basic quality systems that need to function well in order for a company to manage quality well. Some of the elements of this standard—including Monitoring and Measurement of Processes, Control of Monitoring and Measuring Equipment, and Monitoring and Measurement of Product—are directly applicable to the role of the process technician.

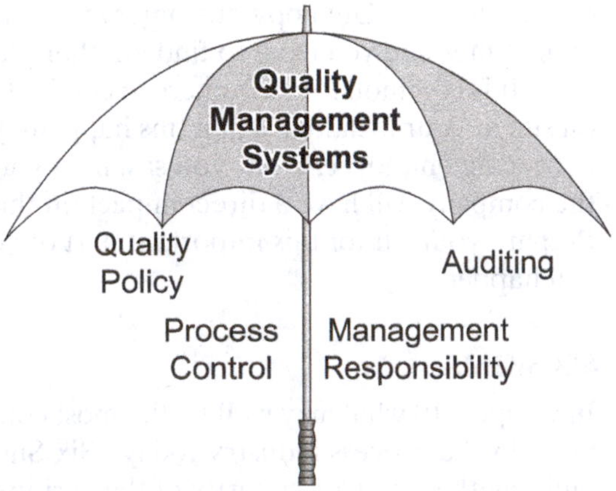

You will be expected to follow the policies and procedures of your company in order to comply with this standard. If your company is ISO 9001 registered, and many process industries companies are, then you will be an active participant in the process.

When the internal and external auditors come around, you will be audited. The auditors will ask you to explain what you do. Knowing what they are looking for will give you the edge in answering their questions. You may even be given the opportunity to serve as an auditor. If so, take the chance and go for it.

The Six Sigma teams talked about in Chapter 10 use many of the tools discussed in other chapters to accomplish their improvements. The control plans documented by these improvement teams often require the improvements to be integrated back into the daily operating instructions (standard operating procedures) in the plant. These documents form the foundation of the ISO 9001 quality management system at the plant—which is where the process technicians get their instructions for how to run the plant. The ISO 9001 documentation system in your plant is the umbrella under which many pieces of the quality puzzle reside.

QUALITY RELIABILITY PLANNING

An often-heard complaint about the quality process is that it is too often focused on problems after the fact. In Chapter 13 the topic of quality reliability planning was presented. This topic is a consolidation of three tools used commonly throughout the process industries under many names.

This process is proactive in nature in that it kicks into gear with a quality design plan before you ever make the first batch of product. In the quality design plan, you

work to document the needs of the customers and your plans to measure your own product to meet the customers' needs. The next phase of this process is the quality control plan in which you document how to control your process in order to achieve the quality necessary to meet the customers' needs. Finally, your process includes a failure mode and effects analysis (FMEA), which is a "what if" kind of tool. In the FMEA, you analyze the probability of failures, the impact of each failure, and your ability to detect each failure before the customer does.

By analyzing your process before something goes wrong, you can implement preventive actions. Although process technicians are not the most common candidates to lead an effort like this, they are most certainly valued participants in these types of efforts due to their hands-on knowledge of running the process. Who better to help the technical staff understand the likelihood of a particular failure than the person on the shop floor?

LEAN

Your collection of quality tools was rounded out in Chapter 14 with a set of improvement methodologies collectively referred to as lean. Most of these tools are applicable at the process technician level. Although the traditional implementation of lean is infused with Japanese buzzwords, this chapter concentrated on the more practical aspects of reducing cycle time, implementing the 5 Ss (a semiformal system for straightening up and organizing the workplace), and driving incremental continuous improvement into the workplace.

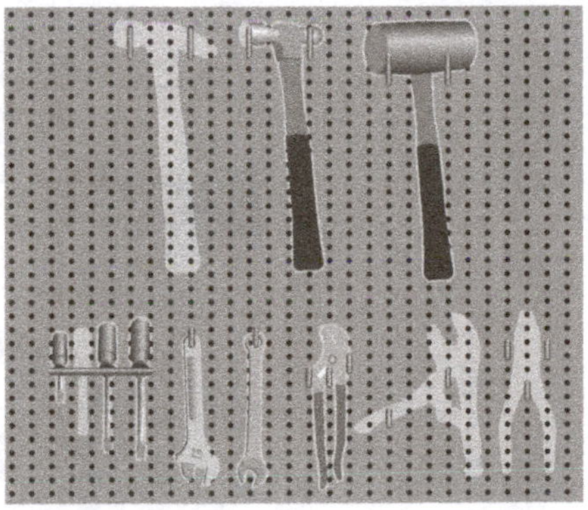

You may recall that some of these tools were as simple as making sure there is a place for everything and everything is in its place. Since the lean concepts have grown in popularity in recent years, you can rest assured you will see these things again.

QUALITY COSTS

Finally, you reached the last content chapter and the bottom line. In the business world the bottom line refers to the profit line on a balance sheet. In Chapter 15 money was

"$ £ ¥ €"
Money:
The Language
of Management

thus discussed from the standpoint of the quality process. Choosing to save the generic study of finance and economics for classes in those topics, this chapter concentrated on how quality impacts the bottom line.

The four types of quality costs and the negative impact that failures can have on the profitability of the company were covered. Also talked about was how describing your improvement opportunities in the language of managers helps you accomplish those opportunities.

There is a cost of poor quality, sometimes called the price of nonconformance. Everything you do in the quality process needs to be focused on improving quality performance so that you reduce waste, which reduces costs, which keeps costs down, profits up, and customers happy. Let us face it—we are in it for the money.

Summary

A lot of ground has been covered just in this one book, which is pretty amazing when you think about it because many of these chapters are the subjects of entire books themselves. What you should walk away with is this:

Quality
Straight Ahead

1. Quality is an important part of your role as a process technician.
2. There are many pieces to the quality puzzle, and these pieces all fit together to help you manage and improve your processes.
3. There are many tools available in your quality toolbox for many different tasks. Applying the right tool for the right job is part of your job.
4. This is not the end of your quality journey; it is just the beginning.

Now go out there and improve yourself and improve your process a little bit every day.

Appendix A

Blank Charts

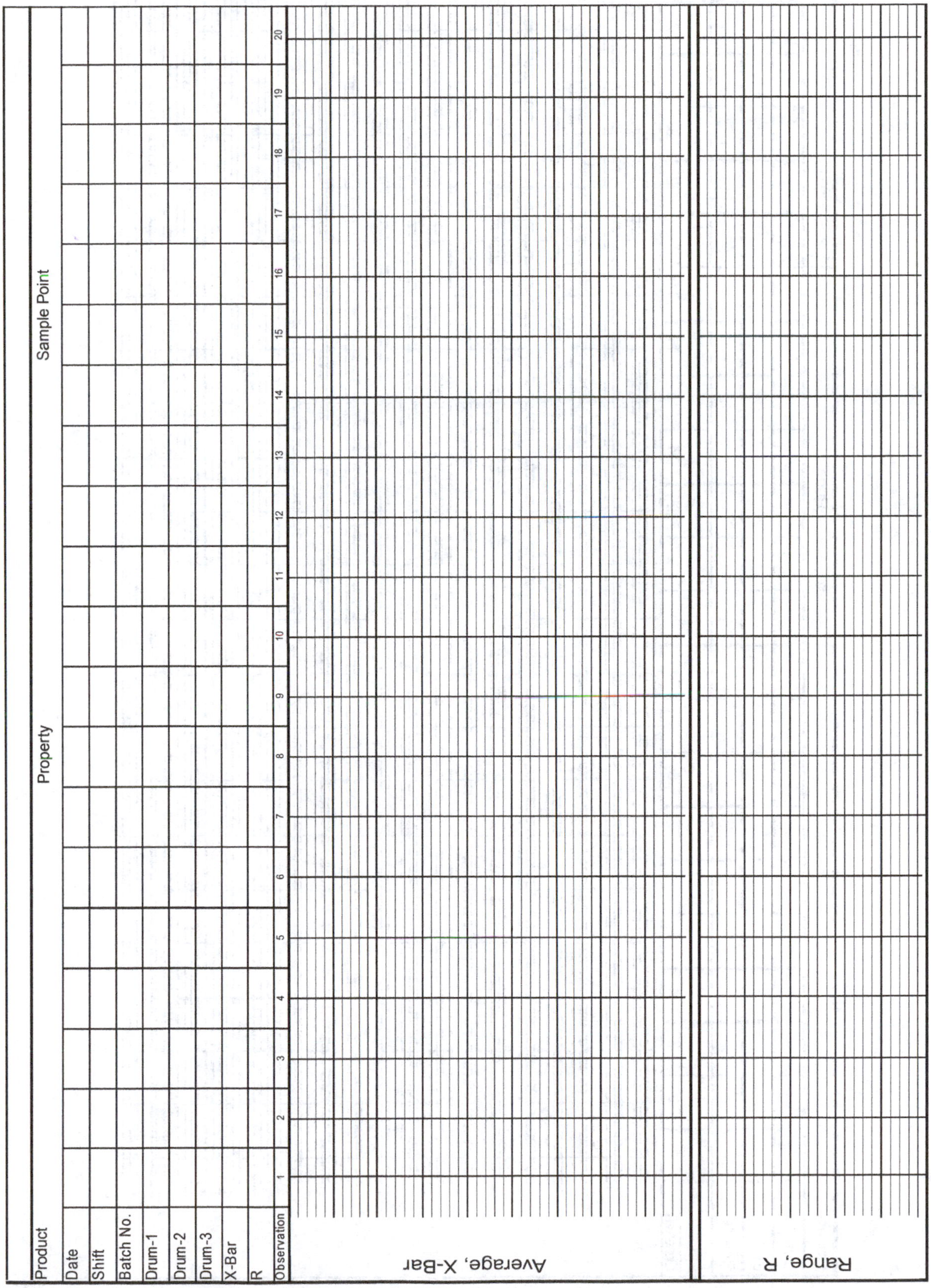

Individuals Control Chart

Product											Property Purity							Sample Point							
Date																									
Batch No.																									
Result																									
MR																									
Observation	1	2	3	4	5	6	7	8	9	10	11	12	13	14	15	16	17	18	19	20	21	22	23	24	25

Zone Chart

Product: Property: Sample Point:

Date																									
Batch No.																									
Results																									
Score																									
Observation	1	2	3	4	5	6	7	8	9	10	11	12	13	14	15	16	17	18	19	20	21	22	23	24	25

Zone	Score
Out of Control	4
Zone C	2
Zone B	1
Zone A	0
Zone A	0
Zone B	1
Zone C	2
Out of Control	4

Attributes Chart C U P NP

Business Unit																									
Month/Yr																									
No.																									
Sample Size																									
No/Sample																									
UCL																									
LCL																									
Observation	1	2	3	4	5	6	7	8	9	10	11	12	13	14	15	16	17	18	19	20	21	22	23	24	25

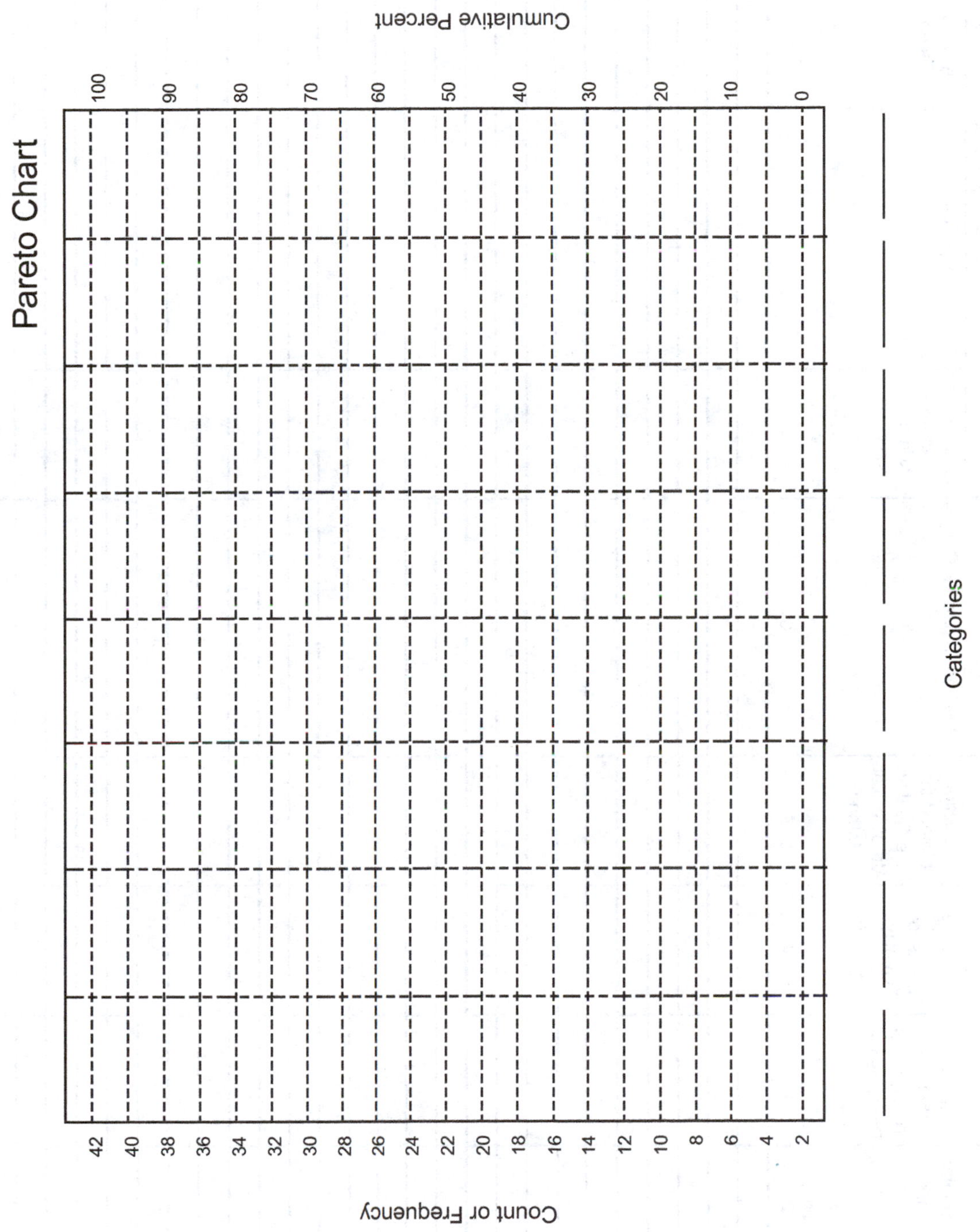
Pareto Chart
Cumulative Percent
100
90
80
70
60
50
40
30
20
10
0
Count or Frequency
42
40
38
36
34
32
30
28
26
24
22
20
18
16
14
12
10
8
6
4
2
Categories

Quality Design Plan for:__________________ Page ____of _____ Date: _____________

						If Method Capability Is Less Than 2.50:			
Customer Performance Attribute 1	Acceptable Limits (If Any Apply) 2	The Quality Attributes That Control Performance 3	Proposed Limits for Quality Attribute 4	Detection Method for the Quality Attribute (Lab Method) 5	Method Capability, Cm 6	Recommended Actions to Improve Precision 7	Resp. Party 8	Actions Taken 9	Date Completed 10

Quality Control Plan for:__________________ Page ____of _____ Date: _____________

Quality Attributes	Sample Plan				Specification Limits for Quality Attribute	Corrective Actions for Non-conformance (Outside Limits)	Controlling Process Parameters	Acceptable Limits for Process Parameters	Process Parameter Control Scheme	Corrective Actions for Non-conformance (Outside Limits)
	Sample Point	Freq.	No. Samples	No. Tests						
1	2	3	4	5	6	7	8	9	10	11

FMEA for:__________________ Page ____of _____ Date: _____________

			Existing Conditions						Resulting Conditions					
Give a Description of the Potential Failure 1	Effect on the Customer Due to This Potential Failure 2	List All Possible Causes for Each Potential Failure 3	O C C 4	S E V 5	D E T 6	R P N 7	Recommended Actions for RPNs >125 8	Resp. Party 9	Actions Taken 10	Date Completed 11	O C C 12	S E V 13	D E T 14	R P N 15

Quality Design Plan for: Date:

Customer Performance Attribute 1	Acceptable Limits (If Any Apply) 2	The Quality Attributes That Control Performance 3	Proposed Limits for Quality Attribute 4	Detection Method for the Quality Attribute (Lab Method) 5	Method Capability, Cm 6	If Method Capability Is Less Than 2.50: Recommended Actions to Improve Precision 7	Resp. Party 8	Actions Taken 9	Date Completed 10

Quality Control Plan for: Date:

Quality Attributes 1	Sample Plan: Sample Point 2	Freq. 3	No. Samples 4	No. Tests 5	Specification Limits for Quality Attribute 6	Corrective Actions for Non-conformance (Outside Limits) 7	Controlling Process Parameters 8	Acceptable Limits for Process Parameters 9	Process Parameter Control Scheme 10	Corrective Actions for Non-conformance (Outside Limits) 11

FMEA for: Date:

			Existing Conditions						Resulting Conditions					
Give a Description of the Potential Failure	Effect on the Customer Due to This Potential Failure	List All Possible Causes for Each Potential Failure	OCC	SEV	DET	RPN	Recommended Actions for RPNs >125	Resp. Party	Actions Taken	Date Completed	OCC	SEV	DET	RPN
1	2	3	4	5	6	7	8	9	10	11	12	13	14	15

Bibliography

Abbott, Wendell H. (1954). *Statistics can be fun.* Chesterfield, OH: Unicorn Grove.

ASQC Chemical and Process Industries Division. (1996). *ISO 9000 guidelines.* Milwaukee, WI: Quality Press.

ASQC Chemical and Process Industries Division. (1999). *Quality assurance for the chemical and process industries.* Milwaukee, WI: Quality Press.

ASQC Quality Audit Division. (1997). *The quality audit handbook*. Milwaukee, WI: Quality Press.

Aubrey, Charles A., III, & P. K. Felkins. (1988). *Teamwork: Involving people in quality and productivity improvement.* Milwaukee, WI: Quality Press.

Bennet, Carl A., & N. L. Franklin. (1954). *Statistical analysis in chemistry and the chemical industry*. Scituate, MA: John Wiley and Sons.

Box, G. E. P. (1979). *Robustness in the strategy of scientific model building*. New York: Academic Press.

Burr, John T. (1989). *SPC tools for operators.* Milwaukee, WI: Quality Press.

Campanella, Jack. (1990). *Principles of quality costs.* Milwaukee, WI: Quality Press.

Cano, Dean L. (1996). *Root cause analysis: Effective problem solving and beyond.* Friendswood, TX: Apollonian Publications.

Cornell, John A. (1984). *Volume 8: How to apply response surface methodology.* Milwaukee, WI: American Society for Quality.

Cornell, John A. (1990). *Volume 5: How to run mixture experiments for product quality*. Milwaukee, WI: American Society for Quality.

Crocker, Douglas C. (1985). *Volume 9: How to use regression analysis in quality control.* Milwaukee, WI: American Society for Quality.

Crosby, Philip. (1979). *Quality is free*. Columbus, OH: McGraw-Hill.

Crosby, Philip. (1984). *Quality without tears*. Columbus, OH: McGraw-Hill.

Deming, W. Edwards. (1986). *Out of the crisis*. Cambridge, MA: MIT Press.

Grant, Eugene L., & R. S. Leavenworth. (1988). *Statistical quality control.* New York: McGraw-Hill.

Guaspari, John. (1985). *I know it when I see it: A modern fable about quality*. New York: American Management Association.

Guaspari, John. (1988). *The customer connection: Quality for the rest of us.* New York: American Management Association.

Gunst, Richard F., & R. L. Mason. (1991). *Volume 14: How to construct fractional factorial experiments.* Milwaukee, WI: American Society for Quality.

Harris, R. Lee. (1991). *The customer is king!* Milwaukee, WI: Quality Press.

Hinton, Tom, & W. Schaeffer. (1994). *Customer focused quality: What to do on Monday morning*. Englewood Cliffs, NJ: Prentice Hall.

Imai, Masaaki. (1986). *Kaizen*. New York: McGraw-Hill.

Ishikawa, Kaoru. (1986). *Guide to quality control*. Shelton, CT: Productivity, Inc.

Juran, Joseph M. (1989). *Leadership for quality, an executive handbook*. New York: Free Press.

Juran, Joseph M., & F. M. Gryna. (1951). *Juran's quality control handbook*. Columbus, OH: McGraw-Hill.

Lamar, Jim, & P. Murphy. (2001). *Quality management in a wellness program.* Lincoln, NE: Writer's Club Press.

McNair, Carol J., W. Mosconi, & T. F. Norris. (1989). *Beyond the bottom line: Measuring world class performance*. Homewood, IL: Business One Irwin.

The memory jogger. (1987). Salem, NH: Goal/QPC.

Mills, Charles A. (1989). *The quality audit*. Milwaukee, WI: Quality Press.

Montgomery, Douglas C. (1984). *Design and analysis of experiments.* New York: John Wiley and Sons.

Peters, Thomas J., & R. H. Waterman, Jr. (1982). *In search of excellence: Lessons from America's best run companies.* New York: Harper & Row.

Peterson, Jim, & R. Smith. (1998). *The 5 S pocket guide.* New York: Productivity Press.

Problem solving and decision making. (1980). Princeton, NJ: Kepner Tregoe.

Process quality management & improvement guidelines. (1987). AT&T.

Pyzkek, Thomas. (1988). *What every engineer should know about quality control.* New York: Marcel Dekker.

Rummler, Geary A., & A. P. Brache. (1991). *Improving performance: How to manage the white space on the organization chart.* San Francisco: Jossey-Bass.

Rust, Roland T., A. J. Zahorik, & T. L. Keiningham. (1994). *Return on quality.* Chicago: Probus Publishing.

Sayer, Natalie J., & B. Williams. (2007). *Lean for dummies.* Hoboken, NJ: Wiley Publishing.

Scholtes, Peter R. (1988). *The team handbook: How to use teams to improve quality.* Madison, WI: Joiner Associates.

Schrock, Edward M. (1950). *Quality control and statistical methods.* New York: Reinhold Publishing.

Sewell, Carl, & P. B. Brown. (1992). *Customers for life: How to turn that one-time buyer into a lifetime customer.* New York: Doubleday Currency.

Shewhart, Walter A. (1986). *Statistical method from the viewpoint of quality control.* Mineola, NY: Dover Publications.

Statistical quality control handbook. (1956). Western Electric Co., Inc.

Tapping, Don. (2003). *The lean pocket guide.* Cedar Rapids, IA: MCS Media, Inc.

The team memory jogger. (1995). Salem, NH: Goal/QPC & Oriel Incorporated.

Trutna, Ledi. (1990). *Teaching statistics using the Roman catapult.* Austin, TX: Texas Instruments.

Wellins, Richard S., W. C. Byham, & J. M. Wilson. (1991). *Empowered teams: Creating self-directed work groups that improve quality, productivity, and participation.* San Francisco: Jossey-Bass.

Wheeler, Donald J., & D. S. Chambers. (1986). *Understanding statistical process control.* Knoxville, TN: Statistical Process Controls.

Glossary

5 Ss A floor-level improvement methodology taken from five Japanese words: Seiri, Seiton, Seiso, Seiketsu, and Shitsuke.

Analysis of variance (ANOVA) A statistical tool that helps you understand the nature of variation in your process.

Appraisal costs The cost of checking performed to attempt to catch the mistakes before they get out the door.

Attributes data Data from a discrete distribution wherein only whole numbers (counts) are possible.

Audit A review of the management system to see whether the management system is effectively implemented.

Auditee A person being audited.

Auditor A person conducting an audit.

Binomial distribution Distribution of discrete data that are made up of data that have only two choices (e.g., pass/fail).

Black belt Team leader on a Six Sigma team that has training in the basic and some advanced quality improvement tools plus training in project management.

Brainstorming A group exercise designed to solicit a large number of ideas in a short amount of time.

C chart Control chart for counts of defects (Poisson distribution) with a constant sample size.

Central tendency Center or middle of a distribution of data; it is measured as the mean, median, or mode.

Champion A manager in the Six Sigma world responsible for the overall effectiveness of the Six Sigma implementation.

Common cause variation Variables in your process that cause the process to vary and are built in and inherent to the process. The variation of your process due to common causes is the variation that is there all the time when things are running normally.

Continuous data Numbers that can be of any value or fraction of a value.

Control chart A graphic depiction of a process that is used to distinguish common cause from special cause variation.

Correlation The mutual relationship of two or more things; the degree to which two or more attributes tend to vary together.

Cost of poor quality (COPQ) Another term commonly used in industry to describe the cost of poor quality, of failing to meet the customers' needs.

Cumulative summation (CUSUM) chart A special type of control chart used to detect small shifts in the process mean.

Design for Six Sigma (DFSS) The application of appropriate Six Sigma tools to new process designs rather than to the improvement of existing processes.

Discrete data Numbers that are not continuous (e.g., whole numbers). Also known as **variables data**.

Documents Instructions (written or electronic) that define how a task is to be accomplished.

Expectations Desires of the customer that might not prevent the product from working as intended but would cause dissatisfaction if not met.

Exponentially weighted moving average (EWMA) chart A special type of control chart that takes previous data into account when evaluating each point; also capable of detecting small shifts in the mean.

External failure costs The cost of mistakes that made it out the door to customers.

Failure mode and effects analysis (FMEA) The third and final step in the quality reliability planning process, in which you examine every possible failure mode for your process in order to understand the consequences of process failures from the customers' perspective.

Fishbone diagram A graphic presentation device whereby ideas (or causes) are grouped into common categories such as manpower, methods, machines, and materials. Also known as an Ishikawa or cause-and-effect diagram.

Green belt Team member on a Six Sigma team that has training in the basic quality improvement tools.

Hidden factory The portion of the factory that produces non-value-added product.

Histogram Bar chart that shows the distribution of a set of data.

Internal failure costs The cost of mistakes made that were caught before they got out the door.

International Organization for Standardization (ISO) Federation made up of over 100 member countries that have agreed to publish common standard methodologies for various aspects of conducting global business.

ISO 9001 The internationally recognized Quality Management System standard.

ISO 14001 The internationally recognized Environmental Management System standard.

ISO registered The recognition granted to a company by an accredited registrar when the registrar has verified that the company has effectively implemented the management system.

Just-in-time (JIT) Process in which the company creates a product for the customer when it needs it.

Kaizen Japanese term for continuous improvement or incremental improvement.

Kanban A signal card used to pull production.

Lower control limit (LCL) Typically 3 standard deviations (σ) below the mean.

Lower specification limit (LSL) The lowest level allowed by the customer for a specific quality measure.

Malcolm Baldrige National Quality Award (MBNQA) A United States–based award established by Congress in 1987 that recognizes not only the effectiveness of the management system but also the company results achieved as a result of the management system.

Management system A collection of activities, usually but not always documented, that are intended to ensure that products and services meet specified needs.

Master black belt Six Sigma expert with training in many advanced quality improvement tools. Is responsible for coaching black belts and providing training to green belts and black belts.

Mean Arithmetic average of a set of data.

Measurement capability index (Cm) The ratio of your specification band to your measurement process variability.

Median Geometric center of a data set.

Mode Number that occurs most frequently in a data set.

Moving range (MR) Difference between two consecutive points.

Muda Japanese term for traditional waste.

Mura Japanese term for unevenness.

Muri Japanese term for overburden.

Nested design A statistical tool used to segregate total process variation into separate components. Also referred to as a component analysis of variance.

Nonconformance Evidence of deviation from documented requirements.

Normal distribution Distribution of continuous data that is bell shaped. The data in a normal distribution are distributed such that approximately 68% of the data fall within 1 standard deviation of the mean, approximately 95% of the data fall within 2 standard deviations of the mean, and approximately 99.7% of the data fall within 3 standard deviations of the mean. Also referred to as a Boltzman distribution.

NP chart Control chart for counts of defectives (binomial distribution) with a constant sample size.

Out of control Points outside the 3 sigma (σ) control limits.

P chart Control chart for counts of defectives (binomial distribution) with a variable sample size.

Performance model The mathematical model that predicts how your process will perform based on the input variables.

Plan-do-check-act (PDCA) cycle A cycle for continuous improvement that implements the following steps: planning, doing, checking, and acting.

Poisson distribution Distribution of discrete data that is made up of counting data. In this case only whole numbers are present.

Prevention costs The costs of improving processes so that they do not have to be checked to catch failures because everything was done right the first time.

Price of nonconformance (PONC) A term coined by Philip Crosby used to describe the costs of poor performance and failing to meet the customers' needs.

Process capability index (Cpk) The ratio of the customer's specification range to your process variability taking the average (process center) into account; this index assumes the data have been "cleansed" of special causes.

Process performance index (Ppk) The same calculation as the Cpk but without the assumption that the data have been cleansed of special causes.

Process potential index (Cp) The simple ratio of the customer's specification range to your normal process variability without taking the average (process center) into account.

Quality The performance, products, and services that consistently meet or exceed the expectations of the customer by doing the right job, the right way, the first time, every time.

Quality circle A group of people, typically organized by logical structure within a company, who work together for a common goal.

Quality control plan (QCP) The second step in the quality reliability planning process. The specific process control handles are specified to ensure that your process is capable of meeting the needs of the customer. Process capability is captured at this stage.

Quality design plan (QDP) The first step in the quality reliability planning process. The needs of the customer are translated into specific requirements (specifications), and the measurement techniques are specified to ensure that those needs are met.

Quality reliability planning (QRP) A procedure designed to ensure that when a new product is being introduced, proper communications occur between all of the functional groups and that a system is in place that will allow the product to meet the customers' expectations consistently. The procedure consists of the quality design plan, the quality control plan, and the failure mode and effects analysis.

Range (R) Difference between the high and low values in a group of numbers.

Records Evidence that a task has been accomplished.

Regression The statistical analysis that generates a mathematical representation of the correlation between two or more variables.

Requirements The criteria that describe the traits that a product or service must possess in order to function as intended. Performance characteristics of your product.

Risk priority number (RPN) A calculated number within the failure mode and effects analysis process that helps you to prioritize the actions that need to be taken to minimize the risks of process failures and consequences to the customer.

Root cause analysis (RCA) The analysis of data to determine the true root cause of a problem. Also called root cause investigation.

Shifts Runs of 8 points above or below the average.

σ Lowercase Greek letter sigma; used in mathematics to denote the standard deviation.

Σ Uppercase Greek letter sigma; used in mathematics to indicate a summation.

Special cause variation Event-related items in the process that cause the process to vary outside of its normal operation. These causes are in addition to the common cause variation sources. Also referred to as assignable cause variation.

Specifications The subset of the requirements that the customer puts into the transaction contract. May also be the translation of the requirement into a measurable output for the purpose of tracking compliance.

Spread The broadness of the distribution of the data set; it is measured as the difference between the highest and lowest values in the data set or by using the standard deviation.

Stable process State of your process when only common cause variation is present; also stated as the absence of special or assignable causes. A stable process is, by definition, predictable.

Standard deviation Statistical measure of the spread of a set of data. Often referred to as sigma (σ), which is always used to denote the standard deviation.

Statistical process control (SPC) The application of statistical techniques to the process that generates a product.

Statistical quality control (SQC) The application of statistical techniques to the output (quality) of a process.

Synergy The phenomenon in which the total effect of a whole is greater than the sum of its individual parts.

Takt time A German word that means beat. In quality it is the available production time divided by the customer demand.

Team A small group of people, with complementary skills, committed to a common set of goals and tasks.

Total quality management (TQM) The pulling together of many different components of quality into a single program (sometimes shown as TQC—total quality control).

Trends Runs of 8 points going up or going down.

U chart Control chart for counts of defects (Poisson distribution) with a variable sample size.

Unstable process State of your process when both common cause and special cause variation are present. An unstable process is, by definition, unpredictable.

Upper control limit (UCL) Typically 3 standard deviations (σ) above the mean.

Upper specification limit (USL) The highest level allowed by the customer for a specific quality measure.

Value stream map (VSM) A high-level flowchart of a process. Included within each step of the process is the amount of time and what value it adds from the customer's perspective.

Variables data Data from a distribution wherein any number or fraction of a number is possible.

x-bar ($\overline{x}$) Used in mathematics to indicate the mean.

x-bar and R ($\overline{X}R$) A control chart used when data are grouped together.

Index